INNOVATIONS IN BRIDGE ENGINEERING TECHNOLOGY

BALKEMA – Proceedings and Monographs in Engineering, Water and Earth Sciences

Innovations in Bridge Engineering Technology

Edited by

Khaled M. Mahmoud
Bridge Technology Consulting
New York City, USA

Taylor & Francis
Taylor & Francis Group
LONDON / LEIDEN / NEW YORK / PHILADELPHIA / SINGAPORE

Cover photo information
Front Cover:
I-87 Twin Arches over the Mohawk River, New York, USA
Photo courtesy of New York State Department of Transportation and CHAS H. Sells, New York

Back Cover (From the Top):
Royal Bridge, Palm Beach, Florida, USA
Rendering courtesy of H2L2 Architects, New York

I-87 Twin Arches over the Mohawk River, New York, USA
Photo courtesy of New York State Department of Transportation and CHAS H. Sells, New York

Alexander Hamilton Bridge, New York City, USA
Photo courtesy of Jacobs Edwards & Kelcey, New York

I-87 Twin Arches over the Mohawk River, New York, USA
Photo courtesy of New York State Department of Transportation and CHAS H. Sells, New York

Cover Design:
Khaled M. Mahmoud

Taylor & Francis is an imprint of the Taylor & Francis Group, an informa business

Typeset by Charon Tec Ltd (A Macmillan Company), Chennai, India
Printed and bound in Great Britain by Antony Rowe (A CPI-group Company), Chippenham, Wiltshire

Published by: Taylor & Francis/Balkema
P.O. Box 447, 2300 AK Leiden, The Netherlands
e-mail: Pub.NL@tandf.co.uk
www.balkema.nl, www.taylorandfrancis.co.uk, www.crcpress.com

ISBN13: 978-0-415-45337-0

ISBN (eBook): 978-0-203-93849-2

Table of Contents

Preface

In the last few years, remarkable technological advances have been achieved in bridge engineering technology. These cover a wide spectrum of issues, ranging from design, maintenance, and rehabilitation methodologies to material and monitoring innovations.

Within an international framework of exchanging the state-of-the-art in the field of bridge engineering, the Fourth New York City Bridge Conference was held on August 27–28, 2007. This book contains a selected number of papers that were presented at the conference. These papers are valuable contributions to the body of knowledge in bridge engineering technology. The Fourth New York City Bridge Conference was distinguished for its global impact. Bridge engineering experts presented papers from Belgium, Canada, Croatia, England, France, Germany, Italy, Japan, Lebanon, Northern Ireland, Scotland, Switzerland, Taiwan and Turkey. These, along with a list of prominent bridge engineering professionals from the United States, will assure the archival quality of this book.

The proceedings lead off with a paper by Forde and Ohtsu on the "International state of practice in the inspection of grouted duct post-tensioned concrete bridge beams and decks". Post-tensioned concrete bridges have been used for both rail and road bridges for some forty years. Problems were first noticed with road bridges due to the use of de-icing salts. However, there have been collapses of post-tensioned concrete railway bridges. Following these series of collapses, the Highways Agency in the United Kingdom enforced a ban on the construction of post-tensioned bridges with metallic tendon ducts. The moratorium was lifted with the introduction of plastic tendon ducts. The paper focuses on the analysis of impact-echo NDT of concrete beams with plastic tendon ducts – using both a conventional frequency domain analysis and using the Japanese SIBIE (Stack Imaging of Spectral Amplitudes Based on Impact-Echo) method. The authors show that the SIBIE method of analysis exhibits great promise in testing these previously difficult to inspect plastic tendon ducts, using impact-echo. Segmental bridges are commonly used in the last thirty years as part of modern roads and highways in Croatia. During regular and preventive bridge inspections some defects were detected. In "Durability of concrete segmental bridges" Banic et al present typical types of concrete segmental bridges built in Croatia, types of construction and condition state of several types of segmental bridges. Condition of those structures was determined during bridge inspections conducted in the last five years. Characteristic damages of segmental bridges, rehabilitation procedures applied in several cases and also some improvements on new structures are presented. In "Cyclic tests of precast segmental unbonded post-tensioned concrete bridge piers", Ou et al present an experimental study on the seismic performance of precast segmental unbonded post-tensioned concrete bridge piers. The pier specimen consists of a foundation, four hollow column segments and 5.7-meter high pier cap. The prestressing tendons are located inside the hollow core of the pier column and hence are unbonded with the surrounding concrete. In the first pier specimen, no mild steel reinforcement is extended across the column segment joints. In the other two specimens, longitudinal mild steel bars, also referred to as energy dissipation bars or ED bars, anchored at the foundation and extended up to the pier cap, are added to enhance the seismic energy dissipation. The test results showed that all the pier specimens exhibited satisfactory ductile behavior. The hysteretic energy dissipation, lateral strength and residual drift upon unloading of the specimens increased with the increase of the amount of ED bars. The Jamestown-Verrazzano Bridge over Narragansett Bay, Rhode Island, USA, features a 1,509 meter-long prestressed segmental box girder main bridge with 23 spans varying in length from 33 meter to 194 meter, and a 732 meter-long trestle structure. The bridge was open to traffic in 1992 and a baseline inspection was conducted in 1999. The scope of the baseline inspection included analysis, load rating, and comparison of creep deflections based upon as-built shop drawings, and casting and stressing schedules versus field-surveyed conditions. During subsequent inspections nondestructive and destructive methods were used to investigate the post-tensioning ducts for the presence of voids. Over 28,000 meters of nondestructive impact-echo (sonic/ultrasonic) measurements were taken on the concrete top slab, webs and bottom slab containing the tendons to evaluate the grouted tendon ducts for voids. Of the approximately 1,520 tendon ducts tested, 7.5% or 114 tendon ducts were determined to have voids. Void lengths ranged from 0.3 meter to over 94 meters. In most cases the tendons were grout covered but some of the tendons were exposed and exhibited corrosion. At the time of construction, grouting methods were not always fully effective. Currently, voids in post-tensioned ducts are an issue for a number of bridge owners. Other repairs include use of epoxy-injection with CFRP reinforcement system for the cracked webs of the segmental box girder pier tables. In their paper "Inspection and rehabilitation of Jamestown-Verrazzano segmental concrete bridge", Abrahams et al discuss these findings and repairs that are currently underway.

Cable-supported bridges are notable for their aesthetic appeal and ability to link long spans. Many of the issues associated with these structures require thorough studies prior to construction. In "Ultimate capacity of suspension bridges with arbitrary imperfect towers", Inoue investigates the difference of ultimate capacity of suspension bridge due to the imperfection of towers. The measurements of tower deviations from the ideal position for constructed suspension bridges, mainly in Japan, have been studied and the tendencies of imperfection have been classified into different types. The effect of tower imperfection for the ultimate capacity has been investigated by 1-1/2 order analyses using 2-D bridge model. Four types of imperfect tower with the different imperfect shapes were modeled at the freestanding position. Finally, the difference of ultimate capacity among the imperfect models is summarized and some remarks are offered for more reliable and economical bridge in the future. The new trend in design of footbridges in Turkey is to utilize cables. Some of these bridges have fake cables while others partially rely on the cable system. These steel composite bridges typically constructed over highways span about 40 to 60 meters. It was observed that the bridges with fake cables can be substantially heavier than the ones with functional cables. In "Cable supported footbridge analysis with construction staging", Caner studies the importance of tensioning sequence of cables and impact of construction staging on the design forces at superstructure to have economical designs. A case study is illustrated as an example design. Locked coil cable assemblies are used in cable supported road bridges (e.g. as suspenders in suspension bridges and hangers in arch bridges) and a large variety of pedestrian and cycle bridges. Despite of lots of installations all over the world and recent product enhancements, locked coil cable assemblies are not so well known in the USA. Recently, increased demand for the product has been observed. In "Locked coil cable assemblies for bridges", Bechtold et al introduce an overview about present and past applications of locked coil cables.

Isolation bearings have become a standard tool for engineers designing bridges in seismic regions. However, the added complication of cold weather has raised concerns with rubber isolators and their performance in northern regions of the United States. As a result, bridge designers are migrating towards sliding isolation bearings (SIB) in these regions. SIB have been proven to be cost effective and high damping devices on numerous projects to date. Watson describes the research that led to the development of SIB. In addition several case histories will be reviewed in an effort to demonstrate SIB capabilities in low temperature environments. Deformations on the order of 11 mm in the masonry plates of installed lead rubber isolation bearings were observed in a highway bridge. Of the more than 400 isolators in the project, approximately 30 showed deformations greater than 2 mm. Due to the cost, accessibility issues and traffic impacts of removing and replacing the isolators, the Owner agreed to accept laboratory testing as a means to determine which effects, if any, the deformation had on the properties of the lead core isolation bearings. Jacak and Pezzotti present the "Results of tests performed on lead-rubber seismic isolators with deformed masonry plates". New bearings were manufactured in accordance with the original project requirements. The new bearings were first tested to establish baseline properties and validate their compliance with the contract documents. Subsequently, the isolators were deformed in the lab to achieve a similar deformation as that observed in the structure. The bearings were then tested in the deformed condition and the results compared to the baseline properties. Soil-Structure interaction may play a major role in the seismic response of a bridge structure. Specifically, a significant reduction in soil stiffness and strength may result in permanent displacement of the abutments and foundations, thus imposing important kinematic conditions to bridge structure. In "Humboldt Bay Middle Channel Bridge: 3D bridge-foundations-ground system", Trombetti et al show the effects of this behavior referring to the Humboldt Bay Middle Channel Bridge, in California, USA. The Finite Element model and nonlinear solution strategy are built in the open-source software plat-form OpenSees. The 3D nature of bridge response imposes significant computational challenges. The soil is modeled as a nonlinear material with a Von Mises multi-surface kinematic plasticity model so as to reproduce elasto-plastic shear response. The results obtained using 1978 Tabas earthquake record show that changes in properties of the superficial soil layers dictate significantly different time histories of dynamic excitation at the various support points of the bridge (piers and abutments).

Bridge design methodologies have made significant strides due to technological advances in construction, fabrication and testing techniques. The $210 million Florida Avenue Bridge project is being designed to provide reliable access between St. Bernard and Orleans parishes over the Inner Harbor Navigational Canal (IHNC) in New Orleans, Louisiana. The project includes a five-span high-level bridge over the IHNC with a 143-meter center span. Bridge type studies were completed to determine the most viable structure type. Both cast-in-place segmental concrete box girder and steel plate girder alternates were selected for final design. In "Design of Florida Avenue Bridge over the Inner Harbor Canal", Nelson presents the design of the segmental concrete alternate. The superstructure consists of a variable depth twin-cell box girder that is supported by voided box column piers and steel HP piles. The bridge will be built with form travelers using the balanced cantilever method of construction. Heat curving is widely used for fabricating curved steel bridge I-girders. Curving is accomplished by asymmetric heating of the flanges of the straight girder. Heat is applied along the girder length continuously or intermittently with the heated width varying from $^1/_{12}$ to ¼ of the flange width depending on the curvature. Curvature develops after the girder cools to ambient conditions. Current practice limits the maximum temperature to 620°C for conventional Grades 250 and 345 steels. The "Guide for Highway Bridge Fabrication with HPS 485W Steel" recommends investigating heat curving of HPS 485W at 705°C. In their paper "Heat curving HPS 485W bridge I-girders", Gergess and Sen evaluate the validity of the 705°C temperature using non-linear finite element analysis. Other fabrication issues relating to heat curving stiffened and hybrid girders are also addressed. Results show that the maximum temperature can be somewhat lower. Stiffeners may reduce the curvature by up to 10% while hybrid girders with top and bottom flanges made of different steel grades require different heating profiles. In "Testing of a novel flexible concrete arch system", Taylor et al describe the testing of a flexible masonry concrete arch system which requires no centering in the construction phase or steel reinforcement in the long-term. The arch is constructed from a 'flat pack' system by use of a polymer reinforcement for supporting the self-weight of the concrete voussoirs and behaves as a masonry arch once in the arch form. The paper outlines the construction of a prototype arch and load testing of the backfilled arch ring. Some comparisons to the results from analysis software have been made. The arch had a load carrying capacity far in excess of the current British Highways Agency design wheel loads.

The Great River Bridge, built in 1939, is located in downtown Westfield, a City in Western Massachusetts, USA. The through truss bridge is a landmark for the City, forming perhaps the most distinctive structure in the downtown area. The project scope was initially limited to the rehabilitation of the 112 meter long, two-span structure. However the project has grown to include the design of two other bridge structures, four landscaped parks, several thousand feet of urban roadway, and two miles of railroad track. In "Westfield Great River Bridge", Ennis presents the different components of the project. The State Route 62 Bridge over Crooked Fork Creek in Morgan County, Tennessee, USA, was originally designed in 1940. After more than six decades of service to the citizens of this rural community, the bridge had become structurally deficient and functionally obsolete. The Tennessee Department of Transportation (TDOT) decided an extensive rehabilitation was needed to address the structural problems and improve its functionality. The project entailed a complete replacement of the original superstructure as well as repair and modification of the existing substructure units. In "Renewing the Crooked Fork Creek Bridge", Wilson describes the project which was accomplished without construction within the channel of the creek. To avoid the need for a lengthy detour, construction activities were phased and traffic control designed so that one lane of traffic could be maintained across the bridge throughout the duration of the project. The aging highway bridge continuously renewed while accommodating traffic flow. The traveling public demands that this rehabilitation and replacement to be done more quickly to reduce congestion and improve safety. Conventional bridge reconstruction is typically on the critical path because of the sequential, labor-intensive processes of completing the foundation, the substructure, the superstructure infrastructure in the United States is being subjected to increasing traffic volumes and must be components, railings, and other accessories. Bridge systems can allow components to be fabricated off site and moved into place quickly while maintaining traffic flow. Depending on the specific site conditions, the use of prefabricated bridge systems can minimize traffic disruption, improve work-zone safety, minimize impact to the environment, improve constructability, increase quality, and lower life-cycle costs. In "Rapid delivery! New Jersey overnights bridge rehabilitation for Trenton Bridges" Capers and Cheng discuss the adopted approach to the replacement of the superstructures of two structurally deficient bridges carrying a freeway section of Route US 1 through the capitol city of Trenton, New Jersey, USA. The Route 70 over Manasquan River Bridge replacement project utilized an innovative precast substructure solution on a project requiring difficult coordination of highway and marine traffic, environmental constraints and community involvement. The 7.6-meter high, 220.7-meter long bridge, which is supported on five architecturally treated precast High Performance Concrete (HPC) in-water piers, crosses a navigable waterway in the coastal region of the State of New Jersey, USA. The precast pier column and cap components were fabricated offsite, delivered via barges and trucks and assembled using post-tensioning. Pier foundations were constructed at the waterline within precast concrete cofferdam shells, which provided pile driving templates, served as architecturally treated formwork for the footings and eliminated construction of traditional cofferdams. Yermack presents the details of the project in his paper "Accelerated construction of precast concrete piers on the Route 70 over Manasquan River Bridge replacement project". Stainless steel reinforcing has been used in numerous structures throughout North America. Recent advances in concrete technology have provided structural designers with materials which can easily last over 100 years, and the life of many concrete structures today is limited by the reinforcing. Improvements in the life of the reinforcing can be translated directly into extended life of the structure. Current projections by several transportation agencies show that the use of solid stainless steel reinforcing bar in bridge decks will more than double the life of the bridge deck. While solid stainless steel reinforcing bar can increase the cost of the bridge deck by as much as 12% (compared to carbon steel reinforcing), the economic value of the longer life outweighs the initial higher cost. In "Improving tomorrow's infrastructure: extending the life of concrete structures with solid stainless steel reinforcing bar", Schnell and Bergmann discuss corrosion resistance and cost saving offered by the use of stainless reinforcing.

The paper "Use of structural health monitoring techniques for a forensic study of bridge accidents", by Yun, presents an overview of a real-time web-based continuous monitoring system for the Vincent Thomas Bridge. An effective multi-thread bridge monitoring system architecture is shown. Using the bridge monitoring system, the bridge response to earthquakes, bridge-ship collision and ambient vibration was measured, and the bridge modal frequencies were successfully determined with vibration-based identification methods. In "Bridge Management and Inspection System for Montgomery County, Maryland", Shaffer and Schellhase cover an overview of the county's needs and the solutions that have been developed to significantly improve both the inspection and management processes. Electronic forms were created to meet the county's requirements, the most rigorous in the state of Maryland, USA, and thus allowing for entry of all information from the inspection. Inspection of bridge decks generally relies on visual inspection and use of basic non-destructive testing techniques. Assessment typically involves comparison of observed condition with pre-defined condition states. Current condition states require little quantitative data and must apply across many different material types and bridge elements. Use of these types of subjective techniques may lead to uncertain assessment of structure condition. This is particularly true when comparing different structures or structures assessed by different personnel. One improvement that may be considered to reduce the uncertainty or subjectivity of the current process is the introduction of quantitative measures within the condition states. In "Objective condition states for concrete bridge deck assessment", Knight discusses condition states developed for assessment of concrete cast-in-place bridge decks. The proposed condition states include basic quantitative information and address specific forms of deterioration consistently identified during inspection. The Brooklyn Bridge stands as a monument of bridge engineering, and while easily accessible to the public, access for structural inspection is difficult. As part of the 2006 biennial inspection of the Bridge, a detailed masonry inspection of the Manhattan and Brooklyn towers was conducted using rope access techniques to examine areas previously investigated only through remote visual methods. In "The 2006 rope access inspection of the Brooklyn Bridge towers: a new view of an old bridge", Schmidt discusses the access methods employed for a detailed inspection of the bridge tower masonry. Challenges included performing this work without adding anchors to the towers, registration of inspection findings on a massive masonry structure in a repeatable format, and providing tactile inspection access in stone overhangs, beneath steel walkways and within recesses.

Bridge structures stand as landmarks of aesthetics and monument of engineering ingenuity. On the theme of historic bridges, Melewski et al take the reader along a "Walkway over The Hudson (historic bridge to northeast recreational destination)". The paper discusses the historic significance of the Poughkeepsie Railroad Bridge, which was opened in 1888. It was the longest bridge in the world when the first train crossed it. As the first bridge constructed across the Hudson River between New York City and Albany, the bridge had an enormous impact on transportation throughout the Northeast United States. After a long history of ownership and uses, the bridge suffered damage from a fire in 1974 that rendered it unusable for railroad traffic. A comprehensive study has begun to certify structural integrity and to produce a plan to establish it as a public park and walkway, as well as a bridge engineering educational resource. The paper provides a brief historic overview, discusses the objectives of the comprehensive study and the findings of the late 2006 underwater inspections. In "Aesthetics and durability aspects in the realization of small and medium span arch bridges" Siviero and Zanchettin present the importance of function and its harmony with the surrounding environment. The authors discuss the value of aesthetics within the context of long lasting durability. In January of 1886, an American contractor, Union Bridge Company of New York City, won an international competition to design and build a two-track steel railroad bridge of approximately 3,000 feet in length over the Hawkesbury River in New South Wales, about 30 miles north of Sydney, Australia. At the time it was the biggest public works project in the southern hemisphere. In his paper "Hawkesbury Railway Bridge near Sydney, Australia", Gandhi gives a background of this project; details of 14 designs submitted by different contestants from England, France, Australia, and the US; construction methods; key individuals involved in this project; difficulties encountered during construction; and its successful completion. The bridge was completed in 34 months and opened to traffic with great fanfare in May 1889. It linked the north and south regions of Australia. The bridge was strengthened several times and ultimately replaced in 1946. Projects responsive to the purpose and needs defined by the stakeholders generate greater participation and ownership in the project. Involving the community early builds support and minimizes resistance that sometimes develops in response to change. Officials benefit by helping the public anticipate construction with the knowledge that any inconvenience will be rewarded by a structure that is responsive to the specific needs of the community. In "Historic bridge replacement: a collaborative approach to context sensitive design", Piotrowski and Chamberlin present architect's experience in the replacement of the Royal Park Bascule Bridge, Florida, USA.

The contributions of an outstanding body of technical experts from all over the world ensure the archival value of this set of proceedings. The presented material in this volume reflects state-of-the-art innovations in bridge engineering technology. The editor thanks the authors and expresses a special note of gratitude to the reviewers. This volume is a result of the sacrifice of time and effort, dedication and collective wisdom of all contributors.

Khaled M. Mahmoud, PhD, PE
Chairman of Bridge Engineering Association
www.bridgeengineer.org

President of Bridge Technology Consulting
www.kmbtc.com
New York City, USA

New York City, August 2007

1 Concrete segmental & post tensioned bridges

Chapter 1

International state of practice in the inspection of grouted duct post-tensioned concrete bridge beams and decks

M.C. Forde
University of Edinburgh, Edinburgh, Scotland, UK

M. Ohtsu
Kumamoto University, Japan

ABSTRACT: Post-Tensioned concrete bridges have been used for both rail and road bridges for some 40 years. Problems were first noticed with road bridges due to the use of de-icing salts. However there have been collapses of post-tensioned concrete railway bridges. Following this series of collapses, the Highways Agency enforced a ban on the construction of post-tensioned bridges with metallic tendon ducts. The moratorium was lifted with the introduction of plastic tendon ducts.

This paper will focus on the analysis of impact-echo NDT of concrete beams with plastic tendon ducts – using both a conventional frequency domain analysis and also using the Japanese SIBIE (Stack Imaging of Spectral Amplitudes Based on Impact-Echo) method. It will be shown that the SIBIE method of analysis exhibits great promise in testing these previously difficult to inspect plastic tendon ducts, using impact-echo.

1 INTRODUCTION

Post-Tensioned concrete bridges have been constructed in the UK since 1947. In the case of highways, a major issue has arisen with the grouting of the Post-Tensioned tendon ducts. If water, chlorides and oxygen infuse into these ducts then the tendon corrodes reducing the strength, ultimately leading to structural collapse. The collapse mechanism is brittle and little or no warning may be given.

Railway bridges are less vulnerable than highway bridges as the latter are not normally subjected to de-icing salt. None-the-less, railway bridges remain vulnerable (Woodward & Williams, 1998), albeit the time scale to failure may be longer.

2 APPROACHES TO INSPECTION OF P-T CONCRETE BRIDGES

A number of different approaches have been developed for the inspection of P-T concrete bridge beams. These are summarised in Table 1.

2.1 *Drilling and inspection by borescope*

Drilling and inspection by borescope (Stain & Dixon, 1993) remains the most widely used technique in the UK. This procedure is seen as definitive at the point of inspection. GPR is often used to detect the location of the metallic tendon ducts. Drilling into the duct is undertaken with great care with an automatic cut-out switch. Finally a borescope is used to inspect the void and the state of the tendons. Sometimes a gas test may be used to estimate the volume of the void. Apart from being slow, risky and expensive.

Table 1. Methods of detecting voids grouted ducts of P-T concrete bridges (Highways Agency, 2007).

Investigation method	Cost of method	Metal ducts	Plastic ducts	Effectiveness of technique
Visual Inspection	Low	No	No	Technique if ineffective as bridges rarely show distress before catastrophic failure.
Load Test	Relatively high	No	No	Ineffective procedure and dangerous as the structure could fail before any meaningful deflection response is obtained.
Stress/strain measurement	Relatively high	No	No	Generally ineffective as Cavell (1997) has shown that post tensioned bridge strain variations due to loss of pre-stressing can be similar to variations resulting from temperature gradients throughout the year. Thus this technique is not sensitive to the defects in post tensioned bridges.
Impulse radar	Intermediate	No	Yes	Effective with non metallic liners such as in the joints of segmental bridges and in the newer post tensioned bridges. Radar will not penetrate post tensioned metal ducts.
Impact echo	Intermediate	Yes	Maybe	Potentially useful in identifying voiding in non metallic and metallic post tensioned ducts. Essential to ensure that impact frequency is sufficiently high to identify the defect.
Manual drilling of tendon duct with visual inspection using endoscope	Intermediate	Yes	No	Statistically limited and potentially dangerous if the tendons themselves are drilled. Advantage is that a direct physical observation can be made.
Radiography	High	Yes	Yes	High powered radiographic techniques give good image of voiding but requires closure of the bridge and may not be used in urban areas due to risk of radiation.
Ultrasonic tomography	Intermediate	Yes	Yes	Promising technique that could identify voids by producing a 2-D or 3-D image of the beam cross-section.

2.2 *Radiography*

Radiography or high energy X-ray was popular in France in the 1960s and 1970s. This procedure has fallen out of favour due to both the expense and the risk to human health.

2.3 *Ultrasonic tomography*

Ultrasonic tomography of concrete was researched at the University of Edinburgh (Martin et al., 2001). The word "tomography" is derived from the Greek "tomos" meaning a slice and involves reconstructing a section of an object using measurements taken from outside the object. The tomographic imaging method uses ultrasonic pulse velocity information taken through a section to develop a two or three-dimensional reconstruction of the velocity distribution in that section – Figures 1 and 2.

Tomographic modelling is only really effective when one can gain access to 4 sides of a beam – not normally possible with post-tensioned bridge beams.

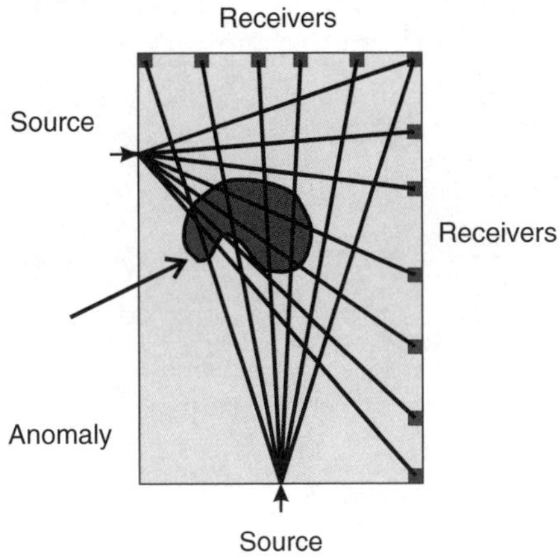

Figure 1. Tomographic paths.

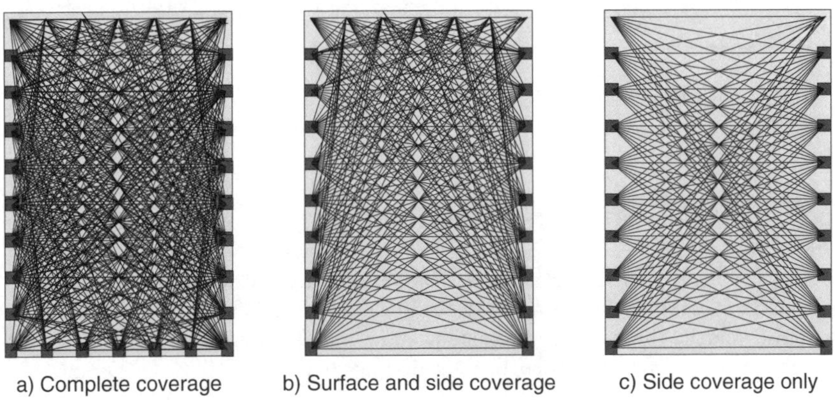

Figure 2. Ray path coverage for different transducer arrangements.

2.4 *Impact echo*

Impact echo was developed and promoted in the USA (Sansalone & Streett, 1997). The Impact Echo Test (IE) can be summarised as:

- The Impact-Echo (IE) technique utilises impact generated stress waves which propagate through the medium reflecting at boundaries or flaws.
- These reflections are detected by a transducer next to the impact.
- A voltage vs. time trace is recorded from the surface displacements.
- The dominant resonant frequencies are deduced by carrying out a Fast Fourier Transform of this trace.
- The peak frequencies are then related to their corresponding depth in the medium.

A schematic of the test is procedure is shown in Figure 3 (Sansalone & Streett, 1997).

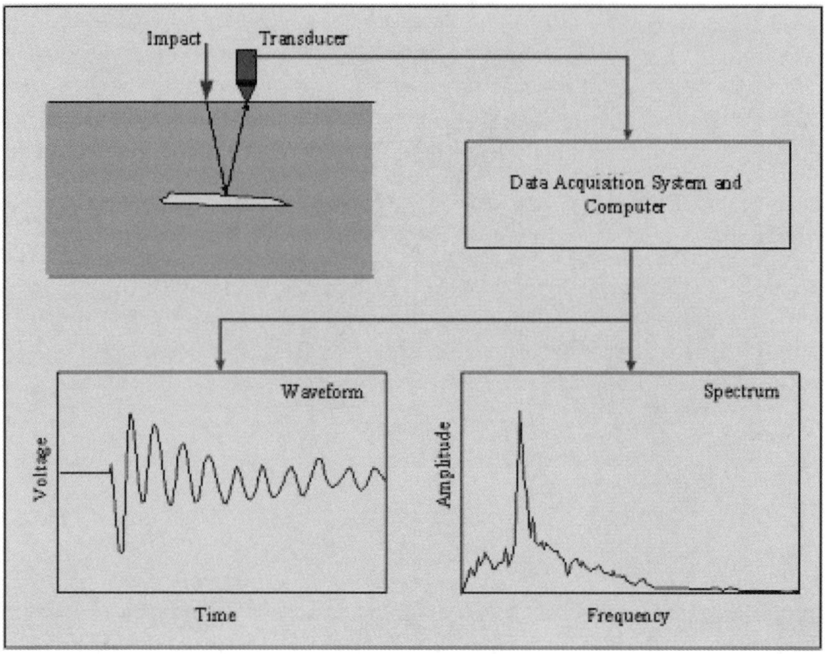

Figure 3. Schematic showing the transfer from a test to a voltage vs. time trace and finally shown in the frequency domain.

3 IMPACT ECHO EXPERIMENTS AT THE UNIVERSITY OF EDINBURGH

A series simulated post-tensioned concrete beams were constructed at the University of Edinburgh.

The standard method of analysis of impact echo data is given below.

If the pulse velocity through the specimen is known and the time to the arrival of a reflection from within the specimen is measured, then the distance to the target is:

$$2d = V_p t \tag{1}$$

where: d = Depth to target
V_p = P-wave velocity
t = time to reflection

As the test object becomes more complex, the time trace of the receiver becomes difficult to analyse. It is usually much more straightforward to analyse data from these tests in the frequency domain by carrying out a Fourier transform on the data.

The appropriate ball bearing was chosen for the impact-echo test by comparing the required resolution and thus the required wavelength (higher resolution = short wavelength = smaller ball bearing) and the depth of penetration needed (greater penetration = longer wavelength = larger ball bearing). The appropriate ball bearing to use on the beams was the 10 mm diameter ball bearing (Martin, 1997).

The velocity of the concrete was calculated by using one transducer, and the equation in Figure 8.

The velocity is measured by impacting the surface and recording the frequency over an area of solid concrete. To simulate testing on site the side of the beam was tested rather than the top as the top would not be accessible on site. The frequency of the rear wall (fT) was found to be the same

Figure 4. Edinburgh test beam.

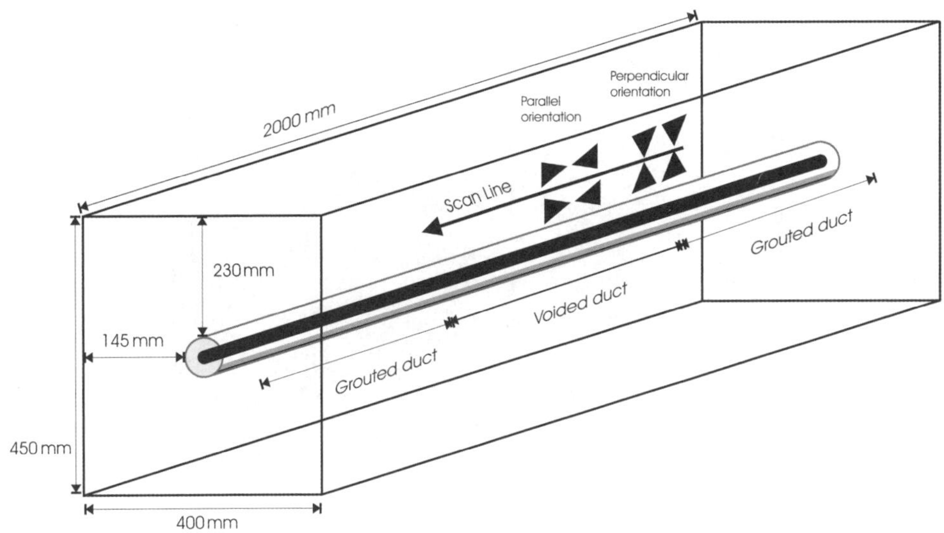

Figure 5. Schematic drawing of the Edinburgh beam.

on all the beams tested, as all the beams where cast from the same batch of concrete. The fT was found to be 4.9 kHz and thus the velocity through the concrete (Cp) was found to be $4083ms^{-1}$.

From this initial calculation it was then possible to calculate the expected position of the fvoid and the fsteel:

fvoid = 13 kHz,
fsteel = 6 kHz

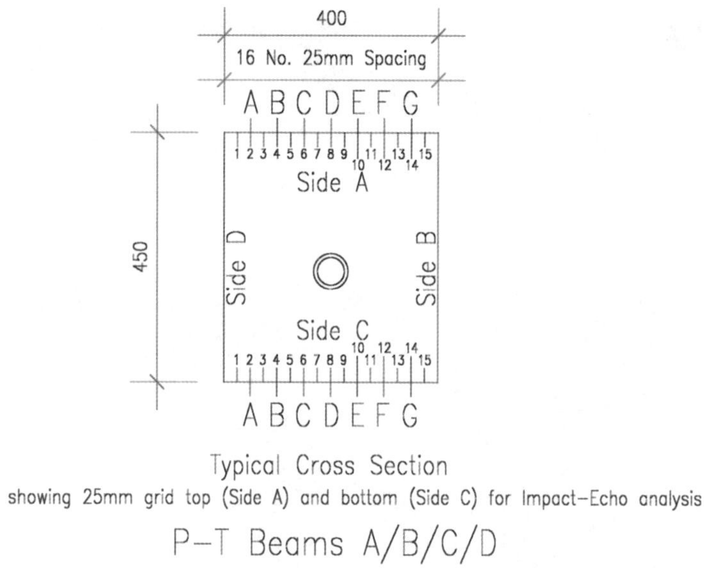

Figure 6. P-T beam section with Resonant Frequency Analysis (I-E) measurement points.

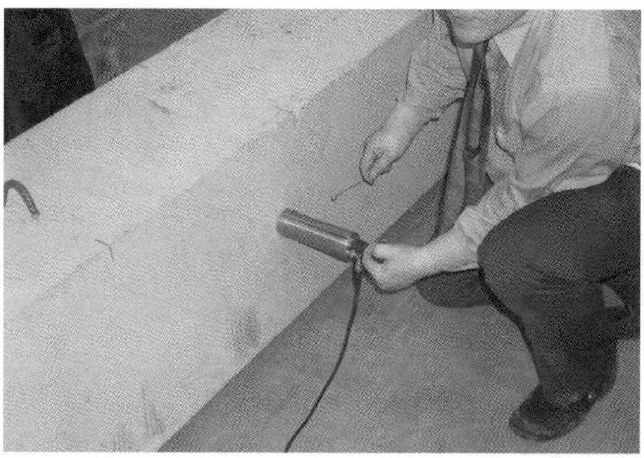

Figure 7. Showing the impact-echo test taking place. (Colombo et al, 2002).

Figure 9 shows the frequency response from a voided duct whilst figure 10 shows the frequency response from a fully grouted duct. These results where taken at the middle of the concrete beam to reduce any end effects. It can be seen that from Figure 9 the initial peak (fT) has moved forward from 4.9 (plain concrete), and there is a peak with a higher frequency, these are typical of a voided duct. In figure 10 the initial fT has not moved forward and there is a pulse at 6 kHz.

The same readings were then taken on a beam with a voided plastic tendon duct – figure 11. It can be seen that this data is very difficult to interpret compared to Figure 9 (Muldoon et al, 2007).

As a result of this difficulty, Kumamoto University developed a new innovation in the interpretation of impact echo test data – the SIBIE method.

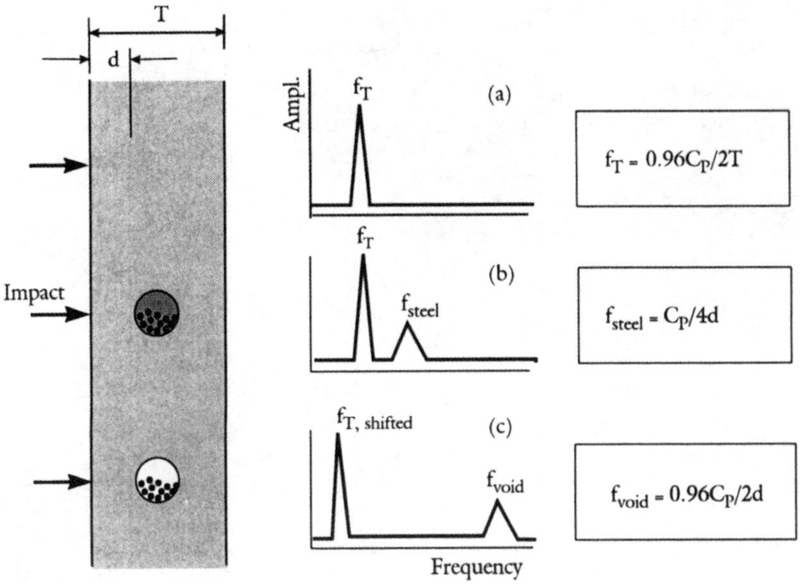

Figure 8. Schematic diagram showing three types pf response for a plate containing post-tensioning ducts: a) solid plate, b) grouted duct; and c) duct containing a void. (after Sansalone and Streett, 1997) (Axes: ordinate = signal amplitude [volts]; abscissa = frequency [kHz])

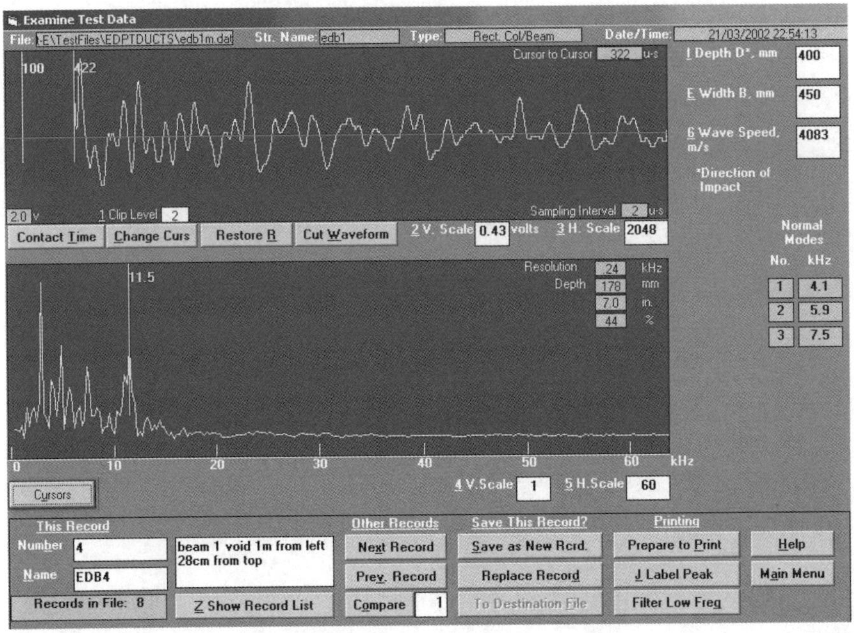

Figure 9. Result of an impact echo test over an un-grouted (voided) tendon duct.
Notes: (a) Top half of figure is a time domain plot: ordinate = amplitude [units = +volts]; abscissa = time [units = μs]; (b) Bottom half of figure is a frequency domain plot: ordinate = amplitude [volts]; abscissa = frequency [kHz]

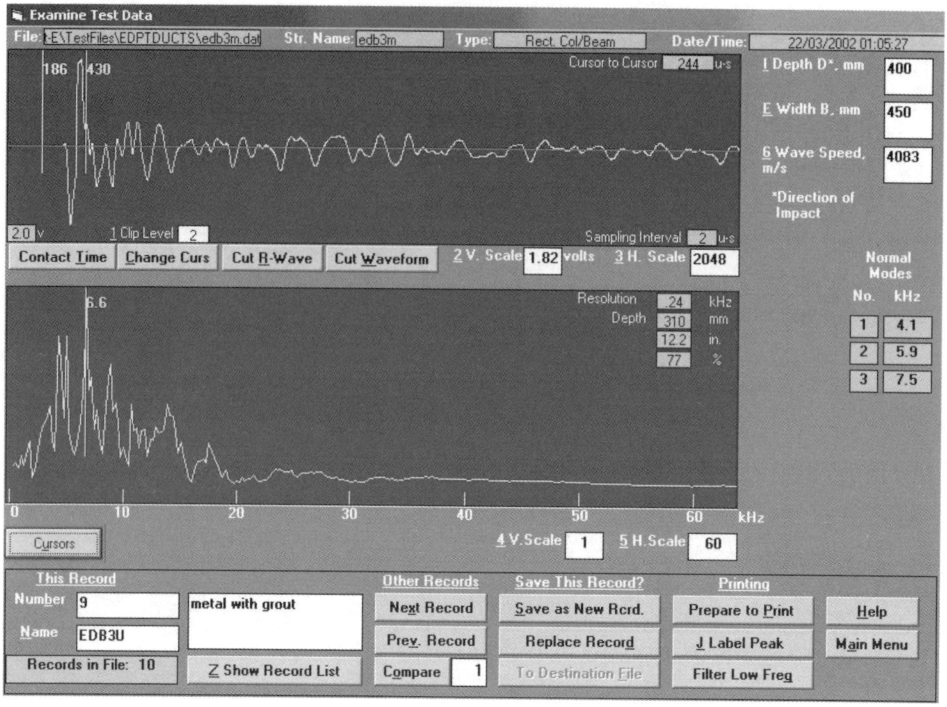

Figure 10. Result of an impact echo test over a grouted tendon duct.
Notes: (a) Top half of figure is a time domain plot: ordinate = amplitude [units = +volts]; abscissa = time [units = μs]; (b) Bottom half of figure is a frequency domain plot: ordinate = amplitude [volts]; abscissa = frequency [kHz]

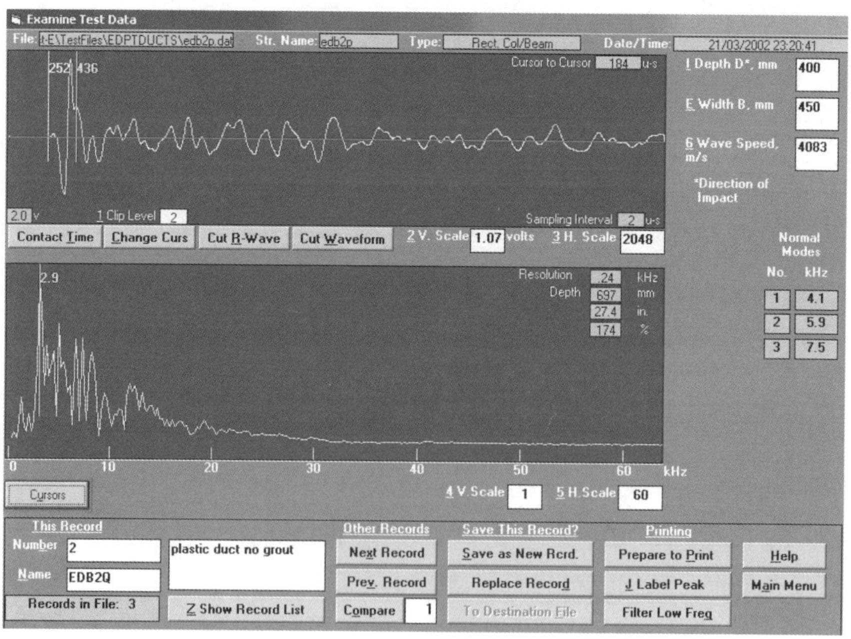

Figure 11. Result of an impact echo test over a fully voided plastic duct.

4 SIBIE METHOD

The SIBIE (Stack Imaging of Spectral Amplitudes Based on Impact-Echo) procedure is an imaging technique for analysing waveforms in the frequency domain (Ohtsu & Watanabe, 2002). In the procedure, first, a cross-section of concrete is divided into square elements as shown in Figure 12. Then, resonance frequencies due to reflections at each element are computed. The travel distance from the input location to the output through the element is calculated as:

$$R = r_1 + r_2. \tag{2}$$

Resonance frequencies due to reflections at each element are calculated from:

$$f_2' = \frac{C_p}{r_2}, \quad \text{and} \quad f_R = \frac{C_p}{R}. \tag{3}$$

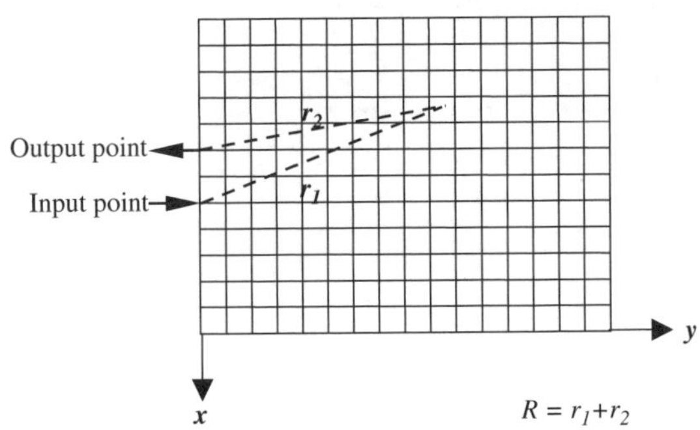

Figure 12. Spectral imaging model.

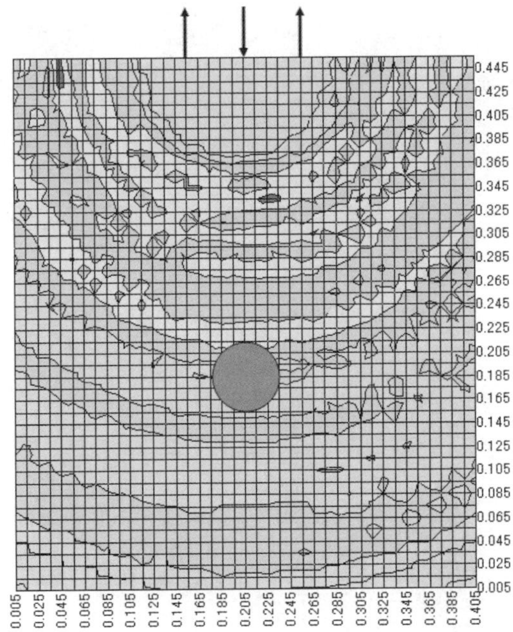

Figure 13. Fully grouted plastic duct beam: distance between the impactor and the receiver = 100 mm.

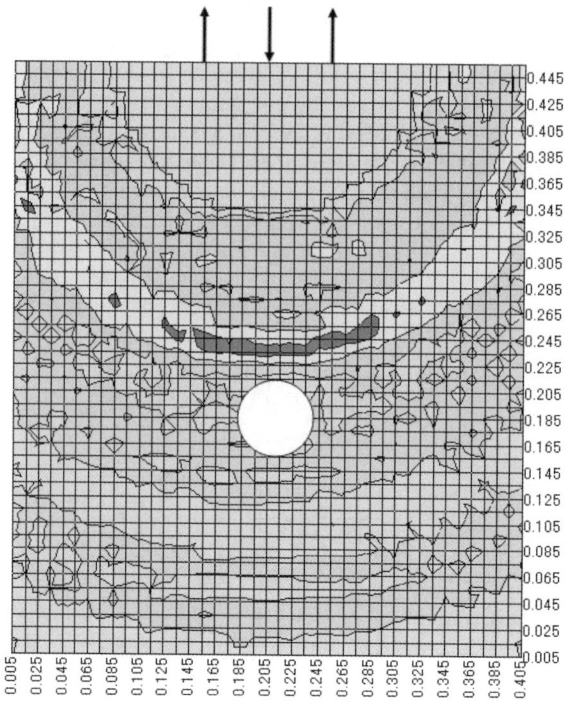

Figure 14. Ungrounted plastic duct beam: distance between the impactor and the receiver = 100 mm.

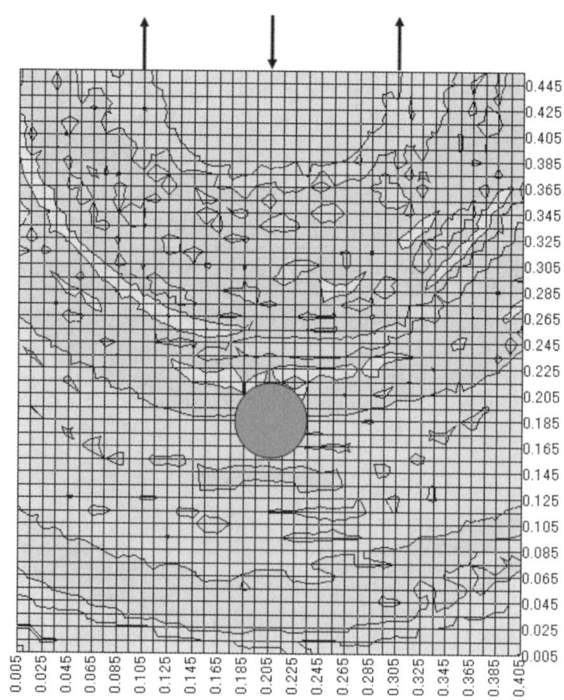

Figure 15. Fully grouted plastic duct beam: distance between the impactor and the receiver = 200 mm.

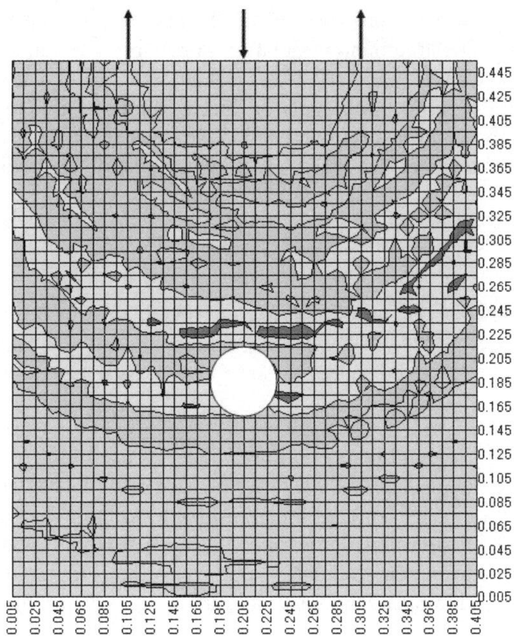

Figure 16. Ungrouted plastic duct beam: distance between the impactor and the receiver = 200 mm.

Spectral amplitudes corresponding to these two resonance frequencies in the frequency spectrum are summed up at each mesh. Thus, reflection intensity is estimated as a stack image at each element. The minimum size of the square mesh Δx for the SIBIE analysis should be approximately:

$$\Delta x = C_p \Delta t/2, \tag{4}$$

where C_p is the velocity of P-wave and Δt is the sampling time of a recorded wave.

Using this technique, the beams at the University of Edinburgh re-tested.

The cases of the distance between the impactor and the receiver = 100 mm.

5 DISCUSSION OF SIBIE DATA

Issues arise in the validation of these techniques as acceptability inevitably depends on the consistency of the results. If these techniques can be proved to be consistently accurate they would be under great demand because of the time and money they could save.

Impact-Echo already offers answers to many structural questions including the assessment of the grout condition in metal ducts. However in recent years, plastic ducts have become common place and more research needs to be undertaken with regard to the grout condition within plastic Post-Tensioned ducts.

The SIBIE method with a distance between the impactor and the receiver = 100 mm, clearly gives the best and least ambiguous results in this set of experiments.

6 THE INTERNATIONAL PERSPECTIVE

6.1 *United Kingdom*

The UK took the problem of tendon corrosion in grouted duct Post-Tensioned concrete bridge beams very seriously when it imposed a moratorium on any further construction (DTP, 1992). This moratorium was lifted once new construction procedures were introduced using plastic ducts (Concrete Society, 1996). The argument was that these new bridges could in the fullness of time be examined using radar (Concrete Society, 1997; Bungey et al, 1997; Giannopoulos et al, 2002). Meanwhile, for the older P-T bridges with metallic ducts – radar was used to identify the ducts with judicious drilling prior to visual inspection. The UK strategy is pragmatic, but expensive, risky and of low statistical significance. The latest UK thinking is summarised in BA86/06 (Highways Agency, 2006).

6.2 *France*

Historically France used to use the high energy X-ray techniques, which can give excellent results but have high health risks in urban areas.

6.3 *Germany*

Germany has used the drilling and inspection technique, as used in the UK. More recently the BAM group (Algernon & Wiggenhauser, 2007) have focused on high level off site signal processing of impact-echo. They have also used radar (not applicable to metallic ducts) and shear wave ultrasonic arrays to enable data fusion. Their general conclusions are that for data fusion, robotic positional accuracy is needed to overlay the data. Contrary to the findings of Sansalone & Streett, (1997), the BAM group find impact-echo unsuccessful on full scale bridges.

6.4 *USA*

In the USA, considerable confidence is shown in the impact-echo technique on post-tensioned concrete bridge beams. However, US companies have the most experience of using this technique. Early US work is summarised in ACI 228.2R-98 (ACI, 1998). The update to this ACI 228 document is due in 2008.

6.4 *Japan*

Japan has focused on refining and developing the impact-echo test interpretation using the SIBIE technique. The procedure looks very promising and is licensed to Japanese industry. The technique has not been adopted in Europe or North America to date.

6.5 *Discussion*

There is still no international standard for the inspection of grouted duct Post-Tensioned bridge beams. However significant, if relatively slow progress is being made towards an internationally acceptable and common approach.

7 CONCLUSIONS

(1) it has been shown that there is a demand for NDT inspection of grouted duct post-Tensioned bridge beams.
(2) It has been shown that impact echo testing of beams with metallic ducts, which are voided, can be inspected using regular impact echo. The USA is more confident of this strategy than Germany.
(3) Sansalone and Streett demonstrated the difficulty of using impact echo to test P-T beams with plastic ducts – confirmed in this paper.
(4) The SIBIE method offers a considerable gain in the interpretation of impact-echo test data from P-T beams with plastic ducts. Japan leads the world in this aspect of testing.

ACKNOWLEDGEMENTS

The authors acknowledge the provision of facilities by the University of Edinburgh, Scotland and Kumamoto University, Japan. They also gratefully acknowledge the funding provided by EPSRC; The Highways Agency, London; and Network Rail Infrastructure Limited. The contributions of many PG and UG students are gratefully acknowledged.

REFERENCES

ACI Technical Report 228.2R-98 (1998) *Nondestructive Test Methods for Evaluation of Concrete in Structures*, ACI, Farmington Hills, MI, USA, p. 62.

Algernon, D. & Wiggenhauser, H. (2007) Impact-Echo Data Analysis Based on the Hilbert-Huang Transform, TRB 2007 Annual Meeting CD-ROM.

Bungey, J.H., Millard S.G. & Shaw, M.R. (1997) Radar assessment of Post-tensioned concrete. *Structural Faults + Repair-97*, Engineering Technics Press, Vol 1, 331–339.

Cavell, D.G. (1997) *Assessment of deteriorating post-tensioned concrete bridges*, PhD thesis, University of Sheffield.

Colombo, S., Giannopoulos, A. & Forde, M.C. (2002) Accuracy of radar testing of masonry arch bridges, *IABMAS'02*, July 2002, UPC, Barcelona, 252–253 + CD-Rom.

Concrete Society Technical Report TR47 (1996), *Durable Bonded Post-Tensioned Concrete Bridges*, The Concrete Society, Slough, UK, p. 64.

Concrete Society Technical Report TR48 (1997), *Guidance on Radar Testing of Concrete Structures*, The Concrete Society, Slough, UK, p. 88.

DTP (1992) *Press Notice No. 260*. Published by DTP, London 1992.

Giannopoulos, A., Macintyre, P., Rodgers, S. & Forde, M.C. (2002) GPR detection of voids in post-tensioned concrete bridge beams, 9th Int Conf, *GPR-2002*, Santa Barbara, CA, 29th Apr–2 May 2002.

Highways Agency (2007) BA86/06: Advice notes on the non-destructive testing of highway structures, *Design Manual for Roads and Bridges*, Vol 3 Section 1, Part 7, p. 247.

Martin, J., (1997), *Non-destructive Testing of Metal Ducted Post-tensioned Bridge Beams using Sonic Impact-Echo Techniques*, PhD Thesis, The University of Edinburgh.

Martin, J., Broughton, K.J., Giannopoulos, Hardy, MSA & Forde, M.C. (2001) Ultrasonic Tomography of Grouted Duct P-T R.C. Bridge Beams, *NDT&E International*, Elsevier Science, 2001, 34, 107–113.

Muldoon, R., Chalker, A., Forde, M.C., Ohtsu, M. & Kunisue, F. (2007) Identifying Voids in Plastic Ducts in Post-Tensioning Prestressed Concrete Members by Resonant Frequency of Impact Echo, SIBIE and Tomography, *Construction & Building Materials*, Vol 21, No. 3, 527–537.

Ohtsu, M. & Watanabe, T. (2002), Stack imaging of spectral amplitudes based on impact-echo for flaw detection, *NDT&E International*, Vol.35, 189–196.

Sansalone, M.J. & Streettt, W.B. 1997, *Impact-Echo: Non-Destructive Evaluation of Concrete and Masonry*, Bullbrier Press, Jersey Shore, PA 17740, 339p., ISBN: 0-9612610-6-4.

Stain, R.T. & Dixon, S. (1993) Inspection of cables in post-tensioning bridge – what techniques are available. Proc. 5th International Conference on *Structural Faults & Repair – 93*, Engineering Technics Press, Edinburgh, UK, 1993. Vol. 1, 297–306.

Woodward, R.J. & Williams, F.W. (1988) Collapse of Ynys-y-Gwas Bridge, West Glamorgan, *Proc. Instn Civ Engrs, Part 1*: Vol 84, August, 635–669.

Chapter 2

Durability of concrete segmental bridges

D.I. Banic, Z. Banic & D. Tkalcic
Civil Engineering Institute of Croatia, Zagreb, Croatia

ABSTRACT: Segmental bridges are most common type road structures built in last thirty years as part of modern roads and highways in Croatia. Fast construction of new highways in almost 10 years was possible only with segmental concrete bridges. Cross section of most bridges consists of prefabricated girders different geometrical shapes such as I, T, H, U and box beams, bulb tee and SAN girders. During regular and preventive bridge inspections some defects were recorded. Deterioration of bridge equipment, drainage, expansion joints, pedestrian lanes, etc., caused significant deterioration of main girders and slabs. Paper shows typical types of concrete segmental bridges built in Croatia, types of construction and condition state of several types of segmental bridges. Condition of those structures was determined during bridge inspections conducted in last five years. Characteristic damages of segmental bridges, rehabilitation procedure applied in several cases and also some improvements on new structures are presented.

1 INTRODUCTION

Period afterward War for Independence was characterized with intensive work on developing adequate road network. Political and economic arguments demanded very fast betterment of existed structures and building new structures. An efficient transportation network was recognized as project of paramount importance for both economic and social development of the country. The construction gained speed and project proved very successful as Croatian motorway network was doubled in just five years. Development of motorway network required construction of large number of motorway structures such as tunnels, bridges, viaducts, overpasses, underpasses, wildlife crossings, etc. due to topographic conditions.

There are five road owners those posses almost 1.200 kilometers of highways and 8000 kilometers roads.

Most of the bridges are short and medium span structures and many of them are similar. Fast construction and very short deadlines prefer the maximum unification and standardization of structures, construction procedure and details. This is a reason why prefabricated concrete bridges make majority of in Croatian Bridge Inventory. In period 1998 to 2006, 700 kilometers of new highways and road network in mountain and maritime areas were constructed and repaired. As part of those highways almost 300 bridges, new viaducts overpasses, underpasses, culverts etc., were built. More than 95 percent of bridges that was designed and constructed were structures defined as simple supported prefabricated beams due to simplest construction but also significant settlements in some areas.

For example, on the whole section of motorway connected Croatian capital Zagreb and main harbor Split on Adriatic coast, there were only five types of bridges and four types of overpasses. Only 10 structures among them were different type of structures or different type of construction excluding two arch bridges. Figure shows motorway network of Republic of Croatia, together with completed and planned motorways.

Table 1 shows number of different type of constructed bridges designed and constructed in period of six years. It can be seen that far most common type of constructed bridge are those consisted of prefabricated girders, almost 500 of this bridges. Figure 1 shows Croatian motorway network with 1019 kilometers of modern roads with further works to complete 1501 km (Radic b). This network

Table 1. The number of bridges built in period of three years.

Year	2001	2002	2003	2004	2005	2006
Arch bridges	–	–	–	–	1	–
Cantilever	–	–	1	1	3	–
Prefabricated Girders	–	70	120	150	90	70
Slab bridge	3	5	10	15	10	8
Culvert	10	20	30	70	50	25
Cable supported	–	–	–	1	–	–

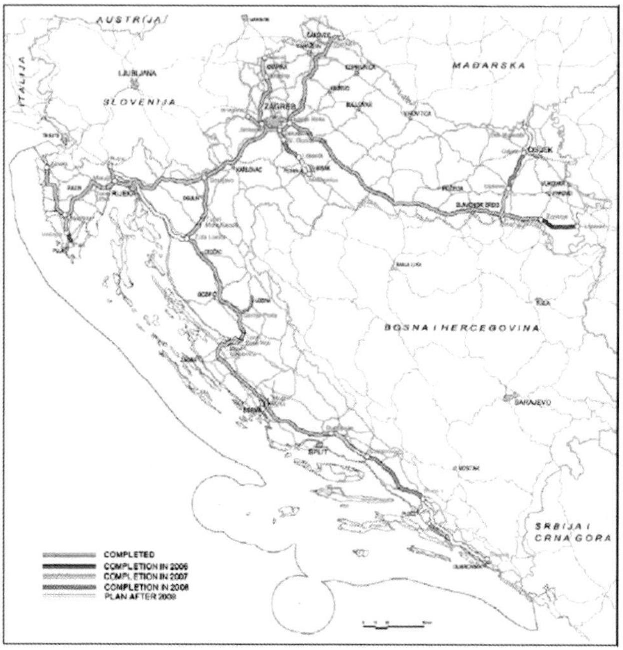

Figure 1. Motorway network in Croatia.

may not seem long but it is worth to mention that Croatia has more kilometers of motorways per 100 000 citizens than UK, Ireland, Greece or even Italy [Radic a].

2 DIFFERENT TYPES OF PREFABRICATED BRIDGES

2.1 *Standardized cross sections*

Crossection of the bridge consists of main girders, that are connected with concrete slab and in some cases cross girders. Figures 2–6 shows crosssections of bridges and viaducts and Figures 7–9 most common crossections of overpasses. Most of the girders have wide upper flange due to lack of any formwork shuttering of the slab between girders (Fig. 3, 4, 7 and 8). In a case of heavy weight girders, spans 40 metres and more, narrow upper flange girders with precast slabs between girders were used. In a course of thirty years, different constructors applied various technology and systems in construction. All those types of girders were produced in great number in very short period of time (U shape girders – cca 2600 pieces in three years, T girders – cca 1800 pieces in four years, SUN girders, 3500 pieces in five years). Girders showed on Figure 6, bulb tee girders, are new in construction work in Croatia but already some advantages were noticed, such as smaller weight, easier transport and assembling, lesser consumption of materials – about 30%, etc.

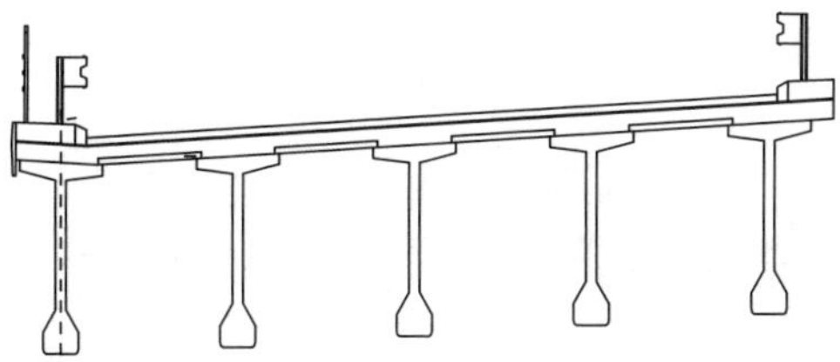

Figure 2. I girders.

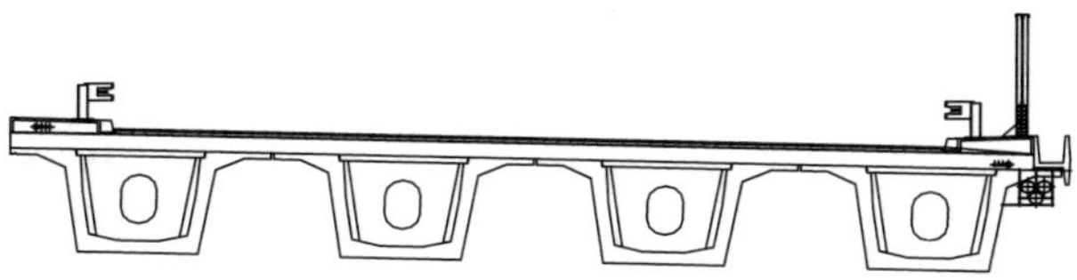

Figure 3. U shape girders.

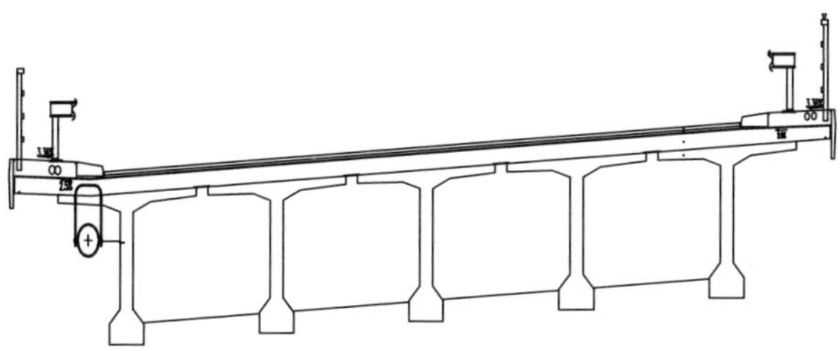

Figure 4. T girders (wide upper flange).

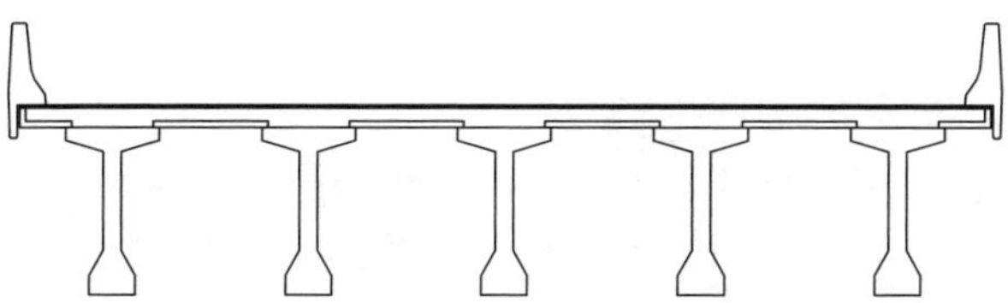

Figure 5. T girders (narrow upper flange).

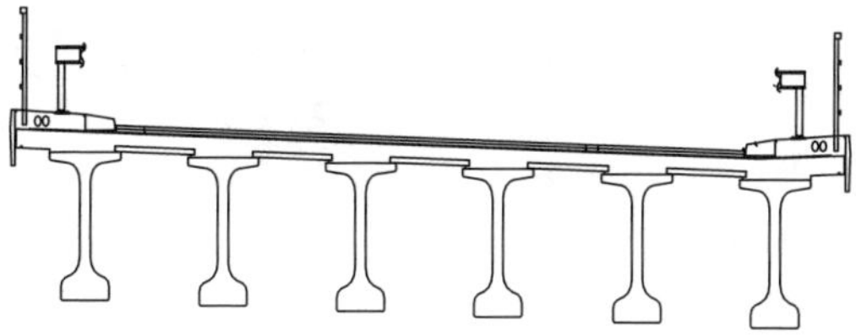

Figure 6. Bulb tee girders.

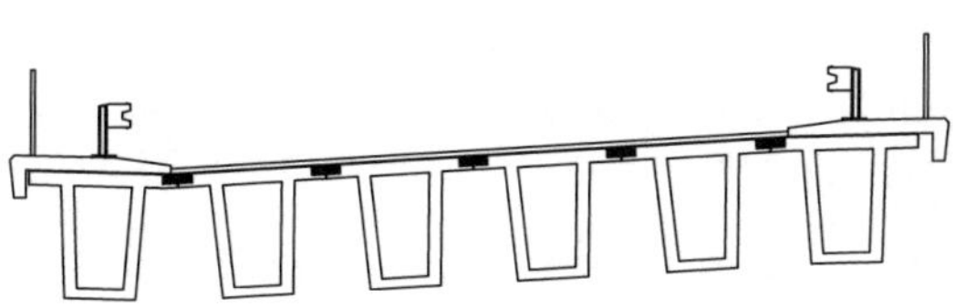

Figure 7. Box girders.

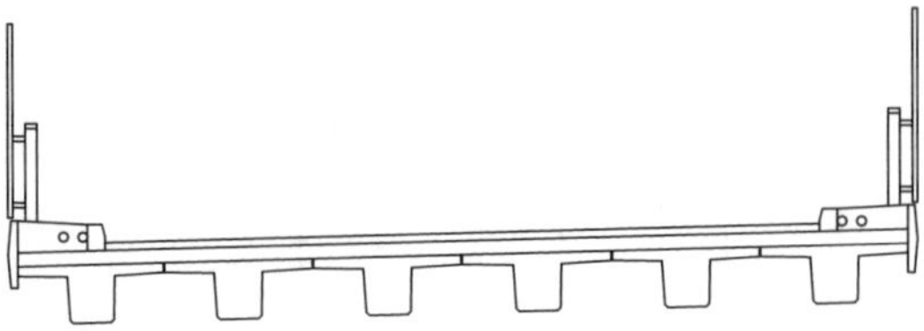

Figure 8. T girders.

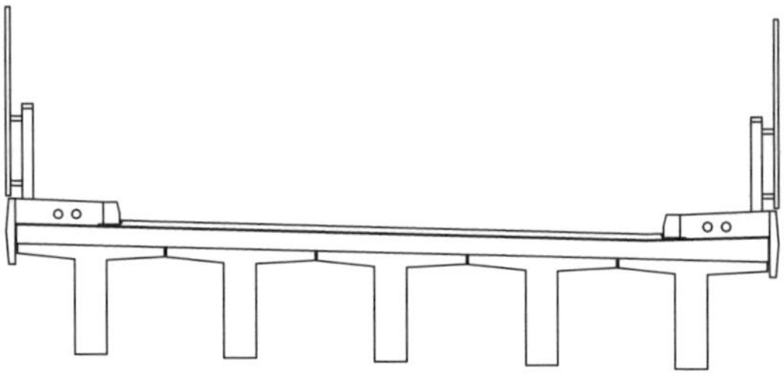

Figure 9. T girders.

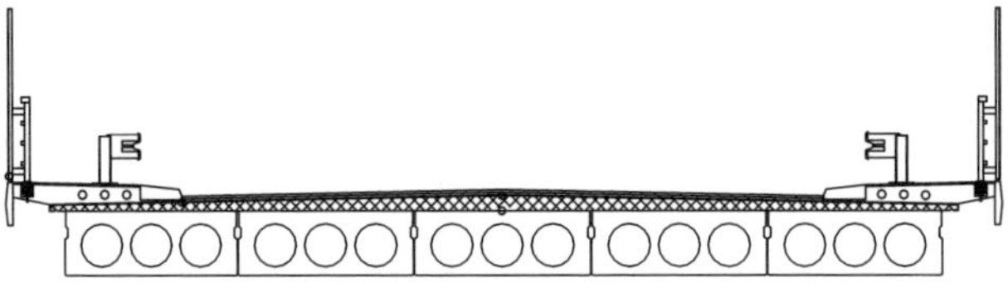

Figure 10. SAN girders.

Figure 11. Precast yard.

Figure 12. Motorway network in Croatia.

2.2 *Construction type*

Speed up construction was able only with production of prefabricated girders in production facilities various contractors or on building sites (Fig. 10). If they were produced in factory, they were transported by the train and then by the trucks to the construction site when they were put in their places either with crane (Fig. 13) on vehicle or with launching truss (Fig. 14) when pies are high. Figures 11 to 14 show typical types of constructions prefabricated girders and cantilever type of construction on Croatian motorways. When everything were going according schedule, each day up to 5 girders have been placed.

Figure 13. Eraction of girders.

Figure 14. Launching truss.

3 BRIDGE CONDITION STATE

3.1 *Bridge inspection*

Regular bridge inspections were not mandatory for bridge owners until five years ago. Inspections were conducted only to assess structural damage and decide what immediate measures have to be undertaken. Today only one bridge owner in Croatia is conducting regular inspections in last ten years.

According Croatian regulation, regular inspection are conducted every 4 to 6 years, depend of bridge condition state determined in previous inspection and bridge environment, maritime or other areas.

Inspections covered the set-up of viaduct elements (bridge equipment, superstructure and substructure, Table 1), inspection of all bridge elements (main and cross girders, deck slab, piers, abutments, pedestrian ways and central reservation, traffic area, drainage system, bearings and expansion joints), making of photo documentation, data processing, damage charting and quantification on the prepared bases. Specimens drilled during the inspection for laboratory testing were analyzed to determine concrete strength, water permeability, gas permeability and chloride content.

These comprehensive actions were undertaken because, in most cases, conducted inspections were first since bridges were released in exploitation.

Every part of bridge was assessed according criteria shown in Table 2 and each bridge was given a final rating at the end of evaluation procedure.

Table 2. Damage classification according HRMOS.

Category	Damage description
0	No damage
1	Minor damage as a consequence of improper construction
2	Minor damage as a consequence of exploitation
3	Damage that affects structural durability, Repairs needed within a certain period of time, preventive measures
4	Damages that will affect structural durability in the near future, immediate repairs
5	Damage that represents a great danger to the safety of the structure, traffic or functionality, restrictions and closing of traffic according to necessity

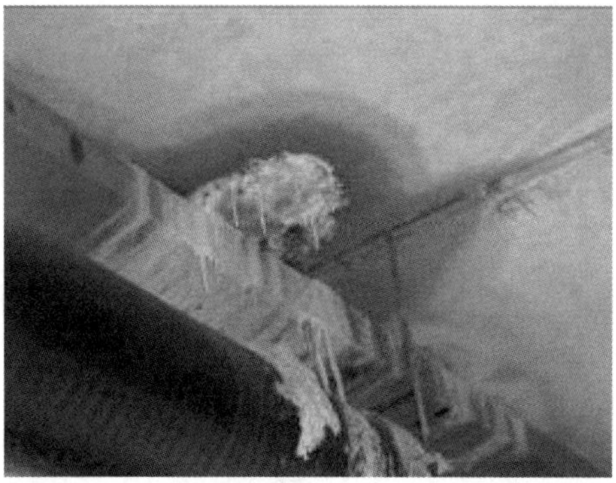

Figure 15. Efflorescence from freezing-and-thawing deterioration.

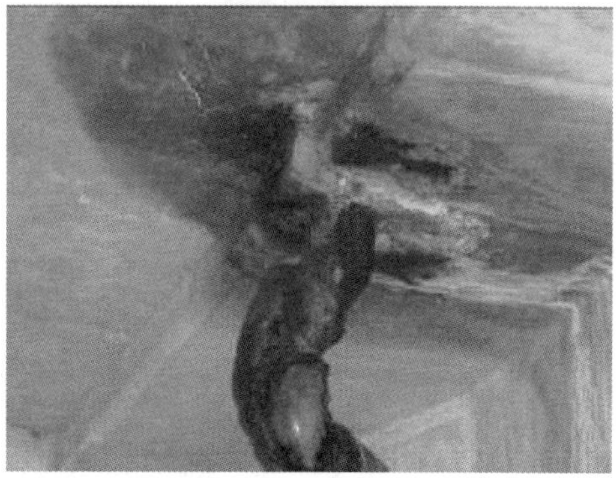

Figure 16. Launching truss.

3.2 *Bridge conditions*

During regular and preventive bridge inspections conducted in last three years some characteristic defects were recorded. Defects were in most cases mostly on bridge equipment, expansion joints, bearings, drainage system etc. Inspected bridges were evaluated with categories 2 and 3 in most cases, bridges showed on Figures 2, 4, 7 and 8, while approximately 5% were evaluated with categories 4 and 5 (Fig. 10).

The most common damages of girders and slabs appeared duo to leakage through water proofing membrane (Fig. 14). Lack of maintenance caused damages on main girders, so some delamination and reinforcement corrosion were visible.

Bridge slabs, on some location, had significant cracks and water leaked through (Fig 17 and 18). The most deteriorated parts of superstructure were those placed under expansion joints because of the failure of waterproofing property of rubber. Rubber parts were not replaced during their service life. It was concluded that prefabricated concrete bridges shown on Figures 2 to 9 are in rather good conditions after almost thirty years only local damages were recorded.

The most damage type of bridge structures were overpasses with main girders called SAN, Figure 10. They were used most frequently as part of the highways. The most deteriorated parts of bridge were girders duo to transversal movement of girders under nonsymmetrical traffic load. This type of loading caused cracking a waterproofing membrane and in this way it was possible leakage and wetting of girders surface.

Damages showed in Figures 19–22 happened in only 25 to 30 years in service. Steel reinforcement corrosion were so severed that about 60 percent of reinforcement missed or has been broken, either in girders or in piers. Concrete covers were delaminated completely in some areas and they

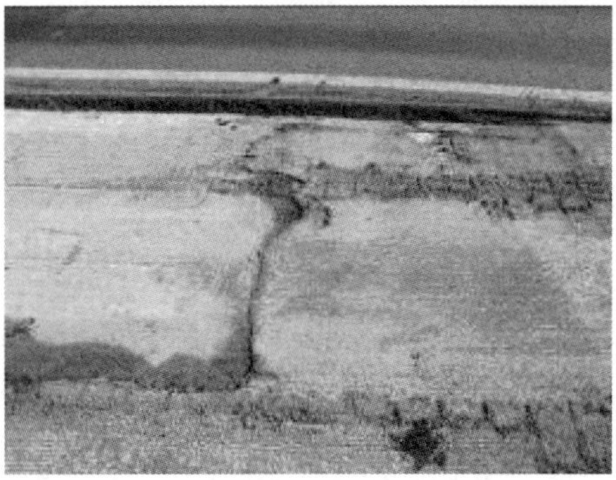

Figure 17. Cracks in slab.

Figure 18. Leakage through Expansion joint.

Figure 19. Cracks in slab.

Figure 20. Broken reinforcement in girder.

Figure 21. Reinforcement corrosion on abutment wall.

Figure 22. Pier reinforcement corrosion.

represented great danger for traffic beneath overpasses (Fig. 19). Prefabricated piers had also severe damages that manifested with significant cracking, delamination and steel corrosion (Fig. 22)

During analysis of findings it was conclude that concrete covers were only 1.5 centimeters that was allowed, at that time, for prefabricated elements.

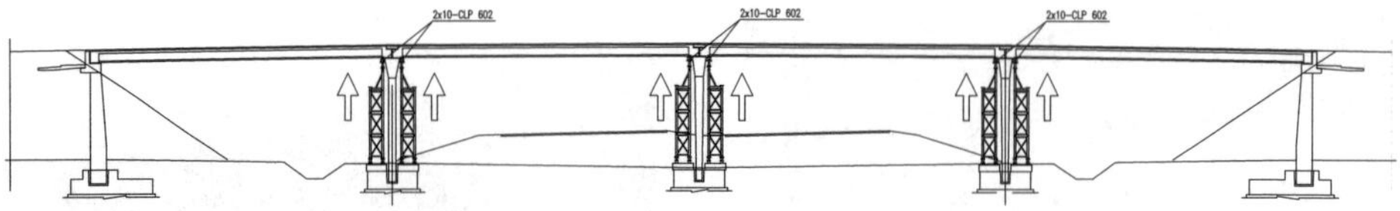

Figure 23. Location of scaffolding.

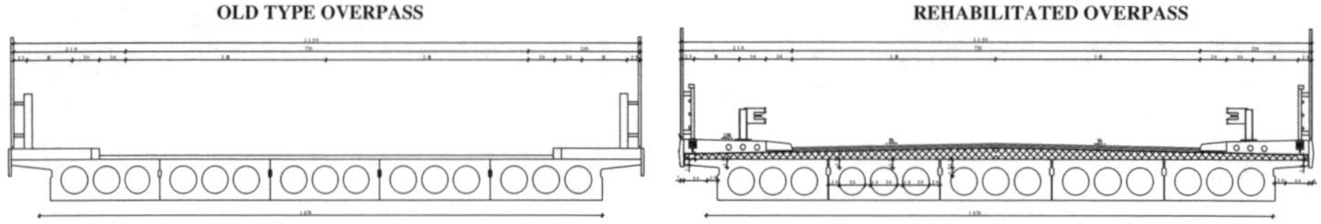

Figure 24. Original and repaired structure.

4 REPAIR METHODS

Rehabilitation of structure was conducted to improve the condition of a structure by restoring and replacing existing components that have been damaged.

Deterioration of non structural parts of bridges and lack of maintenance caused significant deterioration of main girders and slabs. In some cases deterioration were so severe that traffic security and reliability of structure were under minimum requirements. This degree of damages was noticed only 20 to 30 years in service, especially structures wit SAN girders. That is three to four times shorter from planned bridge service life design that amounts 100 years approximately. Some necessary needed measures were conducted. Those measures consisted of completely or partial replacements of structure.

Repairing consists of work on superstructure and work on substructure. Rehabilitation of damaged concrete surface due to freezing and thawing and also piers upgrading was conducted on substructure. Concrete cover of some parts of substructure (piers, pier caps and abutments were removed in layer 5 to 10 centimeters which represented the damaged parts. Since the intensity of traffic increased and since a part of existing pier reinforcement was damaged, additional steel reinforcement to strengthen piers was placed and new cover made.

During process of substructure rehabilitation, overpasses had to be in function through most of the time because most of them are part of important interchanges. Therefore superstructure leaned on scaffolding during repair of substructure (Fig. 23).

Superstructure of those overpasses were in very bad condition so girders were all replaced and bridge deck were added for better transverse load distribution and durability.

Work on superstructure was composed of replacing existing girders, placement of new slab, carriage-way surface, protective facilities, miscellaneous items, road joints and bearings.

In bridge inspection reports, repair measures that would extend service life for 5, 15 and 25 years and their scope were suggested for each bridge owners. Rehabilitation works covered the following: complete repair of bridge equipment, repair of concrete surfaces of girders, piers and abutments and complete replacement of drainage system.

5 CONCLUSION

This article briefly recounts some experience in Croatia regarding use of bridges made of prefabricated girders. Bridges that consist of prefabricated concrete elements manifested very appropriate for fast construction. Condition state of the most of the prefabricated bridges is satisfying except one type of the structure. Bridges that consist of prefabricated elements are also convenient for relatively easy repair even completely replacement of the superstructure. Replacements, conducted on Croatian bridges and overpasses, were performed with non disturbed traffic in reasonably short period.

Concrete segmental bridges are showing given deficiency considering non structural bridge elements that can influenced on durability of structural element. Changes in static system during repair, improvement in details design and especially regular inspection and preventive maintenance can significantly extend service life of bridge and to achieve design age of 100 years.

REFERENCES

Manual for condition evaluation of bridges, 2001, Second edition, American Association of State Highway and Transportation Officials.

Radic, J. 2006a. Croatian achievements in bridge engineering: *Proc. intern. symp., Cavtat, 21-24 May 2006.* Dubrovnik.

Radic J. & co., 2006b. Improvement of bridge structure project: *Proc. intern. symp., Cavtat, 21-24 May 2006.* Dubrovnik.

Chapter 3

Cyclic tests of precast segmental unbonded post-tensioned concrete bridge piers

Y.-C. Ou & G.C. Lee
University at Buffalo, Buffalo, N.Y., USA

P.-H. Wang, M.-S. Tsai & K.-C. Chang
National Taiwan University, Taipei, Taiwan

ABSTRACT: This paper presents an experimental study on the seismic performance of precast segmental unbonded post-tensioned concrete bridge piers. The pier specimen consists of a foundation, four hollow column segments and a pier cap with a height of 5.7 m. The prestressing tendons are located inside the hollow core of the pier column and hence are unbonded with the surrounding concrete. In the first pier specimen, no mild steel reinforcement is extended across the column segment joints. In the other two specimens, longitudinal mild steel bars, also referred to as energy dissipation bars or ED bars, anchored at the foundation and extended up to the pier cap, are added to enhance the seismic energy dissipation. The test results showed that all the pier specimens exhibited satisfactory ductile behavior. The hysteretic energy dissipation, lateral strength and residual drift upon unloading of the specimens increased with the increase of the amount of ED bars.

1 INTRODUCTION

Over the past few years, growing attention has been paid to the investigation, development and application of precast concrete bridge construction for highway bridges. Traditional cast-in-place concrete bridge construction normally causes traffic disruption for a long period of time. Precast concrete bridge construction can offer a viable solution to the problem. It shifts most of the construction activities into precast factories or yards. After adequate concrete strength is obtained, precast concrete products are transported to the construction sites and assembled within a short time, thus reducing traffic disruption. Other advantages of using precast concrete bridge construction as opposed to traditional cast-in-place construction include increasing work zone safety, improving construction quality and reducing environmental impact.

There have been numerous bridge construction projects that successfully used precast concrete construction in superstructures, substructures or both. This research focuses on precast segmental post-tensioned concrete bridge pier construction. Precast segmental bridge pier construction has been used in a number of construction projects in the regions of low seismicity in the U.S. Victory Bridge in New Jersey (NJDOT 2005) and Colorado River Bridge of Hoover Dam Bypass in Nevada (Goodyear et al. 2006) are two recent examples. The use of precast segmental bridge pier construction in the regions of high seismicity is still limited due to the concern regarding the seismic performance of such type of pier construction.

Existing precast segmental bridge piers normally have prestressing steel as the only steel reinforcement across the column segment joints. Longitudinal mild steel reinforcement is usually discontinuous at the column segment joints. Past experimental studies have concluded that this type of precast segmental bridge piers have excellent ductility and minimal residual displacement upon unloading but little hysteretic energy dissipation (Chang et al. 2002, Hewes and Priestley 2002).

The addition of longitudinal mild steel reinforcement across the column segment joints can significantly increase the hysteretic energy dissipation as well as increase the lateral strength of the columns (Chang et al. 2002). Recently, the analytical study by Ou et al. (in press) has shown that by the proper combination of longitudinal mild steel reinforcement, dead load, and unbonded post-tensioning force, precast segmental bridge piers can achieve optimum flag-shape hysteretic behavior with satisfactory energy dissipation and small residual displacement upon unloading.

On the basis the studies by Chang et al. (2002) and Ou et al. (in press), this research designed and constructed three large scale precast segmental unbonded post-tensioned bridge pier specimens. The pier specimens were subject to lateral cyclic loading to investigate their seismic performance.

2 SPECIMEN DESCRIPTION

2.1 *Specimen design*

The specimen has a height of 5.7 m and consists of a foundation, four precast column segments with hollow cross section and a precast pier cap. The size of the foundation is 1.51 m × 2.45 m × 1.2 m (width × width × height). The size of the precast column segment is 0.86 m × 0.86 m × 0.9 m (width × width × height). The wall thickness of the column segment is 0.2 m. Figure 1 shows a typical precast column segment. As shown in the figure, 12 corrugated galvanized steel ducts with diameter of 80 mm are precast with the column segments. Longitudinal mild steel reinforcement, also referred to as energy dissipation (ED) bars, will be inserted through these ducts during the assembling of the column and then pressure grouted. The joint shear friction resistance was calculated to be greater than the maximum shear demand. Thus, no shear key is designed at the segment joint. The pier cap has a span of 3.86 m, a width of 0.86 m and a depth of 0.9 m at the mid span and 0.6 m at two ends of the span. The axial force and the amount of ED bars of the specimens are shown in Table 1. Three pier specimens are designed, namely, C0C, C5C and C8C. The designed dead load is equal to $0.1 f'_c A_g$ and prestressing force equal to $0.075 f'_c A_g$, resulting in a total axial force of $0.175 f'_c A_g$. f'_c is the design concrete compressive strength, i.e. 28 MPa. A_g is the gross cross-section area of the column, i.e. 0.53 m^2. The major design variable is the ED bar ratio, i.e. the ratio of the area of the ED bars to the gross cross-section area of the column. The ED bars are continuous across column segment joints and intended to enhance the hysteretic energy dissipation capability of the columns. The ED bar is made of A706 Grade 60 steel. Specimen C0C has no ED bar to serve as a benchmark. Specimen C5C and C8C have 0.5% and 1% ED bar ratio, respectively.

Each post-tensioning tendon is comprised of two seven-wire strands. Each strain has a nominal diameter of 15.24 mm. The prestressing steel is equivalent to ASTM A416 Grade 270. The tendons

Figure 1. Typical precast column segment.

Table 1. Axial force and ED bar ratio.

Specimen	Dead load (kN)	Prestressing force (kN)	ED bar ratio (%)
COC	1456	1092	0
C5C	1456	1092	0.5
C8C	1456	1092	1.0

Table 2. Material properties.

Material properties	Concrete (MPa)	Grout (MPa)	Prestressing strand (MPa)	ED bar (MPa)
Compressive strength	28	50	N.A.	N.A.
Tensile yield strength	N.A.	N.A.	1670	412
Tensile peak strength	N.A.	N.A.	1860	620

are unbonded with the surrounding concrete to decrease the inelastic straining of the tendons as the columns are push laterally. Considering the advantages in inspection and replacement, external unbonded post-tensioning tendons are adopted instead of internal tendons. The tendons are located inside the hollow core of the column segments and anchored at the top of the pier cap and the bottom recess of the foundation. The material properties are listed in Table 2.

2.2 *Specimen construction*

Considering the height of the specimen, the ED bars are divided into two parts, connected by mechanical couplers when assembled. The lower part of the ED bars are precast with the column segment at the base of the column, protruding from the top and bottom joint interfaces of the segment. The portion of the bars protruding from the bottom joint interface was inserted into the grouted corrugated steel duct precast in the foundation during the installation of the base column segment.

The assembling of the precast column segments began by positioning the bottom of the column segment at the base of the column into the recess on the top of the foundation. After the installation of the segment, the prestressing tendons were inserted through the ducts in the foundation. The other three column segments were then placed sequentially on the top of the previous one. The ED bars were coupled inside the second column segment, which was the one on the top joint interface of the column segment at the base of the column. Figure 2 shows the assembling of the fourth column segment. As can be seen in the figure, the ED bars traveled through the corrugated steel ducts while the tendons through the hollow core of the segment. The pier cap was placed on the top of the fourth column segment, as shown in Figure 3. After the pier cap was installed, the tendons were stressed with hydraulic jacks. The ducts for the ED bars were pressure grouted from the grouting inlets located at the second column segment. Grout was pumped through the grouting inlets to the top of the pier cap. Each batch of grout was monitored with the flow cone test (ASTM C939). Only when the flow of the grout was within 40 to 60 second, the grout was allowed to be pumped. Figure 4 shows the pier specimen prior to testing.

3 TEST SETUP AND LOADING SCHEME

The foundation of the pier specimen was tied down to the strong floor with four steel bars. Two vertical actuators were used to apply the dead load to the specimens. Each vertical actuator generated

Figure 2. Assembling of column segments.

Figure 3. Assembling of pier cap.

Figure 4. Pier specimen prior to testing.

a constant force of 728 kN throughout the tests. One horizontal actuator at one end was mounted on the reaction wall and at the other end attached to the pier cap. The actuator applied the lateral cyclic loading to the pier cap. The cyclic loading was applied under displacement-control to the drift levels of 0.25%, 0.375%, 0.5%, 0.75%, 1.0%, 1.5%, 2.0%, 3.0%, 4.0%, 5% and 6%. Each cycle was repeated twice to allow for the observation of strength degradation under repeated loading with the same amplitude. The drift is defined as the lateral displacement divided by the height of the loading point to the top surface of the foundation, that is, 4030 mm.

4 TEST RESULTS

The test results of the three specimens are listed in Table 3. Figure 5, 6 and 7 show the hysteretic behaviors of the three specimens. The equivalent viscous damping ratio, ζ_{eq}, is defined in Equation 1 and 2.

$$\zeta_{eq} = \frac{E_D}{2\pi K_{eff} d_{max}^2} \tag{1}$$

$$K_{eff} = \frac{f^+ - f^-}{d_{max} - d_{min}} \tag{2}$$

Table 3. Test results.

Specimen	Maximum energy dissipation, ζ_{eq} (%)	Maximum residual drift (%)	Lateral strength (kN)	Yield drift (%)	Failure drift (%)	Ductility
C0C	6	0.2	278	0.33	4.6	15
C5C	16	0.4	363	0.42	6.0	14
C8C	22	2.9	535	0.60	6.0	10

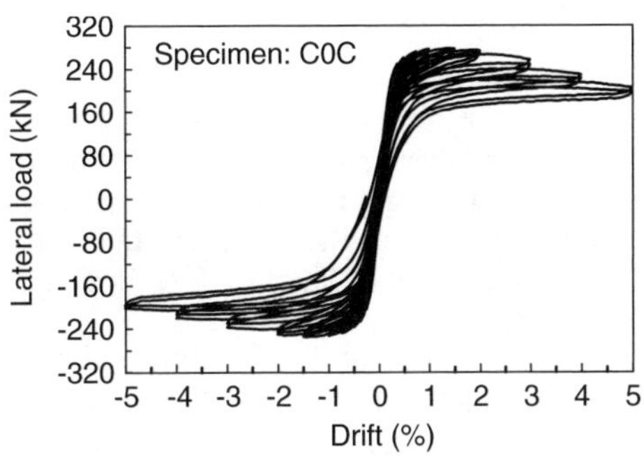

Figure 5. Hysteretic behavior of specimen C0C.

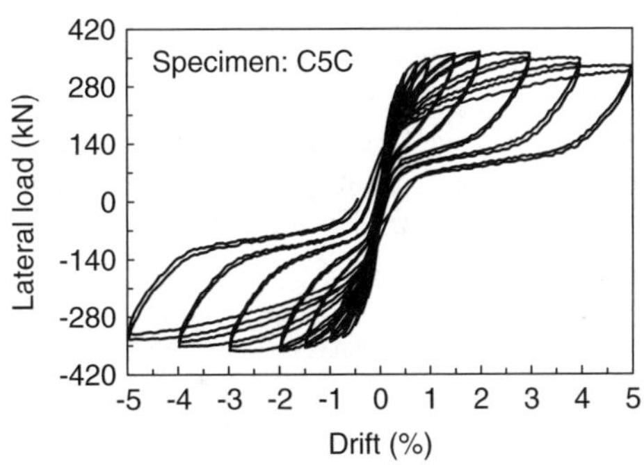

Figure 6. Hysteretic behavior of specimen C5C.

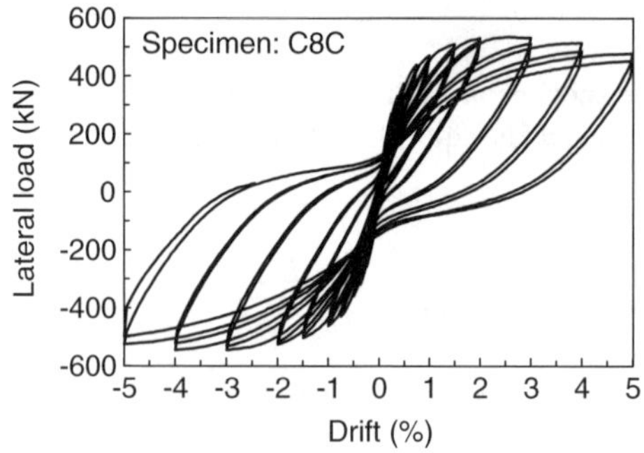

Figure 7. Hysteretic behavior of specimen C8C.

Where E_D = energy dissipation for a cycle of loading, which is equal to the area of the hysteresis loop corresponding to that cycle; K_{eff} = effective stiffness; d_{max} = maximum displacement of the loop; d_{min} = minimum displacement of the loop; f^+ = force at the maximum displacement; and f^- = force at the minimum displacement.

The test result showed that specimen C0C had the smallest hysteretic energy dissipation with a maximum ζ_{eq} of 6% and smallest lateral strength among the three specimens. However, C0C had the best ductility capacity and the smallest residual drift upon unloading. At the end of the test, C0C had a maximum residual drift of only 0.2%. Small residual drift means the column can maintain functionality after a seismic event. It becomes an increasingly favorable characteristic over time in the earthquake research community, since the post-earthquake serviceability of a bridge is important. This is because bridges are often critical for earthquake relief effort to reach the earthquake disaster area. Another advantage of C0C is that it has the fastest construction speed because there is no ED bar in C0C. This characteristic is important since accelerated bridge construction to reduce traffic disruption is one of the motivations of this research.

The maximum hysteretic energy dissipation of specimen C5C in terms of ζ_{eq} was significantly increased to 16% as compared to that of specimen C0C. In the meantime, C5C still had small residual drift upon unloading, with a maximum residual drift of 0.4%, slightly larger than that of C0C. C5C demonstrated a good example of having increased hysteretic energy dissipation while still keeping the residual drift small.

The test results showed that C8C had the highest hysteretic energy dissipation with a maximum ζ_{eq} of 22%. Moreover, C8C had the highest lateral strength because it had the highest amount of ED bars. However, the maximum residual displacement was also significantly increased to a maximum value of 2.9%. Due to its high hysteretic energy dissipation and high lateral strength, C8C is expected to have the smallest size of column cross section among the three types of piers tested under a given design seismic force.

The yield drift and failure drift of the specimens are listed in Table 3. Yield drift is defined as the drift associated with significant softening of the column. The failure drift is defined as the drift at which the lateral strength of the column drops below 80% of the peak lateral strength. Specimen C0C failed due to the P-delta effect while C5C and C8C failed due to the fracture of the ED bars. It is clearly shown that all three piers have satisfactory ductility capacity with ductility factors ranging from 10 to 15. The ductility is defined as the failure drift divided by the yield drift.

5 CONCLUSIONS

Three large scale precast segmental unbonded post-tensioned pier specimens with hollow column segments were designed and tested. The test results showed that all the three pier specimens exhibited satisfactory ductile behavior under lateral cyclic loading. Specimen C0C has no ED bar.

The test results show that C0C has the highest ductility capacity, the smallest residual drift upon unloading and the fastest construction speed among the three specimens. However, C0C has the smallest hysteretic energy dissipation capacity and lateral strength. Specimen C5C has 0.5% ED bar ratio. The hysteretic energy dissipation capacity of C5C is significantly higher than that of C0C. Meanwhile, C5C still has small residual drift upon unloading. Specimen C8C has 1% ED bar ratio. C8C possesses the highest hysteretic energy dissipation and lateral strength. However, the residual displacement of C8C upon unloading is the highest.

REFERENCES

Chang, K. C., Loh, C. H., Chiu, H.S., Hwang, J. S., Cheng, C. B., & Wang, J. C. 2002. *Seismic behavior of precast segmental bridge columns and design methodology for applications in Taiwan*, Taipei, Taiwan: Taiwan Area National Expressway Engineering Bureau.

Goodyear, D., Klamerus, B. & Turton, R. 2006. *Structural Magazine*, January issue: 29–31.

Hewes, J. T., & Priestley, M. J. N. 2002. Seismic Design and Performance of Precast Concrete Segmental Bridge Columns, *Report No. SSRP–2001/25*, San Diego, C.A.: Department of Structural Engineering, University of California.

NJDOT. 2005. Route 35 Victory Bridge Overview. < http://www.state.nj.us/transportation/commuter/roads/rt35victory/>.

Ou, Y.-C., Chiewanichakorn, M., Aref, A. J., & Lee, G. C. 2007. Seismic performance of segmental precast unbonded post-tensioned concrete bridge columns, *Journal of Structural Engineering, ASCE.* (in press).

Chapter 4

Inspection and rehabilitation of Jamestown-Verrazzano segmental concrete bridge

Michael J. Abrahams & Steven Kaufman
PB Americas, Inc., New York City, N.Y., USA

Paul Fisk
NDT Corporation, Worcester, M.A., USA

ABSTRACT: The Jamestown-Verrazzano Bridge over Narragansett Bay, Rhode Island, features a 4,950-foot-long prestressed segmental box girder main bridge with 23 spans varying in length from 109 to 636 feet, and a 2,402-foot-long trestle structure. The bridge was open to traffic in 1992 and a baseline inspection was conducted in 1999. The scope of the baseline inspection included analysis, load rating, and comparison of creep deflections based upon as-built shop drawings, and casting and stressing schedules versus field-surveyed conditions.

During subsequent inspections nondestructive and destructive methods were used to investigate the post-tensioning ducts for the presence of voids. Over 93,000 linear feet of nondestructive impact-echo (sonic/ultrasonic) measurements were taken on the concrete top slab, webs and bottom slab containing the tendons to evaluate the grouted tendon ducts for voids. Of the approximately 1,520 tendon ducts tested, 7.5% or 114 tendon ducts were determined to have voids. Void lengths ranged from one foot to over 314 feet. In most cases the tendons were grout covered but some of the tendons were exposed and exhibited corrosion.

At the time of construction, grouting methods were not always fully effective and voids in post-tensioned ducts are now an issue for a number of bridge owners. Other repairs include use of epoxy-injection with CFRP reinforcement system for the cracked webs of the segmental box girder pier tables. The paper will discuss these findings and repairs that are underway.

1 INTRODUCTION AND DESCRIPTION OF BRIDGE

1.1 *General*

The Jamestown-Verrazzano Bridge is a 7352-foot long structure that carries State Highway 138 in an east-west direction across the West Passage of Narragansett Bay between the towns of North Kingstown and Jamestown, Rhode Island. [See Figure 1] The bridge consists of two portions, designated as the Main Structure and the Trestle Structure, respectively. The Main Structure is further subdivided into three portions: the West Approach Spans, the Main Spans and the East Approach Spans. Construction of the Jamestown-Verrazzano Bridge began in 1985 and was completed in 1992 when the bridge was opened to traffic. The 7-year construction period resulted primarily from problems associated with the driven pile foundations.

1.2 *Main structure*

The Main Structure [See Figures 2 to 4] has 23 spans consisting of the West Approach (Span 1 to Expansion Hinge No. 1 in Span 11), the Main Spans, (Expansion Hinge No. 1 to Expansion Hinge No. 2) and the East Approach (Expansion Hinge No. 2 in Span 16 to Span 23). It begins at Pier 1, which supports Span 29 of the Trestle Structure to the west and Span 1 of the Main Structure to the east, and extends 4950 feet from Pier 1 over Piers 2 through 23 and terminates at the East Abutment located on Conanicut Island. The lengths of the East and West Approach Spans are 1368 and 1818 feet, respectively. The Main Spans are 1764 feet in length and include Piers 13 and 14, which are the only two substructure units that are double-stemmed and integral with the superstructure. The remaining piers receive the superstructure loads through pairs of either fixed or guided expansion pot bearings.

The navigation channel runs beneath Span 13 and has a minimum design clearance envelope at midspan of 135 feet above mean high water (MHW) by 300 feet wide. Five sets of navigation light

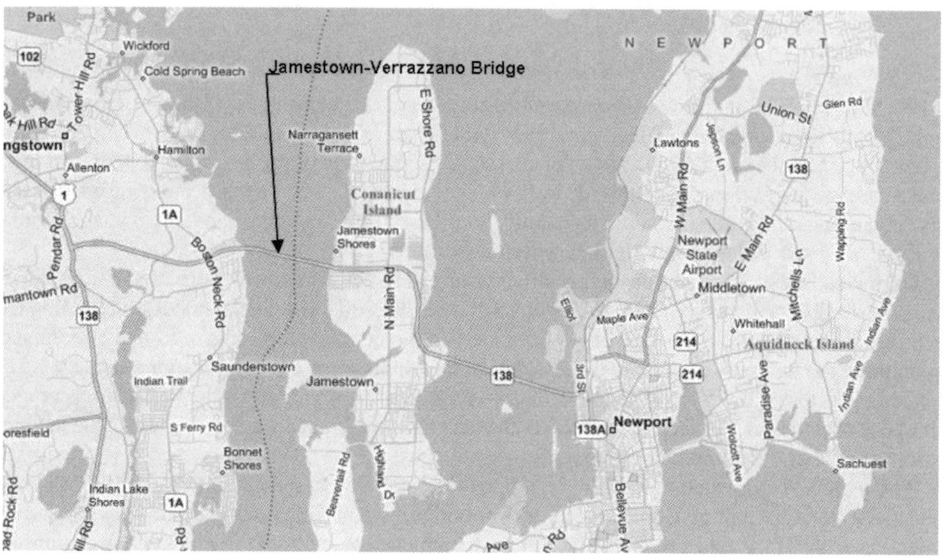

Figure 1. Bridge location.

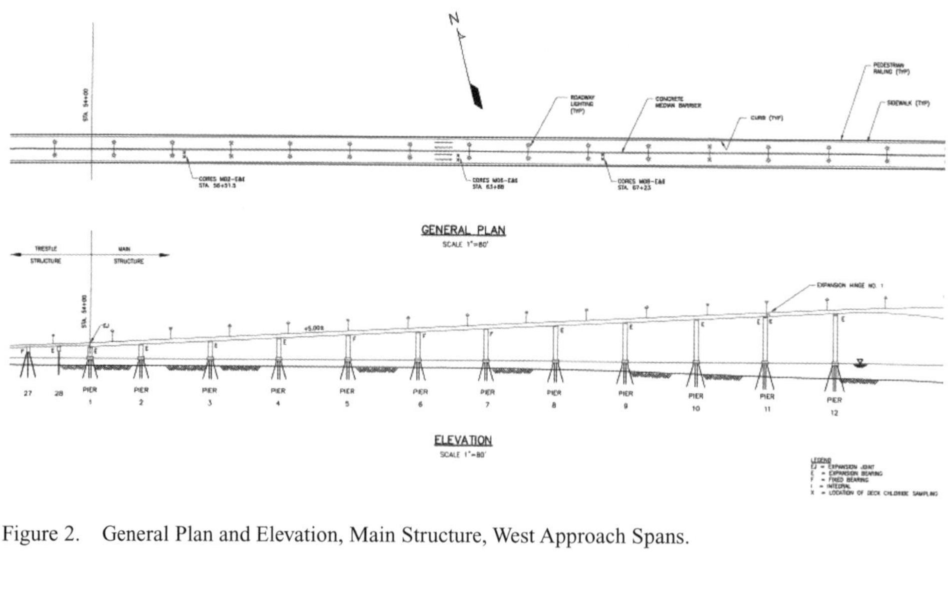

Figure 2. General Plan and Elevation, Main Structure, West Approach Spans.

fixtures are present on each face of this span and consist of one green lamp at midspan, two red lamps at the limits of the clearance envelope, and two additional green lamps located between the red and each outer green lamp.

The West Approach rises toward the Main Spans at a constant 5% grade until reaching approximately the mid-length of Span 11. Similarly, the East Approach descends from the Main Spans starting approximately from the mid-length of Span 15 at a constant 5% grade until reaching the East Abutment. Between the points of vertical curvature and tangency noted above, the structure grade line transitions between the approach spans through a 1600-foot long vertical parabolic curve.

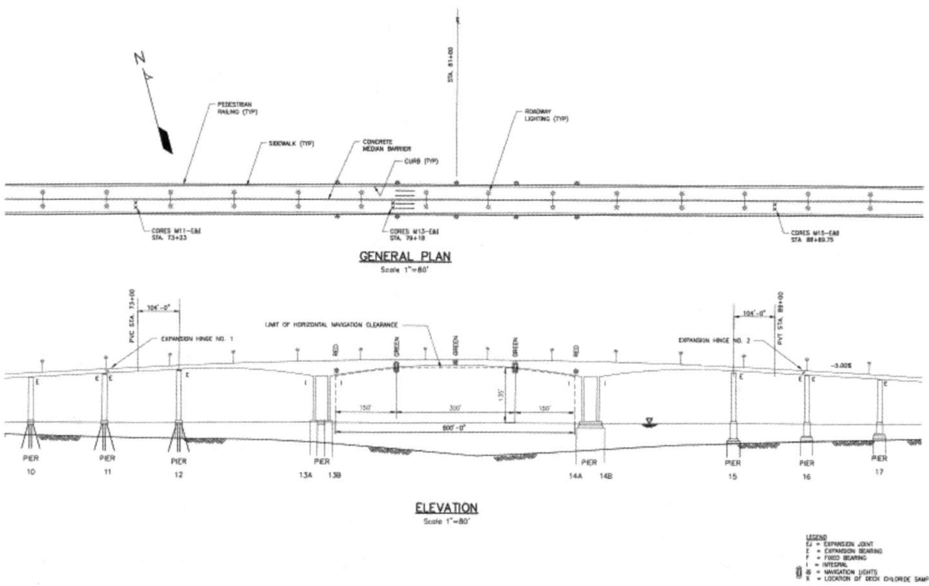

Figure 3. General Plan and Elevation, Main Structure, Main Spans.

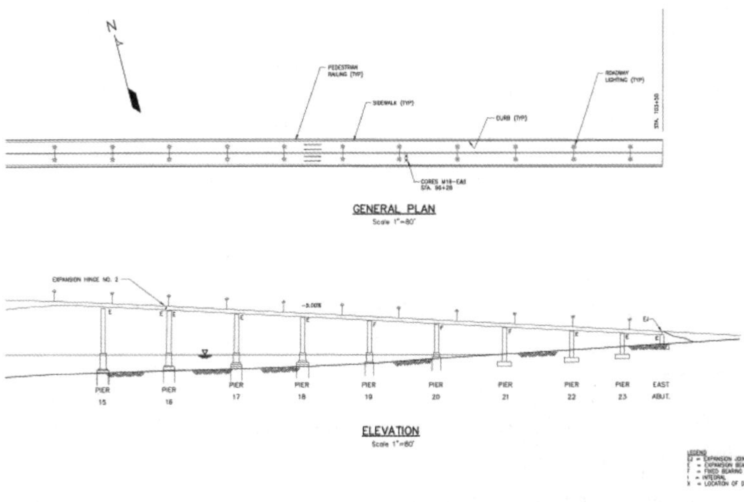

Figure 4. General Plan and Elevation, Main Structure, East Approach Spans.

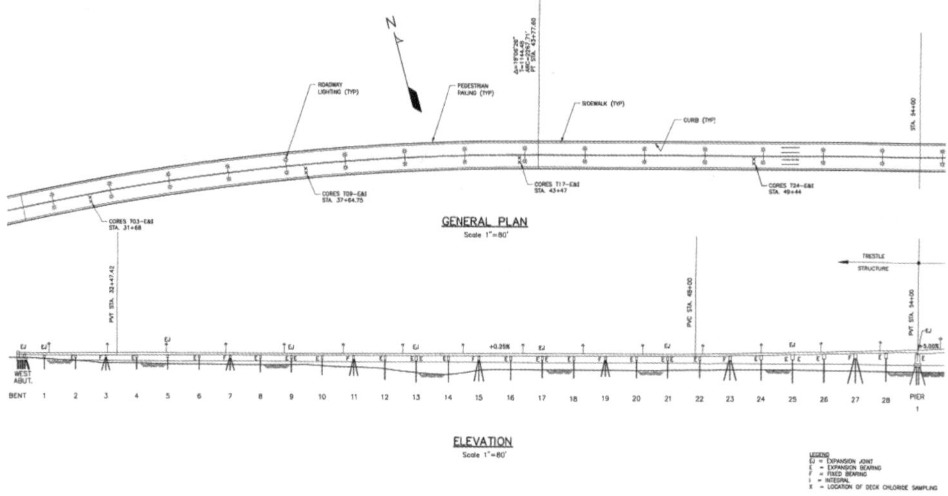

Figure 5. General Plan and Elevation, Trestle structure.

1.3 *Trestle structure*

The Trestle Structure [See Figure 5] is located to the west of the Main Structure. It begins on the mainland in North Kingstown at the West Abutment and extends into the shallow portion of Narragansett Bay for 2405 feet to Pier 1 where the West Approach Spans of the Main Structure begin. The horizontal alignment of the western portion curves as it progresses from the West Abutment into Narragansett Bay turning from an easterly to an east-southeasterly orientation. The horizontal curve reaches the point of tangency near the end of Span 17. The remainder of the structure is tangent through Span 29 at Pier 1. A mild vertical rise of 0.25% is present from the West Abutment until the end of Span 21 where a 600-foot long parabolic vertical curve transitions the structure to a rising grade of 5%, which continues onto the Main Structure.

The Trestle Structure consists of twenty-nine composite concrete girder spans. Span 1 is situated between the West Abutment and Bent 1, and consists of nine simply-supported AASHTO Type IV precast, prestressed girders that are 50'-9" in length. The remaining 28 spans are divided into seven 336-foot long units, each consisting of four 84-foot long spans supported on pile bents. Nine AASHTO Type IV precast concrete girders, initially installed as simply supported and later made continuous for live loads over the four spans, comprise the framing system of each unit.

The West Abutment is constructed from reinforced concrete and is founded on HP12 × 53 piles. The remainder of the structure sits atop reinforced concrete cap beams supported by vertical and battered HP14 × 117 piles jacketed within concrete-filled 36-inch diameter prestressed concrete cylinders. The H-piles and jackets are monolithically connected to the cap beams. There are three bent configurations within each four-span continuous unit:

- A central bent supported by ten battered and two vertical piles. Fixed neoprene bearings sit on the bearing seat of the cap beam.
- Two interior bents – one on each side of the central bent – supported by six vertical piles and two end piles battered normal to the bridge centerline. Expansion neoprene bearings sit upon the bearing seat of the cap beam.
- Two exterior bents – one adjacent to each interior bent – supported by six vertical piles and two end battered piles. Expansion neoprene bearings sit upon the bearing seat of the cap beam. Adjacent four-span units share these bents.

The deck is constructed of concrete reinforced with epoxy-coated rebar, which is used for the full length of the structure. Expansion joints are located between each four-span continuous unit, including at the end of the structure located at Pier 1. At the West Abutment, the superstructure is seated on fixed elastomeric bearing pads.

2 INSPECTION AND ANALYSIS

2.1 *Condition inspection, analysis and design*

In January 1999, PB Americas, Inc. (PB) was engaged by the Department of Transportation of the State of Rhode Island and Providence Plantations (RIDOT) to perform the 1999 Baseline Inspection, which was the first post-construction inspection of the structure, and subsequent biennials were performed in 2001, 2004 and 2006. Portions of these inspections were performed by specialty firms:

- 2001, 2006: Subaqueous inspection of substructure and bottom
- 2004 through 2006: Tendon investigation

In addition, as part of the inspections, the following analyses of the structure were performed:

- Live load rating
- Seismic analyses
- Ship collision analysis
- Scour analysis

Several design contracts to address conditions identified during the inspections and as a result of the aforementioned analyses encompassed the following principal items:

- Replacement of modular expansion joints
- Bearing reinforcement
- Installation of scour monitoring system
- Epoxy injection of cracks in the box girder superstructure
- Reinforcement of pier table webs with carbon fiber-reinforced polymer (CFRP)
- Vacuum grouting of tendons

A few of these items will be discussed within the sections that follow.

3 CONDITION INSPECTION

3.1 *Planning*

The size and complexity of this type of structure typically warrant careful preparation. Thorough preparation of field notes and advance identification of critical structural elements are key to successful execution. On this structure, three different construction methods were utilized, which, in turn, affected the progression of stressing that occurred, the post-tensioning details installed, and the critically stressed locations within these respective spans. This information must be known prior to the inspection to recognize the importance of deficiencies that may be observed.

Inspection of the interior often presents special safety issues:

- Lighting and electrical power are typically needed for inspection equipment. Their absence requires carrying a power source and lighting into the structure along what may be a considerable distance. For this bridge, the Department had added interior lighting and standard 120-volt AC outlets after construction was completed. These proved to be invaluable aids to the work.
- Because there is a significant enclosed volume, confined space issues may need to be addressed if the box is not properly ventilated. This would usually take the form of air testing and, if necessary, opening access hatches for a period of time prior to inspection to vent noxious gas accumulation, which may be accelerated via the use of fans.
- Access may be complicated by the depth of the sections, which, at the piers, may be significant within longer spans. While ladders could be used, they are often cumbersome and slow. Further, the access provisions may limit the size of hardware that can be brought into the box.

Figure 6. UB-60 operation over navigational channel.

Figure 7. Barge and manlift operation.

3.2 *Execution*

The exterior surfaces of the Main Structure located below the deck, including the flanges, webs and bottom slab were inspected using an ASPEN UB-60 under-bridge inspection unit. [See Figure 6] This unit also provided access to the bearings, seats and bifurcated stems of the higher piers that could not be accessed from below. The number in UB-60 designates the maximum horizontal reach of the unit in feet. These units consist of a control basket attached to a set of booms with the capability of extending and rotating. The apparatus is mounted on a counterweighted, rotating platform atop a flatbed truck. Situated adjacent to the curb of the outer shoulder, these vehicles provided complete access to the underside of the superstructure and portions of the substructure. The extended reach of the UB-60 facilitated the inspection of the deep portions of the Main Spans, and the bearings and upper portion of the bearing seats of the Main Structure Approach Spans.

An 80-foot barge-mounted manlift provided access to the lower half of piers while the top surfaces were inspected using the UB-60. [See Figure 7] The tops of the pile caps and the vertical faces of their protective granite panels were accessed via barge. A 60-foot manlift was used to

Figure 8. Interior inspection utilizing 30-foot scaffolding (Not shown to full height).

inspect the East Abutment, the land-based piers and the exterior surfaces below the deck for Spans 23, 22, and most of Span 21.

The interior of the Main Structure twin-celled concrete box was accessed by climbing a 5-foot ladder from the ground to the bifurcation point at the top of Pier 23. This was followed by climbing up existing ladders to the inside of the box through circular portals present in the bottom slab of the superstructure pier segment. Inspection of the interior was conducted by walking through it along the bottom slab. Several hand-held high-intensity halogen lights supplemented existing fluorescent lighting within the box girder. One hundred and twelve-volt AC electrical outlets installed along the faces of the interior webs supplied power for the lamps.

Interior surfaces that could not be reached from the bottom slab of the box were accessed via eight custom-made light-weight aluminum, wheel-mounted scaffolding units. [See Figure 8] These units were purchased as part of the initial inspection and have remained inside the boxes for subsequent inspections. Four of the units are approximately 8 feet in height and facilitated the inspection of the ten-foot deep Approach Spans. The remaining four units are approximately 30 feet high when fully assembled, and provided access to the wall and soffit of the Main Spans.

The latter taller units generally required three to four inspectors to safely assemble, disassemble and move them. Unlike the Approach Spans, which are of uniform height, the sections of the Main Spans vary in depth and width of bottom slab as one travels along its length. This requires continual adjustment of the scaffolding height, width and plumb as it is moved longitudinally through the eastbound and westbound cells of the Main Spans. In addition, access was required within the 30-foot high pair of compartments located between the full-height diaphragms located at Piers 13 and14. To accomplish this, the scaffolding units were completely disassembled from their full height, carried piece-by-piece through the narrow portals of the diaphragms and reassembled to their full height. This was repeated to move the scaffolds through the portal of the other pier diaphragm into the adjacent span.

4 TENDON INVESTIGATION

4.1 *Preliminary investigation*

During the period encompassing the first two bridge inspections, reports were published by the Florida Department of Transportation (FDOT) and others detailing the discovery of voids in the post-tensioned tendon ducts, several with severe deterioration of the tendons, and the subsequent efforts to repair these conditions [Henriksen, 1998; Ghorbanpoor, 2000; Hartt, Venugopalan, 2002; Pielstick, 2002; DeHaven, 2003; Pearson-Kirk, 2004]. During this period, PB had been involved with the discovery of these types of conditions in several segmental structures, and prior to this period in several segmental bridges in the United Kingdom. More recently, the American Segmental Bridge Institute and FDOT have recognized the need for improved grouting techniques and new procedures have been developed to significantly improve grouting hardware, materials and training.

Based on these experiences, it was recommended that an investigation be performed to determine the condition of the internal tendons of the segmental box girder spans (the Main Structure) of the Jamestown-Verrazzano Bridge.

Post-tensioning exists within these spans as draped high-strength steel strand tendons in the webs and straight tendons in the top and bottom slabs that provide longitudinal post-tensioning while straight strand tendons are placed transversely through the top slab of the structure. Additional vertical tendons, consisting primarily of post-tensioned strand tendons, are present in the webs of Main Spans 12 to 14 to provide additional shear resistance. All tendons were to have been grouted.

Since it was not known if problem areas would be found, the extent of the inspection was difficult to estimate. Planning began in early 2004 for a preliminary investigation of several spans that would each be representative of the three modes of construction employed on the Main Structure, which are as follows:

- Spans 1 and 20 through 23 were shored, cast-in-place with Span 1 located within the channel and Spans 20 through 23 at the east end of the bridge on land;
- Spans 2 through 11 and 15 through 19 were precast segmental;
- Spans 12 through 14 were balanced-cantilever cast-in-place

PB collaborated with NDT Corporation, Worchester, Massachusetts, which was instrumental in the development of the methods utilized during the previous investigations of other segmental structures. From October 18 through October 22, 2004, NDT performed the investigation with PB in concurrence with the biennial inspection that was being performed at that time. All work was performed from within the structure with the aid of scaffolding and ladders that were present. The following steps were performed for each tendon investigated to identify and confirm the presence of voids:

- Ground penetrating radar (GPR) using a high resolution 1500 Hz antenna was utilized to locate the centerline of the ducts. The high frequency GPR antenna allowed the technicians of NDT to distinguish the signal representing the ducts from those of the plain reinforcement and marked their location on the surface of the concrete. [See Figure 9.]
- Sonic / ultrasonic frequency (impact echo) detection was then utilized to identify voided tendon ducts by running a four-sensor array just ahead of the impact signal device along the centerline marked out using the GPR. [See Figure 10.]
- Where voids were identified by impact-echo, small diameter holes were carefully drilled through the concrete cover to the surface of the sheet metal duct, which was carefully peeled away to reveal the duct interior. Where a void was confirmed, a bore scope was used to document the size and length of the void, and the present state of the tendon. [See Figure 11.]

In addition, several tendon anchorages were investigated with drilling and borescope inspections. Because the anchorages are located in thicker concrete diaphragms and anchor blocks GPR and impact-echo are ineffective with these components, drilling was solely used to detect the presence of voids. The method followed was very similar to the third step described above in the typical tendon investigation.

Table 1 on the following page summarizes the areas investigated and their observed conditions. The table shows that most of the investigated tendons were in good condition. However, there were concerns that the only two draped web tendons probed in Span 14 both displayed significant voids and evidence of mild corrosion. These findings may be associated with the difficulty in maintaining quality control while working in an open environment required of cast-in-place construction and the particular challenges of variable-depth balanced-cantilever construction.

Figure 9. GPR investigation of box bottom slab using 1500 Hz antenna.

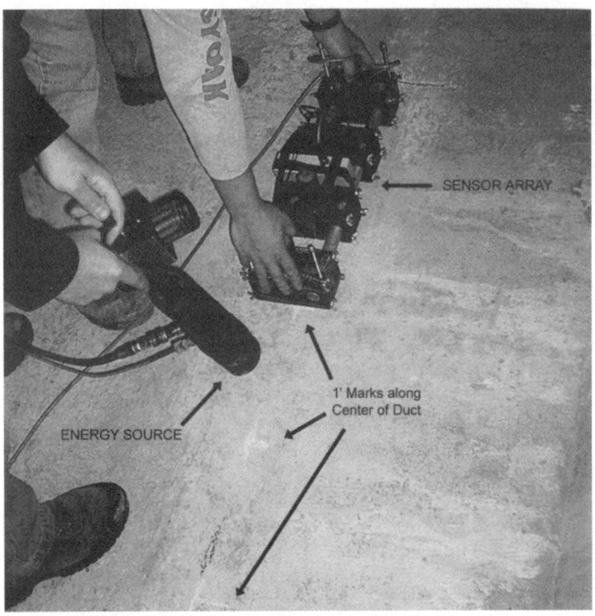

Figure 10. Impact-echo air void detection using 4-sensor array.

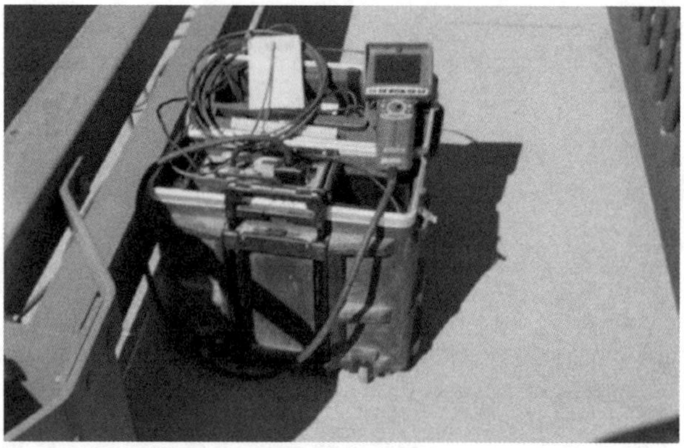

Figure 11. Borescope with video recorder and monitor.

Table 1. Summary of Jamestown-Verrazzano Bridge preliminary tendons investigation.

Span	Tndn no.	Type	Location	Observed condition
14	n/a	4 Short Verticals	North, center & south webs	
	9	Draped	North, web	<V> , <E> , <LC>
	11	Draped	North, web	<V> , <E>
	n/a	5 Transverse	North, web	
19	7	Straight	North web, bottom slab	
	8	Straight	North web, bottom slab	
	9	Straight	North web, bottom slab	
	10	Straight	North web, bottom slab	<V>
	24	Draped	South web	
	25	Draped	North web	
	n/a	5 Transverse	Top Slab	
21	23	Straight	Center web, bottom slab	
	23	Anchor	West end	
	24	Straight	Center, web, bottom slab	
	24	Anchor	West end	
	25	Straight	South web, bottom slab	
	26	Straight	North web, bottom slab	
	n/a	5 Transverse	Top Slab	
22	n/a	Straight	North web, bottom slab	
	n/a	5 Transverse	Top Slab	
23	n/a	Straight	North web, bottom slab	
	n/a	5 Transverse	Top Slab	

Key for observed conditions: <V> Void; <E> Exposed tendon; <LC> Light corrosion.

4.2 *Second phase*

Because of these findings, a second investigation was performed beginning in April 2005 that concentrated on the Main Spans. The proposed scope of work within these spans included:

- All of the continuous draped web tendons;
- One hundred-twenty vertical tendons encompassing 10 per each of the three webs of the 340-foot long clear spans of Spans 12 and 14, and 20 per each of the webs of the 636-foot long clear spans of Span13;
- Six transverse top slab tendons each of Spans 12 and 14, and 8 top slab tendons of Span 13;
- Four longitudinal bottom slab tendons within each span.

In addition, a sample of each tendon type was selected for investigation within the remaining spans Main Structure to assure that potential voids in them would not be overlooked. To maintain satisfactory quality assurance, the size of the sample was statistically determined using ANSI/ASQC Z1.4-1993 entitled "Sampling Procedures and Tables for Inspection by Attributes" [American Society for Quality, 1993] . The resulting sample size represented at least five percent of the total number of each tendon type, with the exception of the draped web tendons of the Main Spans, which represented 100 percent of these tendons. Investigation methods were the same utilized during the preliminary inspection. Included in the scope of work was the option of having a specialty contractor follow on the findings of the investigation to perform repair grouting of the tendons should this be warranted by the findings.

The investigation commenced at the end of March 2005 with work concentrated within the Main Spans. While some additional voids were immediately identified within the draped web tendons, investigation of the vertical tendons soon revealed another concern. Within the Main Spans, vertical tendons are present within each web for the entire length of each span. Their lengths closely match the lengths of the webs, which vary from about 10 feet at the points furthest from Main Piers 13 and 14 to about 30 feet at the piers and within the pier tables. It was at the latter locations that voids ranging from a few feet to three-quarters of the full length of the tendons were present within the top of almost all the tendons investigated at these locations. Of particular concern was the discovery of water and measurable tendon corrosion present at the bottom of one of these voids. (It should be noted that similar problems had been observed by PB in the vertical tendons of another bridge, which had vertical precast pier segments with vertical tendons.) In June 2005, soon after the confirmation of these conditions, the scope of work was revised. Because of the quality of the transverse deck tendons within the Approach Spans - none presented voids throughout their entire length up to this time – their number was halved and top slab longitudinal tendons within these spans were completely omitted to permit the investigation to focus on the Main Span vertical tendons at and within 30 feet of the pier table.

It was during this increased focus on the longest vertical tendons that an undocumented construction modification was discovered. Two adjacent vertical tendons located within the pier table center web of Pier 14 were fabricated from $1^3/_8$-inch Dywidag threaded rod instead of the $6\,^1/_2$-inch strands shown on the shop drawings. In addition, these tendons had voids measuring 20 and 28 feet respectively and surface corrosion present throughout their exposed length.

In August 2005, a contract to grout eleven vertical tendon voids located at Piers 13 and 14 was provided to RIDOT. The contract was let under the auspices of an existing contract underway at the bridge in September 2005 to DSI, Incorporated. Quality of the work was maintained by NDT, Inc and PB, during which time the adequacy of the grouting repair was confirmed utilizing the same methods as the tendon investigation.

4.3 *Third and fourth phases*

The findings from the second phase of the investigation identified deficiencies in a number of the vertical tendons, in particular those at locations with the highest shear stresses within the Main Spans. To address these findings, the scope of the investigation was again modified to include:

- All of the Main Span vertical web tendons;
- A sampling of the top anchorages Main Span vertical web tendons accessed from the bridge roadway;
- All remaining web tendons within the Main Structure;
- Draped tendons of the Main Spans that could only be accessed from the exterior of structure.

This phase of the investigation revealed more voids in the vertical tendon ducts, particularly within the top half of the longer ones. In addition, periodic minor voids were recorded within longitudinal web and bottom slab tendon ducts of many spans. Of particular concern during this phase, was the discovery of voids within the longitudinal tendons of the Main Spans. As result of this finding, in September 2006, a fourth phase that encompassed these tendons was added to the project

Table 2. Summary of Jamestown-Verrazzano Bridge tendons investigation.

Construction type	Tendon type (Location)	Number of tendons tested	Number of voided tendons	Observed condition
Cast-In-Place	Transverse (Top Slab)	20	None	
Spans 1, 20 & 23	Draped (Web)	24	1	
	Longitudinal (Bottom Slab)	12	None	
Precast	Transverse (Top Slab)	60	None	
Spans 2-11 & 15-19	Draped (Web)	138	15	
	Longitudinal (Bottom Slab)	227	13	
Balanced Cantilever	Longitudinal (Top Slab)	108	34	<E> *
Spans 12, 13, & 14	Transverse (Top Slab)	20	None	
	Vertical (Web)	864	31	<E> <LC>
	Draped (Web)	33	14	<E> <LC>
	Longitudinal (Bottom Slab)	2		

Key for Observed Conditions: <V> Void; <E> Exposed tendon; <LC> Light corrosion; * Location of 314-foot long void.

scope. This latter task, and the tendon investigation, concluded in December 2006 with the identification of several voids measuring several hundred feet in length within these longitudinal tendons. As a result of this latter investigation, the total volume of voids recorded for these tendons were the highest for any tendon type within the structure, followed by the vertical tendons. Those with the best findings were the transverse deck tendons, which presented no voids for any of the tendons examined. The results of the nondestructive tendon duct testing are summarized by Span Construction and tendon type in Table 2.

5 LIVE LOAD ANALYSIS

5.1 *Long-term behavior of segmental structures*

One characteristic that distinguishes both segmental concrete design and construction is the need to recognize and account for the long-term deformations and changes in the material properties experienced by these structures. These changes occur as soon as each segment is cast and/or erected and influence the behavior of the structure particularly during the initial five to ten years of the structure's life.

Deformations take two predominant forms: creep and shrinkage. Shrinkage is the reduction in volume of a concrete mass that occurs as it loses free water to either evaporation or hydration with the constituents of the cement. Creep is the plastic deformation experienced by this mass in response to stress, typically compressive. Both of these phenomena are most prevalent during the early life of the structure and their influence rapidly diminishes after the first couple of years. Several references provide methods for approximating these behaviors, in particular the Comité Euro-International du Béton - Fédération Internationale de la Précontrainte Model Code for Concrete Structures [CEB-FIP, 1990].

Over the early years of the structure, especially during construction, these deformations result in significant redistribution of stresses that must be accounted for to accurately predict its behavior, to provide sufficient structural capacity and sometimes to assure its serviceability. For a live load rating, data that must be accrued include the casting date, mix properties and general climate present during the casting of each segment, accurate tendon stressing records, methods of construction utilized, construction loads and any deviations that may have occurred from intended design procedures. The greater the accuracy of this data, the more accurate will be the analysis performed.

For the Jamestown-Verrazzano Bridge RIDOT wisely retained and archived this information. This was critical for several reasons:

- Several different construction methods were employed on different portions of the superstructure: precast, cast-in-place balanced cantilever and cast-in-place shored construction.
- Several delays had occurred that extended the construction to a period of about 7 years.
- Because there are only two expansion hinges present on the Main Structure, the superstructure is continuous for lengths of 1368, 1764 and 1818 feet. This and the prior condition assured the interaction between some portions of the Main Structure that had already undergone significant amount of deformation while others were just being constructed.

These archives were carefully reviewed and the data input into the time-dependent analysis structural model that was developed for this purpose. GT-STRUDL [Georgia Institute of Technology, 1999] and TANGO [Tang, 1993] were each utilized, and their respective outputs from their transverse and longitudinal analyses were post-processed using EXCEL spreadsheets. This effort was later validated by the results of a topographic survey of the deck performed concurrently with the 1999 baseline inspection. The survey results from the 1999 survey records were compared with the survey records during construction. When creep deformation of the piers was accounted for, the changes closely matched the deformations determined by analysis. The greatest deviation between the two occurred at the mid-length of Span 12 over the navigational channel, which, with a clear span of 636 feet, is the longest span of the bridge. The results of the survey versus the analysis were 0.14 feet of creep deformation versus 0.25 feet, respectively, resulting in a difference of 0.11 feet or $1^3/_8$ inches.

Dead and live load ratings were checked by utilizing both allowable stress design (ASD) and load and resistance factor design (LRFD) for the live loads that were typically considered during the period the bridge was designed, which were AASHTO HS-20-44 lane and truck loads, AASHTO H-20-44 lane and truck loads, and the Type 3 and 3S2 legal loads.

The results of this analysis confirmed that: the bridge, in general, had satisfactory capacity to support these live loads. However, it was revealed that the shear capacity of the bridge differed from assumptions made during design. The advantage of current PC capabilities over the computing tools available during the mid-1980s allowed an elastic analysis to be performed without the assumption that all three webs shared dead load equally, which was a standard simplification often made at that time. As a result, it was determined that outer two webs of the bridge carried a higher proportion of the dead load shear than the third interior web with the governing ratings being for shear.

6 CARBON FIBER-REINFORCED POLYMER BOX GIRDER WEB REINFORCEMENT

6.1 *Findings of the 1999 baseline inspection*

The baseline inspection began in June 1999. During the inspection detailed mapping of the cracks present on many of the surfaces of the Main Structure were recorded. Most of the cracks were concentrated on the box girder webs near and at the pier tables, with some of the highest in the vicinity of closure pour placed between the cast-in-place pier tables and the jacked-in-place 169-foot long precast Approach Span segments, which, at the time represented the heaviest such span-by-span construction performed on a segmental concrete structure.

Subsequent to the inspection, as described above, a live load rating was performed that confirmed the presence of high shear stresses at these locations. Further finite element modeling of the closure detail revealed additional stresses incurred by the structure when the jacked spans were released after completion, curing and tendon stressing at the closure. While the live load ratings satisfied AASHTO criteria, these findings, coupled with the cracking observed during inspection, resulted in the recommendation that the webs of the Approach Spans at the pier tables be reinforced

Figure 12. CFRP reinforcement of pier table web. Note 2-inch wide inspection strips between panels.

to assure redundancy of the structure at these locations. In addition to the web reinforcing, cracks throughout the superstructure were first injected with epoxy following the recommendations for allowable crack width prescribed by Table 4.1 in ACI Committee Report 224R entitled "Control of Cracking in Concrete Structures" [American Concrete Institute, 1990].

Carbon fiber-reinforced polymer (CFRP) was selected as the reinforcement material. [See Figure 12.] The choice was made based upon its strength being comparable to other conventional structural materials, its high strength-to-weight ratio that enabled rapid installation by only two individuals, and the ability to install the material using a high strength adhesive without damaging the existing concrete substrate. The CFRP consisted of two layers of Tyfo SCH-41 carbon fiber fabric placed on both sides of each pier table web. Because CFRP is not resistant to UV radiation, a matching protective coating was specified for the second layer of reinforcement placed on the exterior surface of the pier tables, which had the additional benefit of providing a cosmetically favorable appearance to the repair.

Note that a two-inch wide gap was placed between adjacent CFRP panels to permit future inspections to observe portions of the web and determine if further cracking has occurred.

The implemented repair consisted of the following components:

- Epoxy injection of all exterior cracks exceeding 0.012 inches in width and interior cracks exceeding 0.006 inches;
- Smoothing the surface of the concrete and filling in depressions with epoxy grout to provide a smooth continuous substrate for the reinforcement material;
- Application of epoxy adhesive to the surface of the substrate just prior to installation of the CFRP;
- Saturation of the CFRP just prior to its installation;
- Removal of void that might be present between the CFRP and the substrate.

After the 72 hours of curing time had elapsed, quality control was performed by performing pull-off tests at randomly-selected locations and by checking the surfaces of all repaired locations for voids. Before and after conditions where deficiencies were found were documented and photographed.

7 CONCLUSIONS

Segmental bridge technology has become a popular type of construction in the United States, and with its continued maturation, more commonplace. However, relative to the more common forms of long-span bridge design and construction, such the truss, the arch and suspension, segmental concrete construction represents a relatively modern innovation. This construction has produced many ingenious, beautiful and efficient bridge types such as the segmental box girder, the cable-stay, the segmental arch and most recently the extradosed bridge. Given the relative youth of this technique, it will not be surprising to find additional adaptations evolving from this technique.

Nor should it be surprising at this time to discover unanticipated issues that periodically need to be addressed as time reveals which aspects of this construction type still require maturation and refinement. The presence of cracks in the concrete surfaces and incomplete grouting of tendon ducts are two. This investigation discusses these two issues and has provided a possible means of addressing these issues.

ACKNOWLEDGEMENTS

PB gratefully acknowledges the personnel of Department of Transportation of the State of Rhode Island and Providence Plantations who, through their dedicated service to assure the longevity of the State's roads and bridges, and timely and continued participation contributed to our efforts on their behalf. We particularly wish to recognize the following individuals:

Mr. James Capaldi, Former Director
Mr. Edmund Parker, Chief Engineer, Transportation Development
Mr. Kazem Fahoumand, Deputy Chief Engineer – Design, Transportation Development
Mr. David Fish, State Bridge Engineer, Bridge Design Section, Transportation Development
Mr. Rahmat Noorparvar, Project Manager, Bridge Design Section, Transportation Development
Mr. Mazen Alsabe, Former Project Manager, Engineering Section, Highway and Bridge Maintenance

REFERENCES

American Concrete Institute Committee 224. 1990. ACI 224R – Control of Cracking in Concrete Structures.

American Society for Quality. 1993. Sampling Procedures and Tables for Inspection by Attributes.

Comité Euro-International du Béton – Fédération Internationale de la Précontrainte. 1990. Model Code for Concrete Structures.

DeHaven, T. A. October 2003. Grouting of Post-Tensioning Tendons, Overview of Recent Developments. Structures Magazine. Pages 24–26.

Georgia Institute of Technology – CASE Institute. 1999. GT-STRUDL Integrated Software System for General Structural Engineering Design and Finite Element Analysis.

Ghorbanpoor, A. November 14, 2000. Final Report: Condition Assessment of External P-T Tendons in the Mid Bay Bridge.

Hartt, W. H. & Venugopalan, S. April 15, 2002. Final Report: Corrosion Evaluation of Post-Tensioned Tendons on the Mid Bay Bridge in Destin, Florida.

Henriksen, C. F., Knudsen, A. & Braestrup, M. W. October 1998. Cable Corrosion: Undetected?. Concrete International. Volume 20. Issue 10. Pages 65–72.

Pearson-Kirk, D., Collard-Jenkins, S. J., Theryo, T. & Chandra, V. February 2004. Improving the Durability of Segmental Bridges. PB Network. Volume 19. Issue No. 57. Number 1. Pages 23–27, 33.

Pielstick, B. H. 2002. Grouting of Segmental Post Tensioned Bridges in America. Transportation Research Board. Volume 1813. Pages 235–241.

Tang, M. C. 1993. TANGO, For the Analysis of Bridges (Software Manual). DRC International, Ltd., New York, NY.

2 Cable-supported bridges

Chapter 5

Ultimate capacity of suspension bridges with arbitrary imperfect towers

Manabu Inoue
IHI Co., Ltd., Tokyo, Japan

ABSTRACT: In this study, the difference of ultimate capacity of suspension bridge due to the imperfection of tower has been investigated through the past experiences and the case study. At the beginning, the measurements of tower deviations from the ideal position for constructed suspension bridges, mainly in Japan, have been studied and the tendencies of imperfection have been classified into some types. Then the effect of tower imperfection for the ultimate capacity has been investigated by 1-1/2 order analyses using 2-D bridge model. In the analyses, four types of imperfect tower with the different imperfect shapes were modeled at the freestanding. Finally, the difference of ultimate capacity among the imperfect models has been summarized and some remarks were made for the potential of more reliable and economical bridge in the future.

1 INTRODUCTION

Suspension bridges have been constructed using the state-of-the-art technologies and with the possible care. To satisfy the proposed performance, high degrees of accuracy are required during the construction. For example, the requirements for the straightness or the verticality are one of the most severe ones.

In the areas including Japan where there is a potential of big earthquakes with magnitude M8.0–9.0 class such as Great Hanshin-Awaji earthquake in 1995, the towers of suspension bridge are usually composed of steel shell plates to mitigate the inertia. The tower leg is divided into number of blocks with an appropriate height corresponded to some limitations concerned with fabrication, transportation, lifting ability, site condition and so on. As one of solutions to achieve the appropriate straightness among the blocks and verticality of tower, mill to bear connections have been adopted for some bridges and this could minimize the number of bolts.

In reality, several kinds of errors must be inevitable during the fabrication and construction. For the tower, these errors remain as the imperfection at the completion and it is well known that imperfections break symmetry and significantly influence the response of elast-plastic columns. For example, the towers may be subjected to unintended small lateral loads, they may be initially curved rather than perfectly straight, or the axial load may be slightly eccentric. Unlike beams subjected to transverse loads and small axial forces, columns are quite sensitive to imperfections, although not as much as shells.

Thought the weak element in the collapse chain is not usually the tower but hangers for suspension bridges, the tower imperfections are closely related to the robustness and the imperfect sensitivity of the whole bridge and it is of important to keep an appropriate balance among the durability of structural elements such as anchorage, tower, cable, hanger and deck.

The imperfect sensitivity of tower depends on its configuration, section properties, load conditions and so on. However, it is not always clear how the difference of configuration of tower affects the ultimate capacity of the whole bridge. In this paper, the ultimate capacity of steel tower has been investigated through some results of past suspension bridges and the case study.

2 EXAMPLES OF TOWER IMPERFECTION

2.1 *Technical requirement for tower straightness and verticality*

Some specifications mention the allowance value for tower straightness or verticality clearly. For example, for AKASHI bridge the errors in the longitudinal displacement at the tower top within h/5,000 (h stands for the height of tower) was required as the maximum allowance verticality of freestanding tower. The requirements for straightness or the local change of inclination were not specially mentioned in this specification. However, the high accuracy of flatness for each joint was required.

In other specification, only the maximum verticality or deviation to be considered in the analyses is defined and the special attention is required for the tower construction not to lead any deterioration in the performance level.

In the design, safety and sensitivity of the bridge to tower imperfections are verified by using FE-model where the tower imperfection represents as some horizontal deviation at the top or the middle of tower. The design deviation is mainly defined as h/1,000 or h/2,000, some 10–20 times larger than that for the allowance, though these conditions depend on the technical specification in each project. In some papers, the ultimate capacity has been investigated for the bridge model with the tower leaned straightly at the completion. (e.g. Nogami et al, 2001)

Usually the verticality is controlled by keeping the horizontal displacement at each step within some target values. The target value at each step is defined as the function of tower height such as h/5,000 or h/10,000. In reality, it is hard to adjust the verticality at each joint during the block erection. So, for the mill to bear connection, machining is carried out for each joint with special attention at the fabrication phase. In this method, the requirement for the flatness at each joint depends on the design concept. And, the accuracy of tower relies heavily on the result at the fabrication phase. In the trial assembly, more than three leg blocks are lied on the ground with many temporary supports and connected each other to verify the straightness or the verticality of fabricated blocks. It is not usually carried out to trial-assemble all blocks in the same time. Thus, the deviations for tower with all blocks are verified by accumulating these results. In this phase, high quality of verticality has been reported for the past suspension bridges.

However, in the construction phase, the differences of conditions between in trial assembly and in situ cause inevitable errors. In some bridges, re-machining for mill to bear connection of latter block has been carried out to compensate for deviation of constructed leg after measuring the verticality at some height.

2.2 *Examples of constructed towers*

The allowance verticality and the results of maximum longitudinal deviation form the point of ideal line at the top of freestanding tower for some suspension bridges in Japan are shown in Table 1. In some bridges, the configurations of freestanding tower have been measured and reported as reference values. The main target for the accuracy of constructed tower seems to be set to a displacement at the tower top.

The examples of deviation in the longitudinal direction are shown through Figure 1 to Figure 3. Due to some difficulties of measurements, the deviations have been measured at a few points. As is obvious in these figures, the deviation at the tower top is not always maximum value.

Though the number of examples is so limited, the tendencies of deviation at the freestanding can be classified at least to four types, i.e.:

1) Leaning straightly (e.g. 2P-R of INNOSHIMA bridge, Europe-N of 2nd Bosporus bridge)
2) Changing the inclination at some height so as to reach the tower top in the ideal position (e.g. 2P-L and 3P-L of INNOSHIMA bridge)
3) Adjusting the inclination at each step or several times (e.g. 3P-E of AKASHI bridge)
4) Combining among the above mentioned three types (e.g. 2P-E of AKASHI bridge)

Table 1. Allowance deviation for tower verticality.

	Allowance at tower top	Height (m)	Main Span (m)	Max. displacement at tower top
AKASHI	57 mm (1/5,000)	286.7	1,991	39 mm (1/7,389)
MINAMI BISAN SETO	30 mm (1/6,000)	175.7/184.2	1,648	19 mm (1/9,346)
KITA BISAN SETO	30 mm (1/5,000)	165.4/173.3	1,538	10 mm (1/16,535)
SHIMOTSUI SETO	30 mm (1/4,500)	136.9/141.0	1,400	12 mm (1/11,901)
OHNARUTO	13 mm (1/10,000)	128.3	876	6 mm (1/22,514)
OHSHIMA	18 mm (1/5,000)	88.4	840	14 mm (1/6,402)
INNOSHIMA	14 mm (1/10,000)	135.9	770	11 mm (1/12,229)
KANMON	14 mm (1/10,000)	136.0	770	10 mm (1/13,603)
AKINADA	24 mm (1/3,500)	119.5	750	23 mm (1/5,193)
HAKUCHO	19 mm (1/7,000)	129.4	720	14 mm (1/9,585)

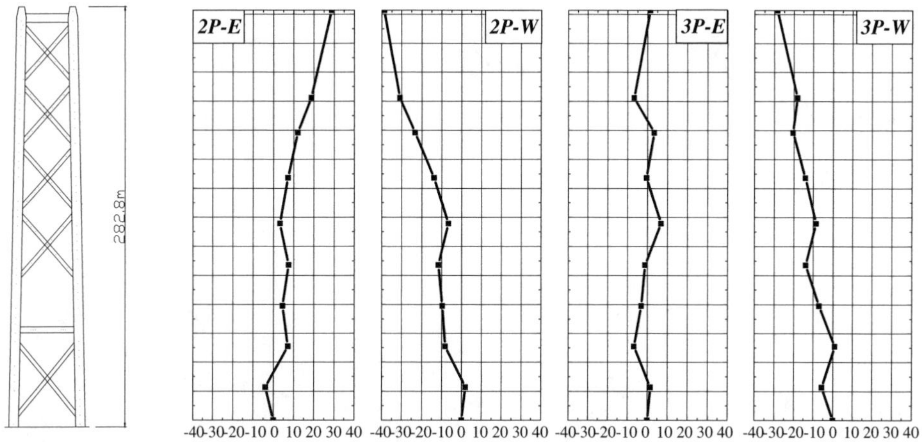

Figure 1. Deviations in the longitudinal direction (Akashi Bridge).

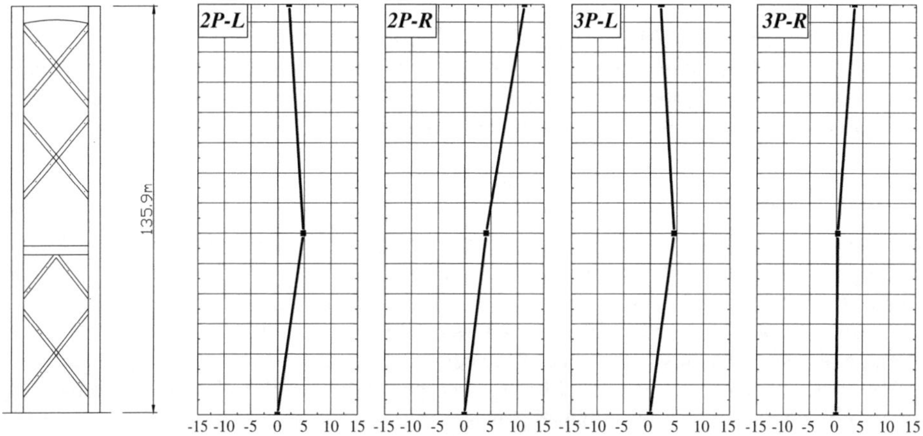

Figure 2. Deviations in the longitudinal direction (Innoshima Bridge).

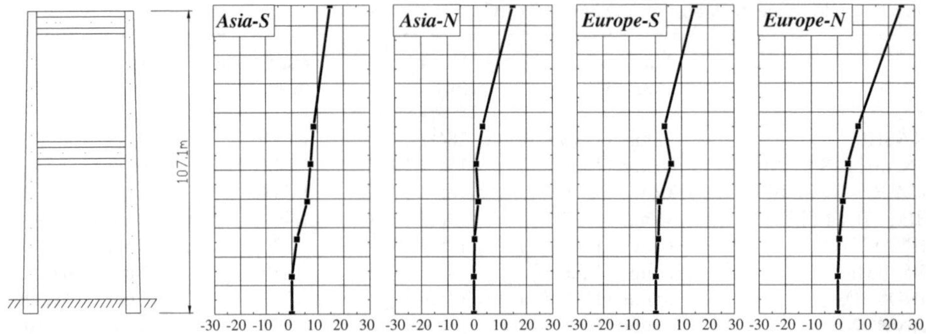

Figure 3. Deviations in the longitudinal direction (Fatih Sultan Mehmet Bridge (2nd Bosporus bridge)).

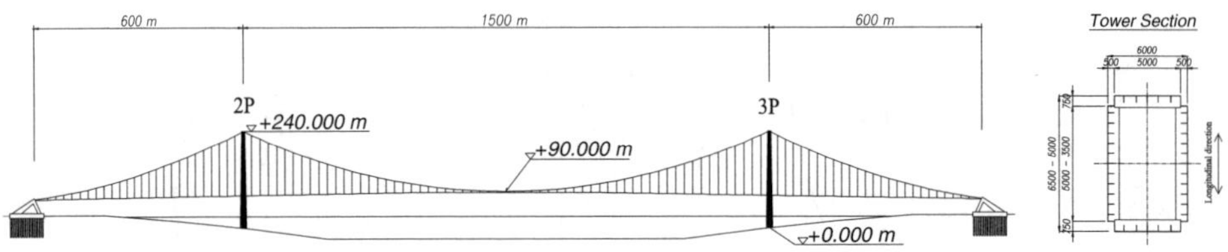

Figure 4. Bridge model.

Table 2. Section properties.

	Area (m^2/bridge)	Moment inertia (m^4/bridge)
Main cable	0.701	–
Tower	2.4–3.5	14.7–50.8 (in the transverse axis)
Deck	0.874	1.560 (in the transverse axis)
Hanger	0.018	–

3 EFFECT OF TOWER IMPERFECTION

The 1-1/2 order analyses have been carried out to oversee how the difference of tower imperfection and shape affects to the ultimate capacity of suspension bridge. In this study, only the imperfections in the longitudinal direction were considered and 2-D global bridge model considering elast-plastic properties only to the tower has been used in the analyses.

3.1 *Bridge model*

The general geometrical layout and section properties are shown in Figure 4 and Table 2. The length in the main span is 1,500 m and the height of tower is 240 m. The material of tower was assumed to be steel with the grade of S355 (EN10025, Minimum yield strength = 355 MPa) for all plates and bi-linear. During the analyses, the tower section was simplified in such a way that the stiffening plates were added to the plate as the equivalent thickness. The residual stresses around the welding of stiffening plates and hardening after the yielding were not considered in the analyses. Also it was assumed on the stiffened panels being stable to its full capacity of the thickness. Materials for other structural elements were assumed as perfectly elastic ones. Thus, the collapse due to the failure of elements except tower was not considered in the analyses.

Four types of tower with the imperfection in the longitudinal direction were used in the analyses. For each imperfect type, the ultimate capacity of tower has been investigated by changing the maximum deviation, the height of inflection point, number of inflection points and so on. During the analyses, only the left side tower (2P) in Figure.4 was assumed to have some imperfection. The imperfect tower models were defined for the freestanding tower as follows, where the deviation is positive when the tower deforms toward the mid centre from the ideal position.

Type-A
The tower is leaned straightly or parabolically, and the deviation at the tower top changes from −240 mm to 240 mm. (correspond to h/1,000)

Type-B
The tower inflects at the height of 60 m with the deviation of −60 mm, and the deviation at the tower top changes from −240 mm to 240 mm.

Type-C
The inflection point is at some height (60, 120 and 180 m), and there is no deviation at the tower top.

Type-D
The inflection point is at the height of 180 m, and the additional one or two inflection points are defined at the heights of 60 m and 120 m.

3.2 *Analyses*

The burdened load on the bridge at the completion under the dead weight was increased incrementally until the tower reaches the ultimate capacity, where the ultimate capacity was defined as those when the burdened load presents deformation of the tower differently from the shape similar to one step before. The burdened load was defined as a multiple of the uniformly distributed live load of HA loading, q = 96.88 kN/m/bridge (8 notional lanes), according to BS 5400-2:2006 in the main span. Although the governing load case depends on the bridge configuration such as a geometrical layout and section properties, this case is the governing one for the tower of this model.

The 1-1/2 order analysis was made in each step using the non-linear analysis program named "Midas/Civil", developed by Midas&CTC Co., Ltd. In the analysis, the resulting member forces and section properties in each step were used in the analysis of the next step and a small increment of the burdened load was applied when the tower came close to the ultimate capacity.

3.3 *Results*

The computation results of configuration and ultimate capacity for tower 2P are shown through Figure 5 to Figure 8. For each model, the bridge becomes unstable due to the tower collapse under some burdened load D+ a*L, where D is the own weight and L is the un-factored distributed live load in the main span. The load multiple factor "a" is used as a representative value for the ultimate capacity in this study. In figures, the values in percentage stand for the ratio of the load multiple factor for the imperfect model to that for the ideal one. Thus, the percentage less than 100% means that the ultimate capacity for the model deteriorates due to the tower imperfection. In figures, a solid line and a dashed one stand for configurations at the freestanding and at the completion, respectively. A dotted straight line shows an ideal position without any imperfection. The difference from a dotted line represents the deviation in that state.

In types-A and -B, the position of tower top moves near the ideal one at the completion because the horizontal displacement at the tower top is mainly governed by the behavior of main cables. The ultimate capacity of tower is in proportion to the deviation at the top of freestanding tower. In type-A-2, the tower shows a parabolic imperfect configuration at the freestanding. Though the configuration of A-2 at the completion is so similar to the ideal one, the ultimate capacity decreases by some 2% and this deterioration is smaller than that for A-1.

In type-B, the ultimate capacity is related to both the absolute maximum deviation at the top of freestanding tower and the configuration at the completion. The ultimate capacity changes in

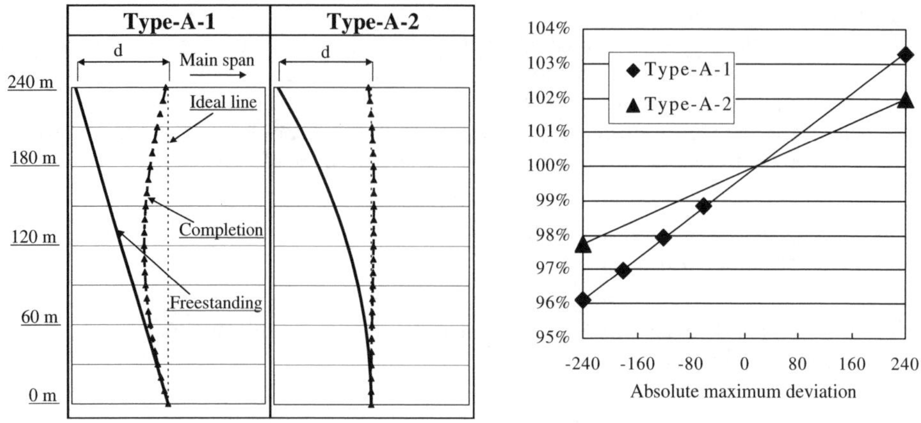

Figure 5. Change of configuration and ultimate capacity for Type-A.

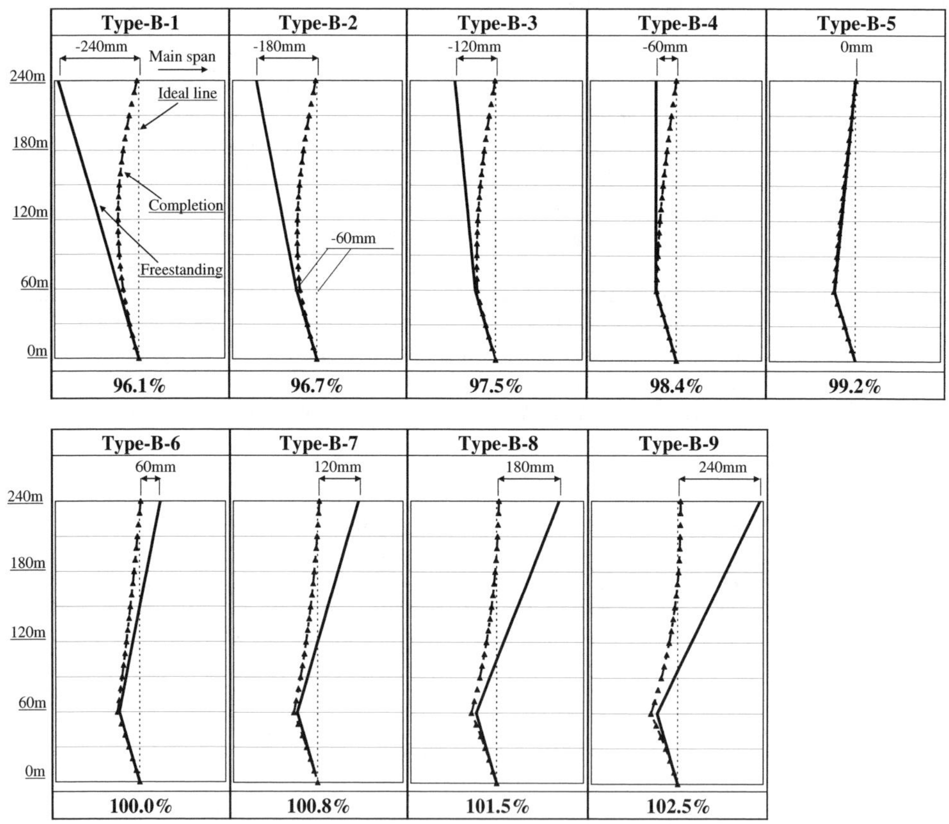

Figure 6. Change of configuration and ultimate capacity for Type-B.

proportion to the absolute maximum deviation at the top of freestanding tower, and it depends on the shape at the completion, i.e. convex or concave in the horizontal axis, whether the ultimate capacity increases or not. Thus, if the tower shape of upper part at the completion, in this case the region from 60 m to 240 m, shows the convex curve in the horizontal axis, the ultimate capacity

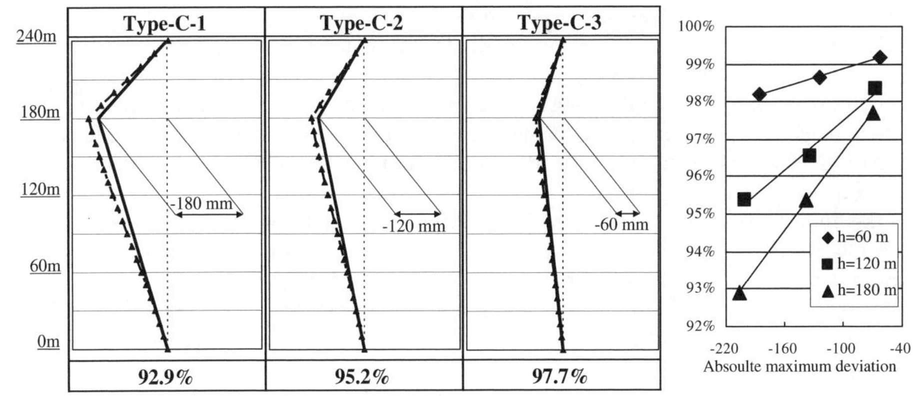

Figure 7. Change of configuration and ultimate capacity for Type-C (h= 180 m).

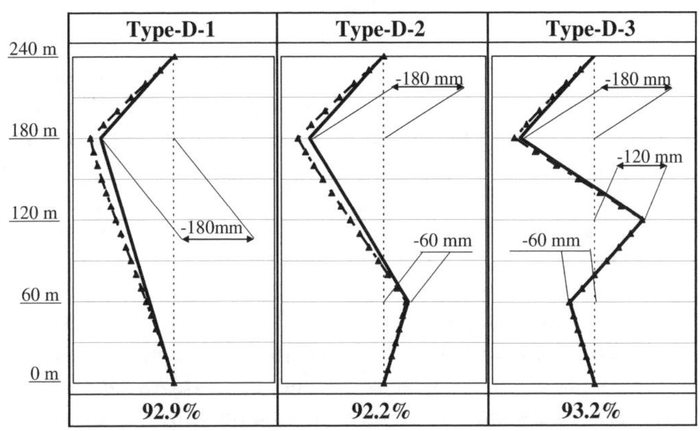

Figure 8. Change of configuration and ultimate capacity for Type-D.

decreases compared with that for the ideal one. Also the critical point appears at the lower height with the incrimination of the deviation at the top of freestanding tower.

In type-C, the difference due to the position of inflection point and the magnitude of deviation are shown in Figure 7. For the imperfect tower with the absolute maximum deviation at the same height, the ultimate capacity changes in proportion to the magnitude of the maximum deviation. If the inflection point appears at the higher position, the incrimination or the deterioration of ultimate capacity for that imperfect model tends to be larger than that for the tower which has the same maximum deviation at the lower position.

In type-D, the effects of the intermediate inflection points have been investigated as shown in Figure 8. In this model, the absolute maximum deviation is positioned at the highest among the inflection points. Though the analyzed cases are so limited, the deviation due to the inflection points at the lower height seems not to affect so much to the ultimate capacity of the tower.

3.4 *Remarks*

Through this case study, the following remarks can be noted:

1) It is effective for dulling the imperfect sensitivity to keep an appropriate accuracy for the verticality of the lowest block or the flatness of basement.

2) If some adjustment to decrease the deviation for latter blocks is carried out by making the inflection point during the tower erection, the inflection point is desirable to be positioned in the lower height.
3) The ultimate capacity is considerably governed by the magnitude of deviation at the highest inflection point. Though adjusting the inclination at some steps is mainly effective for controlling the deviation at the highest inflection point within the target value, it does not so affect to the ultimate capacity.
4) In this model, the inclination toward the anchorage causes the deterioration of the ultimate capacity. Thus, it seems to be adequate that the allowance or target value for the deviation toward the anchorage is more severe than that toward the mid centre.
5) Even if the lower part of tower blocks tends to incline toward the anchorage, the tower can be erected without any deterioration of its performance by making the inflection points and changing the inclination toward the mid centre.
6) If the tower is erected within the conventional target values such as h/5,000 or h/10,000 at the freestanding, the deterioration of ultimate capacity due to the imperfection is negligible.

4 CONCLUSIONS

The effect of tower imperfection in the longitudinal direction for the ultimate capacity has been investigated in this study. The conventional control method of tower construction is surely effective to achieve the appropriate bridge performance. However, this method requires special facilities for machining, a lot of time and much cost.

There are many possibilities of errors which will affect the ultimate capacity of tower, e.g. the deviation of tower in the transverse direction, the difference of imperfection between legs, lengths of cable and hanger, weight of deck and so on. Though these errors were not considered in this study and further investigations are needed, it seems to be feasible to define more flexible requirements or target values, which are some 5–10 times larger than that used in the past suspension bridges.

Finally, the tower construction without higher accuracy requirements but with the well-controlled management can achieve more reliable and economical bridge without any deterioration of the ultimate capacity. The smart and well-controlled method with a consideration of imperfect sensitivity can be useful for the design and the construction especially for super-long suspension bridges in the future.

REFERENCES

Bazant, Z.P. and Cedolin, L., 2003. *Stability of Structures, Dover.*

BSI. 2006, *Steel, concrete and composite bridges- Part2:Specification for loads. BS5400-2.*

European Committee for Standardisation (CEN), 2004, *Hot-rolled products of non-alloy structural steels,* Technical delivery conditions for the thermomechanical rolled weldable fine grain steels. EN10025-4.

Construction report for KANMON Bridge, 1977, Japanese Society of Civil Engineers, (in Japanese).

HSBA. 1985, *Construction report for INNOSHIMA Bridge.* (in Japanese).

HSBA. 1987, *Construction report for OHNARUTO Bridge.* (in Japanese).

HSBE. 1988, *Construction report for SETO Bridge.* (in Japanese).

HIROSHIMA Prefecture, 2000, *Construction report for AKINADA Bridge.* (in Japanese).

KENSETU-TOSYO. 1998. *Erection method for the towers of the suspension bridges. Bridge and foundation engineering Vol.32-No.8*:138–141 (in Japanese).

Nogami, K., Saito, K., Nagai, M. and Fujino, Y., 2001, *Rational allocation of safety factor of cables, hangers and towers in a super long-span suspension bridge, JSCE No.682/I-56*:323–334 (in Japanese).

Yoneyama, M., Koike, T. and Nose, T., 1993, *Fabrication and erection of suspension bridge towers for HAKUCHO bridge for HOKKAIDO development bureau. IHI technical review Vol.33-No.5*:342–346. (in Japanese).

Chapter 6

Cable supported footbridge analysis with construction staging

A. Caner
Middle East Technical University, Civil Engineering Department, Ankara, Turkey

ABSTRACT: The new trend in design of footbridges in Turkey is to utilize the cables. Some of these bridges have fake cables while others partially rely on the cable system. These steel composite bridges typically constructed over the highways span about 40 to 60 meters. It was observed that the bridges with fake cables can be substantially heavier than the ones with functional cables. The focus of this paper is to study importance of tensioning sequence of cables and impact of construction staging on the design forces at superstructure to have economical designs. A case study is illustrated as an example design.

1 INTRODUCTION

Aesthetics of urban pedestrian bridges are mainly governed by the functional, environmental, visual and structural conditions (Yang and Huang 1997). Functional condition includes types of loads, traffic volumes and speeds, and clearance requirements. Environmental condition mostly concerns with the topographic conditions and surrounding environment. Visual conditions should suit the public's taste. Structural condition deals with the serviceability and stability of the bridge. In Ankara, Turkey, the architects' preferences on the shape of pedestrian bridges are either arch or cable-supported type.

Urban pedestrian bridges are more flexible than other type of bridges since the loads are small and bridge widths are narrower (Yang and Huang 1997). Such light structures with small damping values, especially cable supported bridges, are more vulnerable to dynamic forces than other type of foot bridges (Nakamura 2004). Nakamura (2004) and Roberts (2005) have studied vibrations induced by the lateral pedestrian movements over footbridges. London Millennium Bridge was closed for 18 months due to unstable amplitudes of vibration induced by crowded pedestrian crossing and was opened to public use only after a 7 million dollar retrofit. The internal damping of the bridge was increased by the installation of both viscous and tuned mass dampers to eliminate or minimize the vibration problem (Roberts 2005). Similarly, the T-bridge, a cable-stayed bridge in Japan, experienced strong lateral vibration induced by pedestrians (Nakamura 2004). Although this bridge has the same span length as the London Millennium Bridge, it has a heavier superstructure; therefore restricting public access to the bridge was not necessary. The vibration induced displacements were measured to be about 10 mm at the T-bridge, compared to about 60 to 70 mm at the London Millennium Bridge.

Dynamic behavior of a curved cable-stayed bridge, Safti Link Bridge in Singapore, was studied by full-scale testing and analytical methods (Brownjohn et al. 1999). The testing techniques used were ambient vibration testing (AVT) and forced vibration testing (FVT). AVT relies on uncontrollable force inputs such as wind, traffic and other natural sources of vibration. FVT depends on a measurable force input such as an impact hammer. Actual cable tensions were checked by the post-tensioning contractor and were determined to be close to the design values. The cable tension estimations by the vibration measurements (FVT) were found to be mysteriously inaccurate. The AVT testing determined to be more useful than FVT testing due to low signal to noise ratio for vibrations induced by hammer impact.

Simoes and Negrao (2005) worked on the optimum design of cable-stayed footbridges concentrating on minimization of stresses, displacements and cost. Their investigation did not include any type of vibration analysis.

Some of the pedestrian bridges in Ankara, Turkey with span lengths 40 to 60 meters have either fake cable-stays or functional cable-stays. It was observed that these bridges can have sometimes uncomfortable vibration levels due to pedestrian walking or wind. Out of 11 bridges with functional cable-stays under construction, only couple of them has vibration problems believed to be related to improper tensioning of cables, and even in some cases loose cables exist. Vibration characteristics of these bridges are still under investigation. The focus of this paper is to study importance of tensioning sequence of cables and impact of construction staging on the design forces at superstructure to have economical designs.

2 ANALYSIS

2.1 *Description of bridge*

The pylons or tower of the bridge are typically inclined custom made steel "A" type of frames as shown in Figure 1. The back stay cables are anchored to a concrete block at a distance from the bridge. For the bridge with fake cables, the superstructure has usually three longitudinal girders, the middle one being the heaviest section connected to the edge girders with floor beams.

(a)

(b)

Figure 1. Pedestrian bridge (a) with fake cable stays and (b) without cables.

Similar to the fake cable supported footbridge, the bridge with functional ones has the same architectural aesthetics but with a relatively shorter tower and a lighter superstructure. The superstructure has only two edge girders connected with transverse floor beams.

2.2 *Construction staging and cable tension sequencing*

Span length of the superstructure measured from tower connection to the anchor pier connection was taken to be 45 meters. Two edge girders were 3 meters apart from each other and connected with 160 mm deep transverse I shaped floor beams. The cables were selected to be 30 mm diameter bridge ropes with adjustable tension device (turnbuckle) on them. The steel material was selected to have yield strength of 250 MPa. The 80 mm thick reinforced concrete deck had a 50 mm thick finish work on top of it. The erection sequence of the structure is as follows:

- Erect the pylon or tower
- Construct the concrete anchor block for the back stays
- Erect the anchor pier
- Connect the superstructure to the tower and the anchor pier
- Place the cables between the superstructure and the tower, and between the tower and the concrete anchor block
- Place the concrete deck and the concrete tiles
- Tension the cables

In cable tensioning, four different options may be used:

- OPTION 1: Do not tension any of the cables. All cables are non-functional.
- OPTION 2: Only tension back stays to 100 kN to pull back the tower to prevent any sagging of superstructure due to placement of concrete deck. Do not tension cables C1, C2, C3, C4 and C5 shown in Figure 2. Cables C1 to C5 will be additionally stressed by the dead weight of the concrete deck.
- OPTION 3: After placement of the concrete deck, start tensioning the bridge to have the right alignment. Tension all back stays to 100 kN. Start tensioning from the cable closest to the pylon and finish tensioning at the cable closest to the anchor pier without changing the order (tensioning sequence: C1-C2-C3-C4-C5, see Figure 2).

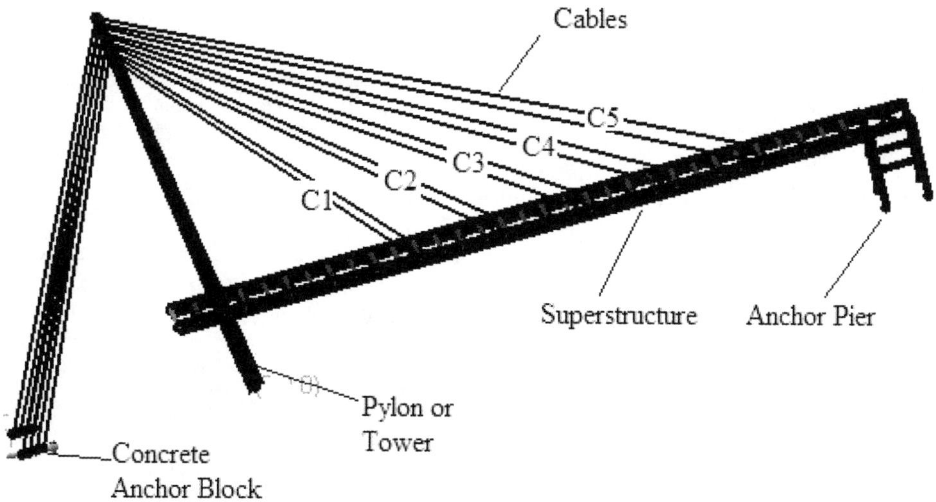

Figure 2. 4-D LARSA Computer Model, fourth dimension being time of cable tensioning.

- OPTION 4: Similar to second option tension all back stays to 100 kN. Tension one set of cable close to the pylon and in the next step tension cables close to the anchor pier and after tensioning both end cables, start tensioning the mid cables (tensioning sequence: C1-C5-C3-C4-C2, see Figure 2).

LARSA structural analysis software was used to model the bridge as shown in Figure 2. In construction staging, full non-linear geometric option was turned on to monitor the change in cable tensions. Deck displacement induced by individual tensioning of cables decreased the initial tension force in the adjacent cables next to the tensioned one.

3 DISCUSSION OF RESULTS

The analysis results for different schemes were tabulated in Tables 1 to 3. Having functional cables, i.e. tensioned cables, helped to reduce the design forces compared to a bridge with fake cables. In comparison, the bridge member design was selected in such a way that at each option the live load deflections of the bridges were almost the same.

Table 1. Cable forces at the initial and final stage without live load.

	Cable Forces (C1 closest to pylon and C5 closest to anchor pier)									
	Cable C1		Cable C2		Cable C3		Cable C4		Cable C5	
Option	I* (kN)	F** (kN)	I* (kN)	F** (kN)	I* (kN)	F** (kN)	I* (kN)	F** (kN)	I* (kN)	F** (kN)
1	0	0	0	0	0	0	0	0	0	0
2	0	65	0	75	0	70	0	53	0	29
3	120	68	110	65	100	59	100	70	130	127
4	100	52	80	80	100	53	100	89	130	100

* Initial cable force
** Final cable force

Table 2 Member forces under dead plus live load.

Option	Edge girder negative moment at tower connection kN-m	Edge girder positive moment kN-m	Tower maximum moment above superstructure level kN-m	Tower base maximum moment kN-m	Tower axial force kN
1	2496	2796	109	2252	505
2	724	524	164	1208	1062
3	676	356	306	1736	1336
4	615	292	279	1593	1206

Table 3. Live load displacement, steel weight of the structure and cost of the steel works.

Option	Live load displacement <90 mm (mm)	Steel weight (ton)	Steel value including labor (USD)
1	90	80	$160,000
2	88	50	$100,000
3	85	50	$100,000
4	86	50	$100,000

Having fake cables at a pedestrian bridge can result in increase in edge girder moment terms up to ten times of a bridge with properly tensioned cables. Since the fake cable-stayed bridge does not rely on the cables, the tower axial forces are about 50% less compared to the functional cable-stay bridges. Cables in tension develop compressive forces in towers, which results in higher tower axial forces.

The steel weight of a fake cable-stayed bridge can be substantially heavier than the one with a functional cable-stayed bridge to satisfy the live load displacement criterion of AASHTO (1997). The steel weight of the option 1 bridge is about 60 percent heavier than the option 2, 3 or 4 bridges. Under its own dead load, the steel profiles of the option 1 bridge could be cambered up to minimize the dead load deflections.

When option 2, 3 and 4 were compared to each other, the importance of cable tensioning sequence could be seen. Even if, these three bridges were identically the same, in option 4 edge girder positive moment is about 80% lower than the ones for option 2. However, no reduction in weight of superstructure is planned when option 4 is used in case contractor selects to use option 2 or 3 as his or her tensioning sequence. Due to cable tensioning sequence, the cable forces in the first tensioned cable can be reduced up to 50% of its initial value. The reductions in cable forces for the last tensioned cables were substantially less and may be ignored in some cases.

4 CONCLUSIONS

- Steel weight and cost of the pedestrian bridge can easily be controlled by the proper tensioning and construction staging.
- The design forces in the same bridge can be varied by 80% by just changing the tensioning sequence of the cables.
- Cable tension forces are very sensitive to the sequence of tensioning. Final tension forces can be reduced up to 50% of their initial values due to the variation in deck displacements during tensioning.

ACKNOWLEDGEMENTS

Author wants to thank to Prof. Dr. Çetin Yılmaz and Assoc. Prof. Dr. Ahmet Türer of Middle East Technical University Civil Engineering Department for their support.

REFERENCES

AASHTO. 1997. *Guide specifications for design of pedestrian bridges*, Washington, DC.

Brownjohn J. M. W., Lee J. and Cheong B. 1999. Dynamic performance of a curved cable-stayed bridge. *Engineering Structures*, 21: 1015–1027.

Nakamura S. 2004. Model for lateral excitation of footbridges by synchronous walking. *ASCE Journal of Structural Engineering*, 130(1): 32–37.

Roberts T. M. 2005. Lateral pedestrian excitation of footbridges. *ASCE Journal of Bridge Engineering*, 10(1): 107–112.

Simoes L. M. and Negrao, J. H. 2005. Reliability-based optimum design of glulam cable-stayed foot bridges. *ASCE Journal of Bridge Engineering*, 10(1): 39–44.

Yang, S. and Huang, D. 1997. Aesthetics considerations for ùrban pedestrian bridge design. *ASCE Journal of Architectural Engineering*. 3(1): 3–8.

Chapter 7

Locked coil cable assemblies for bridges

M. Bechtold & F.E. Rentmeister
BRIDON International GmbH, Gelsenkirchen, Germany

B. Mordue
BRIDON International Ltd., Doncaster, UK

ABSTRACT: Locked coil cable assemblies are used in cable supported road bridges (e.g. as suspenders in suspension bridges and hangers in arch bridges) and a large variety of pedestrian and cycle bridges. Despite of lots of installations all over the world and recent product enhancements, locked coil cable assemblies are not so well known in the USA. In recent times an increase in interest and demand for the product has been observed.

A general introduction to locked coil cable assemblies will be given. After an overview about present and past applications this paper provides detailed information on technical aspects. The elaborations include design aspects for saddles and clamps, installation examples, deck and tower connection examples as well as the latest developments in corrosion protection and sockets.

1 INTRODUCTION

Locked coil cable assemblies are prefabricated tension components for structural applications. They consist of locked coil strands and permanently attached sockets. Locked coil strands are a special type of spiral strand. They are made up using a core of helically spun round wires in several layers onto which covers of helically spun full lock wires in several layers are spun (figures 1 and 2). They are manufactured in the factory, layer by layer. The layers are usually spun in opposite directions to minimize the residual torque.

Locked coil strands (figure 3) have been made from 20 mm (~7/8″) up to 176 mm (~7/8″) in diameter. Their advantage over spiral strand (figure 4) and wire rope (figure 5) include

- better corrosion protection
- higher axial stiffness
- better clamping capabilities
- higher breaking load (only in comparison to wire rope)

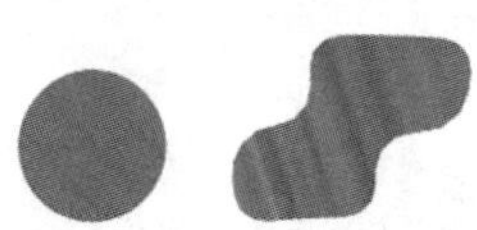

Figure 1. Round and full lock wire for locked coil strand.

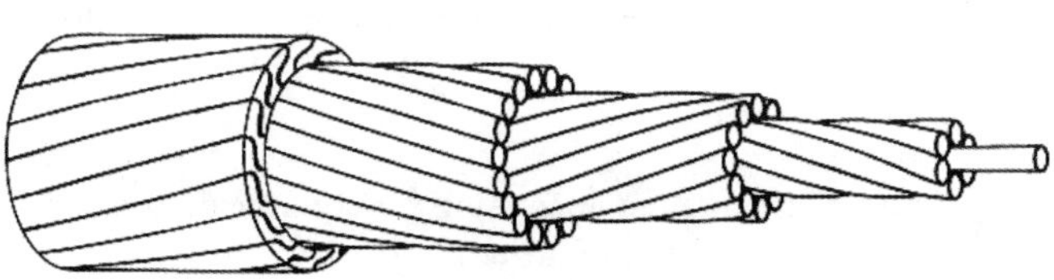

Figure 2. Round wire core with one layer of full lock wires.

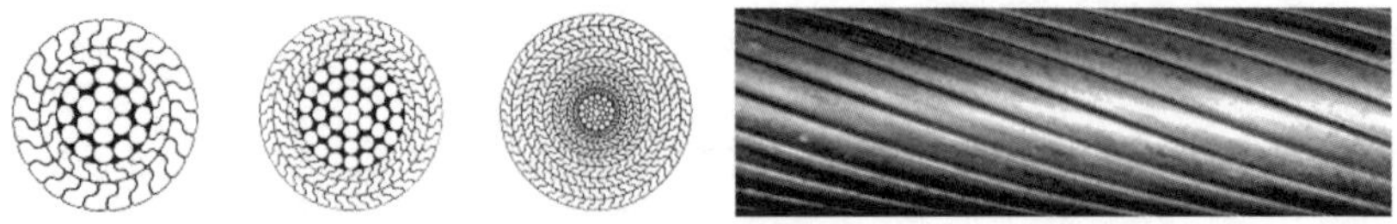

Figure 3. Locked coil strand cross section and side view.

Figure 4. Spiral strand cross section and side view.

Figure 5. Wire rope cross section and side view.

Figure 6. Millennium Bridge London (UK, 2000) and Milwaukee Art Museum Bridge (USA, 2000).

Figure 7. Turtle Bay Sundial Bridge (USA, 2001) and Swansea Sail Bridge (UK, 2003).

2 APPLICATION EXAMPLES

In the past 15 years the world has seen an increasing variety of pedestrian bridges. Locked coil cable assemblies have been used for many of them.

Locked coil cable assemblies are also used as suspenders in vehicular suspension bridges and arch bridges.

Figure 8. Pedestrian Bridge Strasbourg/Kehl (France/Germany, 2003) and River Usk Bridge (UK, 2006).

Figure 9. Fedafjord Bridge (Norway, 2006) and Zaragossa Bridge (Spain, 2007).

Figure 10. Ulvoen Bridge (Norway, 1928).

A remarkable example is the Ulvoen Bridge in Norway from 1928 (figure 10). It has still the original locked coil cable assemblies in service although at the time the wires were not galvanized.

Besides bridges, the focus of this paper, locked coil cable assemblies are also used in roof structures, glass walls and for stayed masts and towers.

3 TECHNICAL PROPERTIES

3.1 *Load carrying capacity*

The wires are made from high carbon steel rod. They get their high strength from the cold deformation process during wire drawing. The wire properties are shown in figure 11.

Locked coil strands are made to the product standard EN 10285-10. Due to the full lock wires locked coil strands with more than 3 layers of full lock wires (~strand diameter >60 mm (~2 3/8″)) have a metallic cross section of minimum 88% compared to the circumscribed circle. This value is higher than the value for spiral strand (~75%) and wire rope (~60%).

The term "minimum breaking load" (MBL) is used to describe the load which will be achieved in a physical breaking load test. Manufacturers of locked coil strand state the MBL of their strand in their brochures.

Shapes	Round 	Full Lock 
Material	High Carbon Steel	
Standards	EN 10264-2	EN 10264-3
Diameter / Height	3.0 – 6.0 mm	3.5 – 7.0 mm
Nominal Strength	1570 N/mm² (230 ksi)	1570 N/mm² (230 ksi) (1470 N/mm² (213 ksi))

Figure 11. Wire properties.

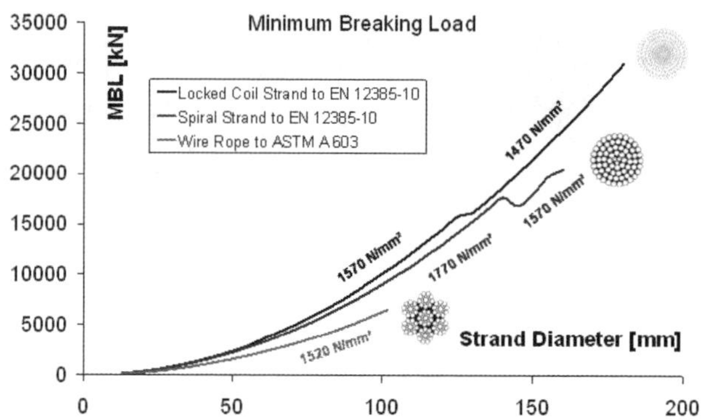

Figure 12. Comparison between the MBL of locked coil strands, spiral strands and wire rope.

Figure 12 shows the comparison between the minimum breaking load of locked coil strands to EN 12385, spiral strands to EN 12385 and wire rope to ASTM A 603.

Sockets are usually designed to develop an ultimate strength greater than the breaking load of the strand.

3.2 *Load elongation characteristic*

The axial stiffness is the product of the elastic strand modulus and the metallic cross sectional area. Locked coil strands have the highest axial stiffness to diameter ratio of all tension members apart from solid steel bars of course.

Additionally to the advantage in metallic cross section described above there is an advantage in elastic strand modulus. The elastic strand modulus is not a material property but a phenomenological property which describes the elongation characteristics of the strand. It considers the material elastic modulus of the wires as well as the geometrical arrangement of the wires in the strand.

The product standard EN 12385 does not give values for the elastic strand modulus but values are given in the design standard Eurocode 3. It gives an elastic strand modulus for locked coil strand of 160000 MPa which seems a bit low in comparison to measured values of ~165000 MPa.

Eurocode 3 gives an elastic strand modulus for spiral strand of 150000 MPA but measured values vary between ~155000 MPa for large diameters (> ~80 mm (~3 1/8″)) and 176000 MPa for small diameters (< ~30 mm (~1 1/8″)).

ASTM A 603 gives an elastic strand modulus of 140000 MPa for wire rope.

Figure 13 shows the comparison between the axial stiffness of locked coil strands to EN 12385, spiral strands to EN 12385 and wire rope to ASTM A 603 under consideration of the realistic elastic strand modulus values.

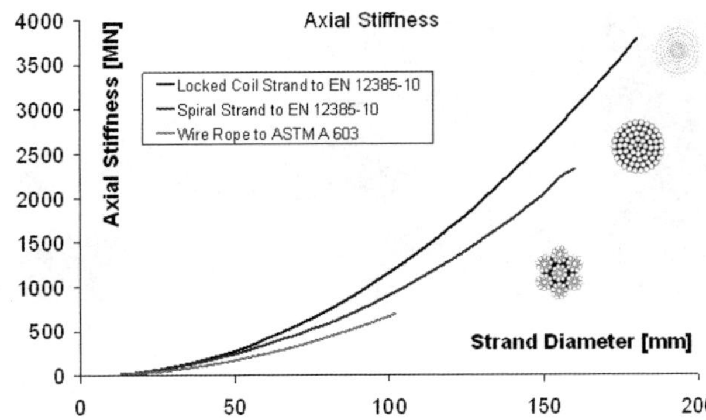

Figure 13. Comparison between the axial stiffness of locked coil strands, spiral strands and wire rope.

Galvanising with Zinc	Galvanising with Galfan®
Hot dip galvanizing process to EN 10244-2	Double hot dip galvanizing process to EN 10244-2 and ASTM A856
99.9% Zn	Alloy of zinc and aluminium at the eutectic ratio of 95% Zn and 5% Al
The ZnO reaction products on the surface are quite easily carried away by rain etc. until complete removal. 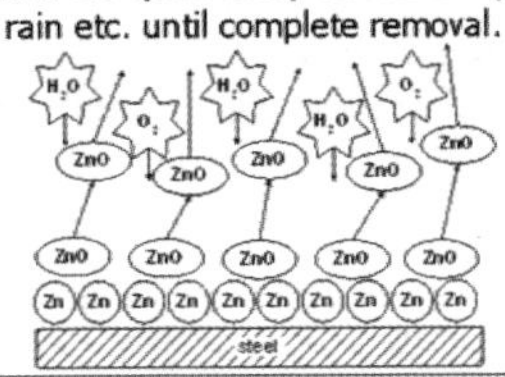	The Al2O3 slows down the deterioration of the protective layer up to the factor 3. 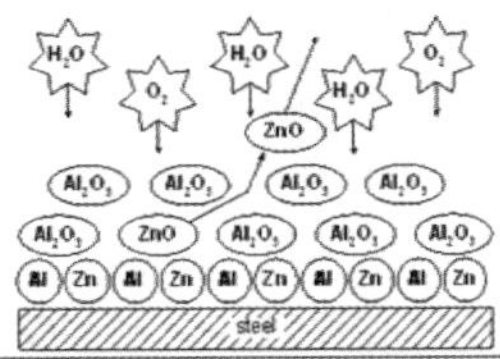
	Better base for additional coatings

Figure 14. Principle functioning of 99.9% zinc and Zn95Al5 galvanizings.

3.3 *Corrosion protection*

3.3.1 *General*

Locked coil strands have three corrosion barriers and an optional fourth one. In addition, potential problems can be addressed during the design phase by preventing localized corrosion points within the strand system. For example, items such as saddles and clamps must be designed to prevent build-up of moisture and it is recommended that the strand manufacturer is consulted at this time.

3.3.2 *Barrier 1*

All wires are usually EN 10264 class A (290 g/m^2) hot dip galvanized in a 99.9% zinc bath. For improved corrosion resistance by approximately factor 3 the wires of the two outer layers are class A hot dip galvanized with the Zn95Al5 double dip galvanizing process. This type of galvanizing is often referred to as Galfan® which is a registered trademark of the International Lead Zinc Research Organization (ILZRO). Although Galfan® was developed already in 1980 it is relatively unknown to the construction industry. Unlike to the class C (915 g/m^2) galvanizing in ASTM A 603 where the wire strength is reduced by 10% and the modulus is reduced by 5% Zn95Al5 galvanized wires maintain their full properties.

The way Zn95Al5 works is that the aluminum oxide that builds up over time sticks better to the surface than zinc oxide (figure 14). Considering this it makes sense to use Zn95Al5 in the two outer layers of a locked coil rope. Zn95Al5 is in use for locked coil cable assemblies in structures since

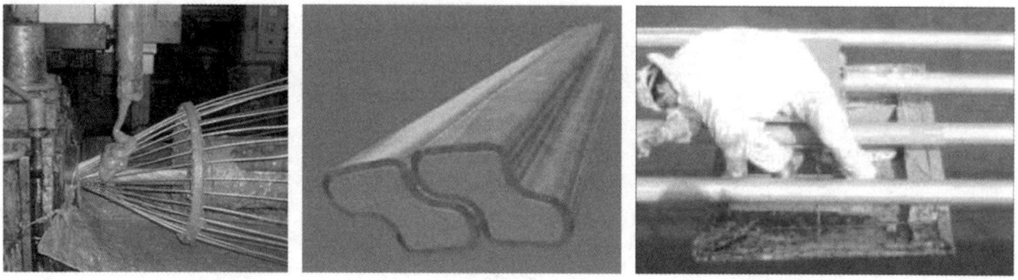

Figure 15. Blocking compound during stranding, interlocking full lock wires and painting of strands.

1991. It is used for spiral strand for mast stays since the 1980s. It has proven its performance multiple times in salt spray (NaCl) and Kesternich (So2) corrosion tests as well as in long term field tests.

ASTM A856/A856M and European wire standard EN 10264 include Zn95Al5 galvanized wires.

3.3.3 *Barrier 2*

During stranding a blocking compound is added to the cable interior (figure 15). By filling up the inter wire gaps and additionally coating the wires inside the strand it prevents the intrusion of corrosive media and is itself a corrosion inhibitor. The blocking compound can be a suspension of aluminum flake incorporated into a hydrocarbon resin carrier or a zinc dust compound.

3.3.4 *Barrier 3*

The full lock wires themselves provide an effective surface barrier against the penetration of corrosive media because of the interlocking of the full lock wires (figure 15).

3.3.5 *Optional barrier 4*

Additional protection which is for example needed for suspenders and hangers of vehicular bridges can be provided by painting (figure 15) or sheathing of locked coil strands.

Strand coating systems are specifically designed to protect against corrosion. Common coating systems include a suspension of aluminum flake incorporated into a hydrocarbon resin carrier, diluted with a solvent for ease of application. Another coating system is a 2 layer polyurethane based two compound paint with an epoxy resin based primer.

Those products do not dry hard like conventional paint systems. Although dry to touch, they remain flexible allowing for the differential wire movement as the underlying cables are tensioned in-service, thus eliminating surface cracking. When selecting the external coating, care should be taken to ensure compatibility with the internal compound.

3.3.6 *Sockets and clamps*

Sockets and clamps need to have the same level of corrosion protection as the strand. The primary corrosion protection of sockets and clamps is provided by applying a zinc coating either using the hot-dip process or by hot metal spraying. As with the strand, additional protection can be obtained by further coating the sockets and clamps with a paint system if required.

3.4 *Fatigue resistance*

The failing of strands submitted to fatigue loading is fundamentally different to that of conventional steel structures where almost the whole fatigue life of the structure is related to the period before the first crack happened. In contrast to this the first wire crack in a strand has no relation to the much later failing of the complete strand. The blocking compound added during spinning and the galvanizing of all wires reduces the inter-wire friction. These measures have improved the fatigue performance in the past few decades. In locked coil strands the interlocking full lock wires avoid the popping out of a broken wire in the outer layer which is an advantage over spiral strand and wire rope. Due to the spiral assembly a broken wire will fully carry load again after approximately three lay lengths ($\sim 3 \times 10 \times$ strand diameter).

If the structure is submitted to considerable fatigue loads, locked coil strands are usually tested with an upper load of 45% of the MBL, a variation in stress of $150\,N/mm^2$ and 2 million load cycles. The remaining breaking load after the test needs to be 81% of the MBL. Recent test results show values of more than 90% of the MBL most of the times.

3.5 *Design aspects*

3.5.1 *Ultimate limit state*

The common value for all design standards is the minimum breaking load (MBL) which is the load that will be achieved in a breaking load test. The MBL is also referred to in some design standards as the characteristic breaking load or as the nominal cable strength.

The design standard Eurocode 3 uses the partial safety factor philosophy which is also known as the load resistance factor design (LRFD). Safety factors are applied to the loads as well as to the materials or structural components respectively. The design resistance of a cable $Z_{R,d}$ subjected to a static load is calculated by dividing the MBL with the partial safety factor of $1.5 * 1.1 = 1.65$. The applied loads are multiplied by safety factors (e.g. 1.35 for dead loads and 1.50 for live loads). The static calculations for different load combinations then results in the design cable tensions $N_{R,d}$. They must be smaller than or equal to the design resistance $Z_{R,d}$.

The design standard ASCE 19 uses the single safety factor philosophy which is also known as the allowable stress design (ASD). This safety philosophy uses a single, overall safety factor. The static calculation for different load combinations results in the cable tensions T. The cable tensions T are multiplied by safety factors (2.0 or 2.2, depending on the load combination) and are required to be smaller than the MBL.

Locked coil strands are designed and made for each particular application. The diameters listed in brochures of manufacturers are just examples and any intermediate diameters can be manufactured. If a range of different diameters is needed, early consultation with the manufacturer will lead to an optimized solution in terms of both product and cost.

3.5.2 *Cable length*

It depends of the structure itself how accurate the cable assembly length needs to be. The following points need to be considered.

In a prestretching procedure the constructional stretch can be taken out of the strand. This is typically done in the factory by loading and unloading the strands up to 10 times between 10% and 50% of the MBL. The strand than behaves elastic and can be marked to the specified length. Load and length at a given temperature need to be taken from the static calculation for the structure.

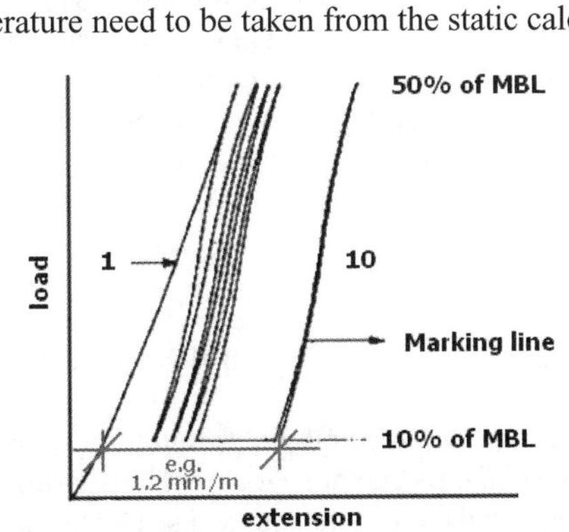

Figure 16. Prestretching of a strand.

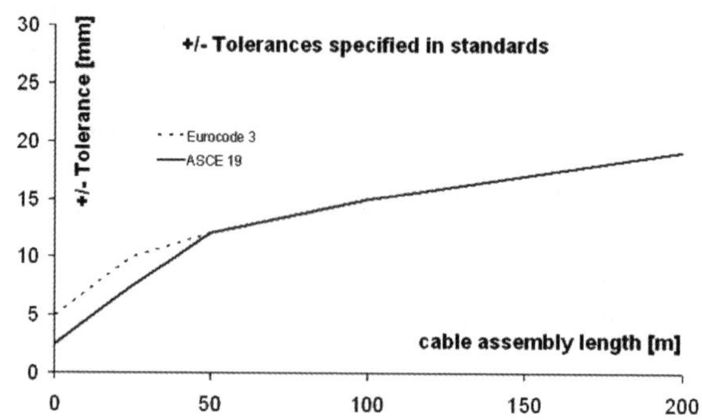

Figure 17. Length accuracy in ASCE 19 and Eurocode 3.

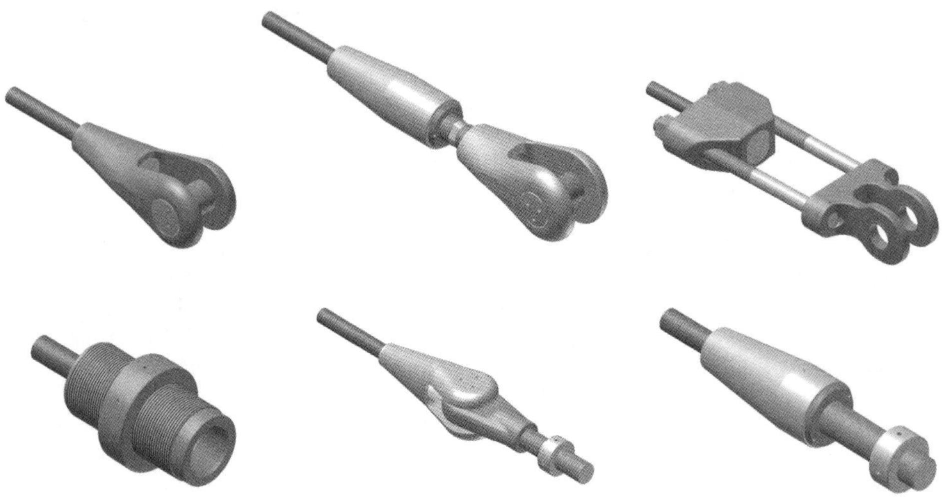

Figure 18. Sockets for locked coil strands.

Even a prestretched strand will show some additional permanent elongation once it is under load in the structure. This phenomenon is usually referred to as strand creep. For a prestretched strand it will be around 0.15 mm/m.

During initial loading of the structure, some socket seating will occur. The magnitude depends on the size and type of the sockets. It will usually be around 1–5 mm per socket.

ASCE 19 and Eurocode 3 call for the following length accuracy for a cable assembly.

There are experience values for all of the above-mentioned effects which can be considered when marking the strand to the required length. For exceptionally length sensitive applications physical tests on cable assemblies can provide the magnitudes for the individual cable assemblies.

3.6 *Architectural socket designs*

Locked coil strands are terminated with poured sockets which are designed to be stronger than the strand. The cable assembly can transmit the MBL of the strand. The most common sockets are shown in figure 18.

The socket design is crucial for architectural pedestrian bridges. Sockets should underline the light and transparent appearance of the structure. Geometrically optimized sockets allow for architecturally pleasing connection details at towers and decks. The use of up to date casting methods and materials with a yield strength of 550 N/mm^2 (80 ksi) and a tensile strength of 700 N/mm^2

Figure 19. Tower connection details of Pedestrian Bridge Strasbourg/Kehl (France/Germany, 2003) and Sale Water Park Pedestrian Bridge in Manchester (UK, 2003).

Figure 20. Deck connection details of Pedestrian Bridge Strasbourg/Kehl (France/Germany, 2003) and Swansea Sail Bridge (UK, 2003).

(102 ksi) and a Charpy value of minimum 27 Joule at −20°C (−4°F) enable the design of slender and safe sockets.

Tower and deck connection details are shown in figures 19 and 20.

3.7 *Clamps and saddles*

With clamps and saddles forces can be introduced into the free length of a strand. This helps by reducing the number of sockets and therefore is architecturally more pleasing and cost saving.

Locked coil strands have a very smooth surface in comparison to spiral strands and wire rope. This allows for a good pressure distribution between the saddles/clamps and the strand surface. The grooves should be lined with a 1 mm soft metal layer (e.g. zinc) to cushion the strand. When cushioned, locked coil strands can be deflected over a radius of 20 times the strand diameter with no loss of efficiency; otherwise 30 times the strand diameter should be applied. When cushioned, transverse pressure from clamps should be a maximum of 100 N/mm^2 (14,5 ksi) for locked coil ropes, 60 N/mm^2 (8,7 ksi) for spiral strand and 40 N/mm^2 (5,8 ksi) for wire rope.

The clamps need to be designed to transmit the force of the attached structural member into the strand. The component of the force which is in the strand's longitudinal direction will be transmitted by friction. Sufficient clamping force from preloaded bolts should be considered to compensate the following effects:

a) Reduction of the diameter if the tension is increased
b) Bedding down of the strand over time
d) Reduction of preload in clamp bolts by external forces
e) Change in temperature

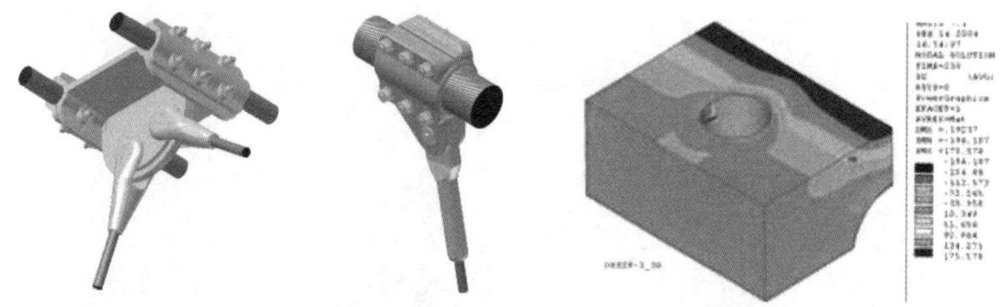

Figure 21. Clamp examples.

Figure 22. Transport of locked coil cable assemblies in coils.

3.8 *Packing, dispatch and installation*

Locked coil cable assemblies should be transported in coils with a diameter of minimum 30 times the strand diameter (figure 22). Coils can be wrapped in polyweave plastic to prevent contamination from dust, sand etc. The sockets can be wrapped to protect against mechanical damage during transit. Smaller diameter coils can be stacked and shrink wrapped on enclosed wooden pallets for added protection.

For a lot of steel contractors steel cable assemblies are not the every day business and therefore assistance of the manufacturer should be asked for. Most manufacturers of locked coil cable assemblies can assist in handling on site, do supervision or even install cable assemblies themselves.

3.9 *Quality assurance*

Locked coil cable assemblies are factory made including the attachment of the sockets. The manufacturing takes place in a fully controlled environment with established methods and machinery. This gives the basis for a very consistent and monitored quality.

Locked coil cable assemblies arrive on site ready for installation. Quality sensitive site work is reduced to a minimum.

Testing starts with the single components wire, socketing material and sockets and it ends with the finished product if a cable assembly type test is required.

4 CONCLUSIONS

Locked coil strands provide additional and improved properties in comparison to spiral strand and wire rope. In combination with aesthetically designed and optimized sockets they are the ideal tension component especially for architectural bridges. For vehicle bridges, their excellent corrosion resistance makes them well suited to use as suspenders on suspension bridges and hangers on arch bridges.

REFERENCES

ASCE 19-96 Structural Applications of Steel Cables for Buildings

ASTM A856/A856M-03 Standard Specification for Zinc-5% Aluminum-Mischmetal Alloy-Coated Carbon Steel Wire

EN 10264:2002 Steel wire and wire products – Steel wire for ropes

EN 12385-10:2003 Steel wire ropes – Safety – Part 10: Spiral ropes for general structural applications

Eurocode 3: Design of steel structures – Part 1.11: Design of structures with tension components (prEN 1993-1.11:2005)

3 Seismic analysis & design

Chapter 8

Sliding isolation bearings in cold weather climates

R.J. Watson
R.J. Watson, Amherst, N.Y., USA

ABSTRACT: Isolation bearings have become a standard tool for engineers designing bridges in seismic regions. However the added complication of cold weather has raised concerns with rubber isolators and their performance in northern regions.

As a result bridge designers are migrating towards Sliding Isolation Bearings (SIB) in these regions. SIB have been proven to be cost effective and high damping devices on numerous projects to date. In addition their outstanding performance in cold temperature testing has proven their efficacy in cold weather regions.

This paper will delve into the research that led to the development of SIB. In addition several case histories will be reviewed in an effort to demonstrate SIB capabilities in low temperature environments.

1 INTRODUCTION

SIB consist of a disc type–high load multirotational bearing coupled with polyurethane displacement control springs referred to as Mass Energy Regulators (MER). The Sliding is accommodated through the use of a stainless steel/PTFE interface which allows multidirectional movement. The MER are designed to control the seismic displacements and restore the bridge back to its original pre-quake position.

Research was conducted at The Multidisciplinary Center for Earthquake Engineering Research (MCEER) which consisted of a detailed study of various sliding surfaces along with component and shake table testing (Fig 1).

The result of this research yielded a cost effective isolation system that offers the following advantages:

1. Significant reduction of seismic forces that are transferred to the bridge substructure.
2. Ability of the designer to direct the seismic loads to those elements that are most capable of resisting them.
3. Ability to accommodate multidirectional non-seismic movement as that of horizontal curved bridges.
4. Use of small movement expansion joints.

In addition this research revealed that SIB were a low profile, cost effective, high damping and versatile device for bridge applications.

To date SIB have been used on over 100 structures worldwide. Engineers designing bridges in cold weather climates have found that SIB can perform the aforementioned features in cold temperatures. Four different bridges in the Ottawa, Ontario Region have been built in recent years using SIB. These case histories will be examined in an effort to determine the benefits of SIB for bridge designers.

Figure 1. SIB Shake table testing at MCEER.

2 CASE HISTORIES

2.1 *Bytown bridges*

The Bytown Bridges are located within the central core of the city of Ottawa. They carry Sussex Drive over two branches of the Rideau River. The existing Bytown Bridges were constructed in 1954 and consisted of 2 similar 3 span precast prestressed concrete T girders.

As a result of a 2001 assessment study commissioned by the City of Ottawa in collaboration with the National Capital Commission, the complete replacement of the Bytown bridges was recommended due to corrosion and seismic issues. The City then contracted Delcan Engineers to design replacement structures. The engineers at Delcan came up with a unique hybrid pre-cast/cast in place concrete bridge design, which incorporated SIB. The use of SIB resulted in the use of 30 rock anchors imbedded 3.0 meters compared to 82 rock anchors imbedded 8.1 meters for the foundation design. This resulted in a substantial savings on the overall structure cost, which easily exceeded the additional cost of SIB [2]. Delcan Engineers saved even more in their design by combining the use of conventional elastomeric bearings at the abutments and SIB at the piers. Additional savings were achieved by connecting the precast prestressed concrete box girders at the piers with a monolithically poured in place concrete diaphragm. This detail reduced the number of isolation bearings required in half (Fig 2).

The engineers at Delcan put together a rigorous set of tests to verify the performance of the SIB. The first requirements was that a prototype bearing be subjected to a 1000 salt spray test to ensure that the SIB can withstand a high chloride road salt environment.

In addition 24 of the 48 SIB required for this project were subjected to quality control testing consisting of proof load testing, seismic testing in accordance with the AASHTO Guide Specification for seismic isolation design [3] and cold temperature testing at +15°C and −34°C. The requirement for this project was that the average force developed over 3 displacement cycles at −34°C was not allowed to be more than 130% of the average force at +15°C (Fig 3).

All of the bearings tested met this requirement. The bridges were completed in 2005 (Fig 4).

Figure 2. Precast girder being lowered onto SIB.

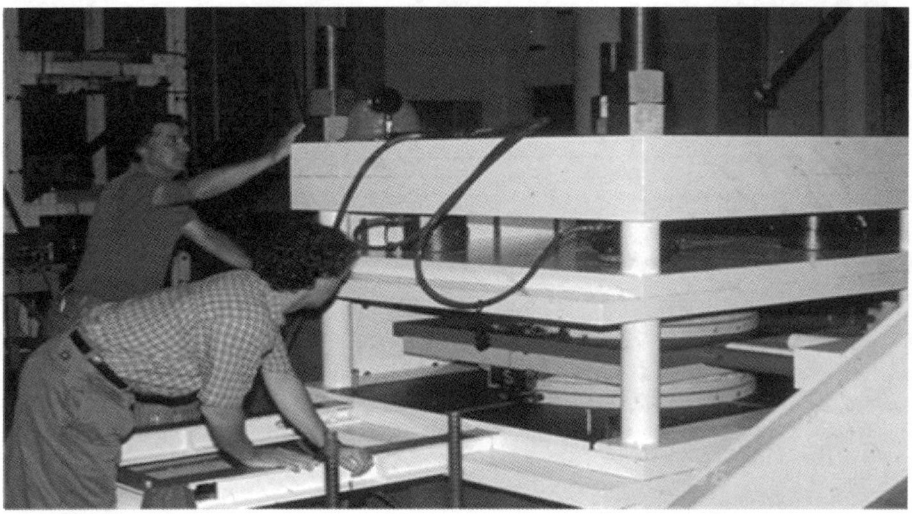

Figure 3. SIB testing.

2.2 *RR 22 over highway 417*

As the Ministry of Transportation, Ontario (MTO) extended the 4 Lane Highway 417 West of Ottawa, new regional road overpasses were required. MTO selected McCormick Rankin to design a 73 meter 2 span overpass for Regional Road 22 to span Highway 417 (Fig 5).

McCormick chose a prestressed concrete box girder design as the most cost effective. They chose the use of SIB to reduce the equivalent static earthquake load from 126.8 Kn/m to 40.9 Kn/m. This resulted in a significant reduction in the amount of piles that were being driven in excess of 50 meters deep at the bridge site. An additional benefit of SIB was the elimination of shear keys at the abutments and high tensile protection ties at the foundations that would have

Figure 4. Bytown bridges.

Figure 5. RR 22 over highway 417.

been required to resist the seismic forces. The 4 abutment bearings had a vertical capacity of 2890 Kn while the 2 pier bearings were designed for 6675 Kn (Fig 6).

2.3 *Pedestrian bridge over hwy 417*

An existing pedestrian Bridge over Highway 417 west of Ottawa did not meet the current Ontario Highway code requirements for clearance. Therefore the City of Ottawa hired Morrison Hirschfield to design a new structure. They chose a 4 span steel truss sitting on 3 reinforced concrete hammerhead piers and two concrete abutments (Fig 7).

The two center spans measure 38.7 meters with the end spans at 25.7 meters each. Since this structure was identified as a critical emergency response path, SIB were evaluated to keep the structure operational after a seismic event. In an effort to save money the engineers came up with a design that enabled them to use conventional elastomeric bearings at the abutments and SIB at the piers to keep the forces below the allowable.

Figure 6. RR 22 pier SIB.

Figure 7. Pedestrian bridge over highway 417.

The design of the 6 SIB required for this project was somewhat unusual in that each bearing contained two 570 Kn capacity polyurethane load and rotational elements. In addition there were 8 mass energy regulators (MER) incorporated into each bearing, which provide recentering features (Fig 8). The SIB were also designed for ± 70 mm of seismic displacement. The other feature of SIB that was incorporated on this project was that of direction specific stiffness. The designer desired a stiffer seismic response in the transverse direction due to structure geometry. This was achieved by designing the transverse MER to be stiffer. The easy way to do this is by shortening the MER although a change in the polyurethane compound can also result in the same effect without resizing the MER.

Figure 8. Load elements and MER on the pedestrian bridge SIB.

Figure 9. King Edward overpass.

2.4 *King Edward Avenue overpass*

When the City of Ottawa decided to upgrade the Ontario approach span to the MacDonald Cartier Bridge, they chose Delcan Engineers to design the new bridge. The New King Edward Avenue Overpass is a curved post tensioned cast in place concrete structure with an overall length of 33 meters. (Fig 9)

The use of SIB on this structure reduced the seismic forces by a factor of 6–7 times when compared to a non-isolated design. For example at one location on the bridge the equivalent static earthquake load was reduced from 355 to 52 Kn/m.

The bearings ranged from 1110 Kn to 4890 Kn in vertical capacity. The SIB were also effective in keeping the seismic displacements down to a manageable 38 mm.

3 CONCLUSIONS

Engineers in cold weather climates have been battling with high seismic forces on their bridge designs due to the latest seismic provisions in current bridge codes coupled with the high forces generated by low temperatures. With the development of SIB, engineers now have a reliable tool to reduce these forces down to a manageable level resulting in a cost effective structure. Now that 4 cold weather bridge projects have now been constructed using SIB, engineers can draw from the experience of these structures for future projects.

REFERENCES

AASHTO Guide Specifications for Seismic Isolation Design, 1999.

Kartoum, A., Constantinou, M.C. & Reinhorn, A.M., 1992. Sliding Isolation System for Bridges: Analytical Study, Earthquake Spectra, Vol. 8 No. 3.

Noureayan, A., Montminy, S. & Watson R.J., 2005. Seismic Design and Construction of a Hybrid Pre-Cast/Cast in Place Concrete Bridge in a Cold Weather Environment, PCI Bridge Conf.

Chapter 9

Results of tests performed on lead-rubber seismic isolators with deformed masonry plates

Mary Jacak
Seismic Energy Products, LP, Alameda, USA

James Pezzotti
Pennoni Associates Inc., Philadelphia, USA

ABSTRACT: Deformations on the order of 11 mm in the masonry plates of installed lead rubber isolation bearings were observed in a highway bridge. Of the more than 400 isolators in the project, approximately 30 showed deformations greater than 2 mm. Due to the cost, accessibility issues and traffic impacts of removing and replacing the isolators, the Owner agreed to accept laboratory testing as a means to determine what effects, if any, the deformation had on the properties of the lead core isolation bearings. New bearings were manufactured in accordance with the original project requirements. The new bearings were first tested to establish baseline properties and validate their compliance with the contract documents. Subsequently, the isolators were deformed in the lab to achieve a similar deformation as that observed in the structure. The bearings were then tested in the deformed condition and the results compared to the baseline properties. One of the deformed bearings was cut in half to permit visual inspection of the bent plates and the inside of the bearing. The test results and visual observations made during and after testing are presented.

1 INTRODUCTION

In October 2005 the isolation bearing manufacturer, Seismic Energy Products, LP (SEP), was notified by the bridge Owner that the masonry plates supporting the Type 3 isolators showed downward deflections at their midpoint ranging from 1 mm to 11 mm (see Figure 1). Approximately 30 bearing assemblies showed deflections greater than 2 mm.

The Contractor informed SEP that axial load consisting of dead and construction loads was placed on the isolator assemblies prior to placement of the non-shrink grout leveling pad. During this initial application of load, the masonry plates were supported solely by four anchor rods (one in each corner of the masonry plates.) The bearing assemblies remained in the deformed position for 6 to 8 months prior to placement of the grout. Since the placement of grout, no additional deflection has been observed.

A series of tests were proposed to determine the effects, if any, on isolator performance in the deformed condition. The tests performed in the deformed condition were to duplicate the maximum field-measured deflection in the masonry plate. The results of the tests performed in the deformed condition will be compared to the original specification requirements and to the data collected from the isolation bearings prior to deformation. Visual inspection of the bearing assemblies will be conducted continuously during testing.

Acceptability of plate geometry and yield state in the bowed plates is beyond the scope of the tests reported herein. The deformed configuration may cause inadequate contact between the anchor rods and the masonry plate, and the masonry and/or load plates may have yielded (American Association of State Highway Transportation Officials *Standard Specifications for Highway Bridges* 1996 with Interims).

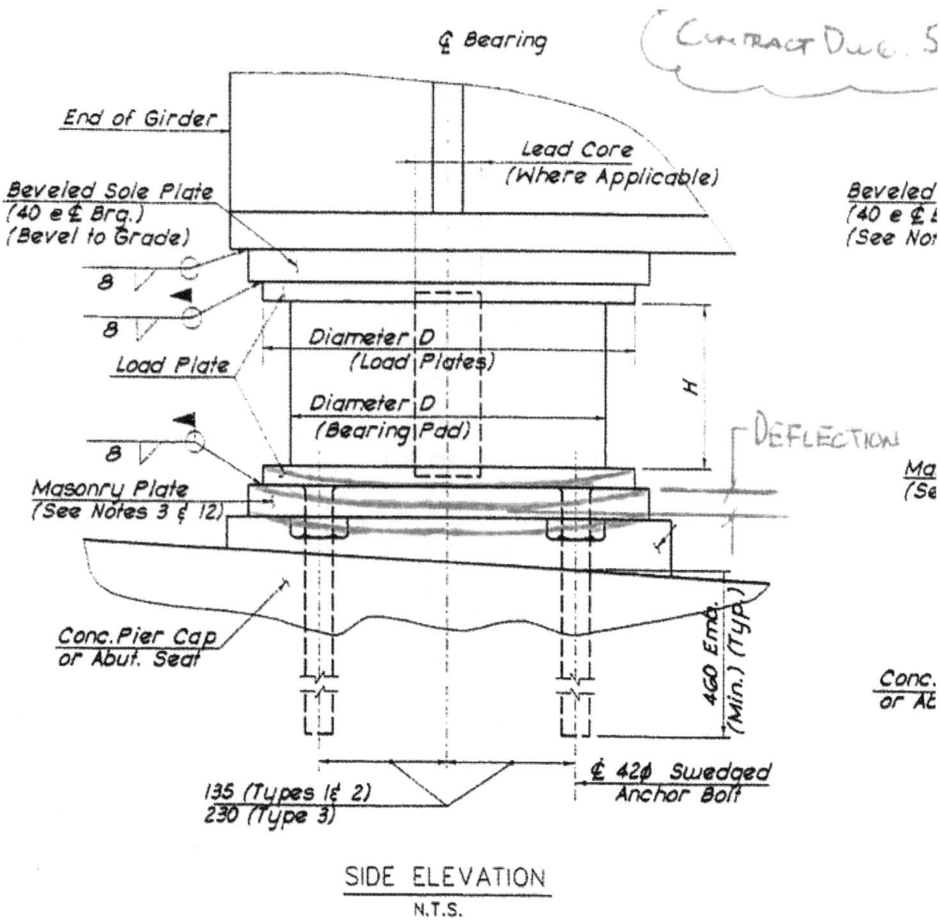

Figure 1. Masonry plate deflections measured in the bridge.

2 TEST PLAN

In order to generate the needed test data without removing isolators from the structure, two new isolator assemblies were fabricated in accordance with the original contract documents and approved shop drawings. The two assemblies consisted of a top load plate, lead-rubber isolator, bottom load plate, and masonry plate. The following tests were to be performed to permit comparison between the undeformed and deformed isolation bearings (American Association of State Highway Transportation Officials *Guide Specifications for Seismic Isolation Design* 1999 & Interims).

The effect of duration in the deformed state on the performance of the bearings is unknown. However, it was not practical to simulate the in-situ condition by maintaining the deformation on the bearings in the laboratory for an extended period of time.

Tests on New Isolators (Undeformed):

A. Perform the 15-hour sustained compression test.
B. Perform a combined compression and shear test on the two isolators in the same manner as for the project to determine the baseline properties of the isolators.
C. Perform a combined compression and shear test for 25 cycles to measure the amount of change in isolator properties over an extended number of cycles.

D. Visually inspect the isolation bearings during and after the tests for irregular bulges, surface cracks and laminate placement faults in accordance with the criteria used on the project.

Tests on Deformed Isolators:

E. Place the isolators in a fixture and apply the axial load (dead plus live) to deform the edges of the plates 11 mm above the bottom surface. Maintain the deformation for a period of 48 hours.
F. Perform the 15-hour sustained compression test.
G. Perform a combined compression and shear test on the isolators in the deformed condition according to the same protocol as for the originally supplied isolators.
H. Perform a combined compression and shear test for 25 cycles to measure the amount of change in isolator properties over an extended number of cycles.
I. Visually inspect the isolation bearings during and after the tests for irregular bulges, surface cracks and laminate placement faults in accordance with the criteria used on the project.
J. Cut one of the isolators to inspect the condition of the interior, load plates and lead core.

3 EVALUATION PLAN

Simultaneous to the proposal of the test plan, criteria for evaluation and acceptance of the test results, and most likely the installed isolation bearings, were developed. This was done to enable prior agreement between the Contractor and Owner as to the course of action and possible outcomes, and to ensure that all necessary tests and measurements would be performed during the testing phase. The following criteria were adopted:

A. Compare the results of both combined compression and shear tests performed in the deflected condition to the corresponding data collected in the undeformed state. If the results from the tests in the deformed state are within 10% of those collected in the undeformed state, it is unlikely that the deformation has caused a change in the bilinear properties of the isolators. If the results are between 10% and 20% of those measured in the undeformed state, the Owner will decide if the changes are acceptable to the performance and longevity of the structure. If the measured properties in the deformed state are greater than 20% different than those measured in the undeformed condition, it will be recommend that the isolators be replaced.
B. Compare the results of the visual inspection between the deformed and undeformed state, and the condition of the cut isolator. If the visual inspection or cut indicates damage to the isolator resulting from the imposed deflection, the extent will be reported for consideration by the Owner.

4 TESTS PERFORMED

On November 16, 2006, the two new (undeformed) isolators were loaded to 942 kips; this load was maintained for 15 hours (Test A). On November 17, 2006, a combined compression and shear test under an axial load of 474 kips for 5 cycles to a displacement of ±2.48 inches was performed (Test B). Test C, a 25-cycle combined compression and shear test, was also performed. The two shear tests provide the baseline properties for comparison with the results of similar tests performed on the deformed isolators. The bearings were visually inspected during and after each test and no signs of defects were observed.

On January 14, 2007, the two isolator assemblies were loaded into the testing fixture and axial load applied until the deflection in the masonry plate reached 11 mm in a manner similar to that of the installed isolators (see Figure 2). The axial load was applied over a period of approximately 1-1/2 hours, and the maximum axial load used was 820 kips (1.3 times the dead plus live load). Upon stabilization of the imposed deformation at a minimum of 11 mm at the midpoint of one edge of the masonry plate, the axial load was reduced to 628 kips, the dead plus live load provided

Figure 2. (a) Deformed isolators in the structure. (b) Deformed isolators in the structure.

in the original contract documents. A steel bolster was placed under each bearing to prevent the deflection in the center of the masonry plate from increasing substantially. The dead plus live load of 628 kips was maintained for 48 hours until the conclusion of Test E at 1:45 PM on January 16, 2007. The deformation at the midpoint of the edges of the masonry plates varied during this test from 11 mm to a maximum of 12 mm. Both isolators were inspected during and after the 48-hour hold for irregular bulges, surface cracks or other signs of distress; none were observed.

On January 16, 2007, after completion of the 48-hour hold, the axial load was increased to 1.5 times the dead plus live load (this test load equals 942 kips) for the 15-hour sustained load test (Test F). During this test, the maximum deformation at the midpoint of the edge of the masonry plates was 11.5 mm on one side and 12.5 mm on the other side. Both isolators were inspected for distress and none was observed.

On January 17, 2007, the two isolators with deformed masonry plates were tested in combined compression and shear. The axial load was 474 kips and the displacement was ±2.48 inches. Test G imposed 5 cycles of shear and Test H imposed 25 cycles of shear. Visual inspection of the isolators during and after the tests showed no sign of distress.

5 TEST RESULTS AND EVALUATION

Results of the tests performed are shown in Tables 1 and 2, and the force-deflection plots can be found in Figures 3 to 6. Table 1 contains the results of the 5-cycle tests and Table 2 presents the results from the 25-cycle tests. Results from the 5-cycle tests are reported in the same manner that was used during fabrication of the original project isolators, namely the average effective stiffness (Keff), post-elastic stiffness (Kr) and energy dissipated per cycle (EDC) over the five cycles of test are calculated and compared to the design values (Buckle et al, 2006). Evaluation of the data reveals that for all three of the measured properties, the deformed isolators meet the requirements of the original specification. It can be seen that the differences in Keff (+2.8%) and Kr (−4.5%) between the deformed and undeformed bearings are within the proposed 10% range for acceptance. The values of EDC in Table 1 and the hysteresis loops presented in Figures 3 and 4 indicate less energy dissipated by the deformed isolators than by the same isolators before the deformation was imposed. The amount of energy dissipated by the deformed bearings, although less than that dissipated by the undeformed bearings, nonetheless exceeds the minimum amount required by the original contract documents.

The results of the 25-cycle tests are reported in Table 2 on a per-cycle basis for each of the imposed 25 cycles. In this case the peak measured force (Fmax) is reported, along with the post-elastic stiffness (Kr) and the energy dissipated per cycle (EDC). The peak measured force (Fmax) is not directly comparable to the contract documents; effective stiffness was not used in evaluating the 25-cycle test due to variability resulting from the method of determining the points between which effective stiffness is calculated. Since the imposed shear displacement is the same in both tests, peak force is a valid measure of stiffness and can be used in lieu of effective stiffness in comparing the properties of the undeformed and deformed isolators, as well as for evaluating the stability of isolator stiffness over 25 cycles of test.

Table 1. Results from Tests B and G (5-cycle combined compression and shear test*).

	Keff (kip/in)	Kr (kip/in)	EDC (kip-in)
Prior to deformation	14.3	8.9	171.0
Deformed condition	14.7	8.5	143.0
Difference or change	2.8%	−4.5%	−16.4%
Design value	14.8	9.4	121.0
Allowable range	12.6–17.0	7.5 min	109 min

*Axial Load = 474 kips, Shear Displacement = 2.48 inches.

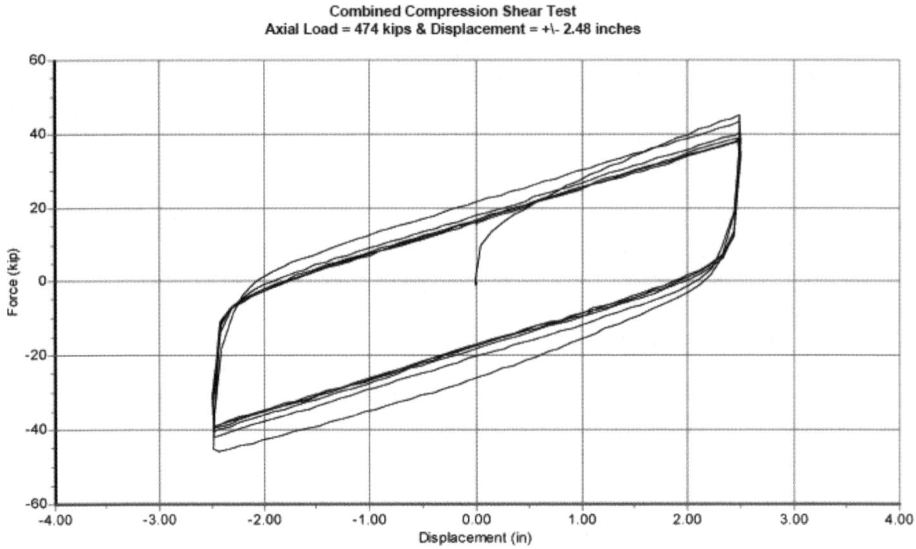

Figure 3. 5-cycle shear test prior to deformation of masonry plate (Test B).

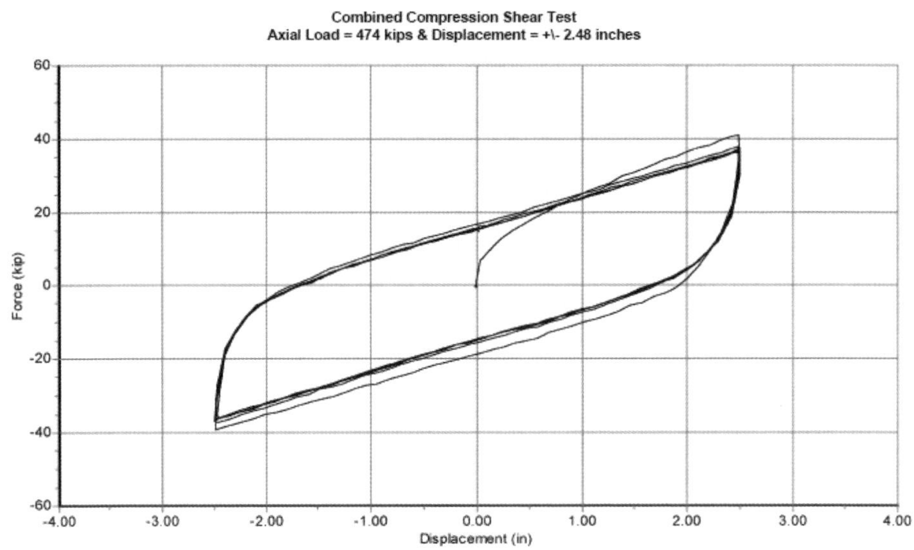

Figure 4. 5-cycle shear test with deformed masonry plate (Test G).

It can be seen from Table 2 that the values of the measured properties are similar between the deformed and undeformed conditions. Both the minimum and average measured Kr and EDC values of the deformed bearings exceed the minimum required values from the original contract documents. The amount of change in the measured stiffness properties (Fmax and Kr) over the 25 cycles is somewhat higher in the deformed bearings than in the undeformed condition; however, the amount of difference is small. The majority of the change in Fmax between cycles occurs in the first 10 cycles or so, although the deformed bearings did not stabilize as quickly as the undeformed bearings (see Figure 5).

Table 2. Results from tests C and H (25-cycle combined compression and shear test*).

	Prior to deformation			Deformed condition			Change		
	Fmax (kip)	Kr (kip/in)	EDC (kip-in)	Fmax (kip)	Kr (kip/in)	EDC (kip-in)	Fmax	Kr	EDC
Cycle # 1	38.4	8.3	163	38.0	8.4	145	−1.04%	1.20%	−11.04%
Cycle # 2	36.6	8.4	156	36.8	8.4	143	0.55%	0.00%	−8.33%
Cycle # 3	36.5	8.0	154	36.5	8.5	141	0.00%	6.25%	−8.44%
Cycle # 4	36.4	8.3	153	36.1	8.5	140	−0.82%	2.41%	−8.50%
Cycle # 5	36.1	8.3	152	36.1	8.4	139	0.00%	1.20%	−8.55%
Cycle # 6	36.3	8.4	152	35.9	8.5	139	−1.10%	1.19%	−8.55%
Cycle # 7	36.1	8.3	151	35.9	8.4	139	−0.55%	1.20%	−7.95%
Cycle # 8	36.1	8.3	151	36.0	8.4	139	−0.28%	1.20%	−7.95%
Cycle # 9	36.1	8.3	151	35.9	8.4	138	−0.55%	1.20%	−8.61%
Cycle # 10	36.0	8.3	150	35.8	8.4	137	−0.56%	1.20%	−8.67%
Cycle # 11	36.0	8.3	150	35.8	8.4	138	−0.56%	1.20%	−8.00%
Cycle # 12	36.1	8.3	151	35.8	8.4	138	−0.83%	1.20%	−8.61%
Cycle # 13	36.0	8.3	150	35.6	8.4	138	−1.11%	1.20%	−8.00%
Cycle # 14	35.9	8.3	150	35.8	8.4	137	−0.28%	1.20%	−8.67%
Cycle # 15	36.1	8.3	150	35.8	8.4	137	−0.83%	1.20%	−8.67%
Cycle # 16	36.0	8.3	149	35.6	8.4	137	−1.11%	1.20%	−8.05%
Cycle # 17	36.0	8.3	149	35.5	8.4	137	−1.39%	1.20%	−8.05%
Cycle # 18	36.0	8.3	149	35.5	8.4	137	−1.39%	1.20%	−8.05%
Cycle # 19	35.8	8.3	149	35.5	8.4	136	−0.84%	1.20%	−8.72%
Cycle # 20	36.0	8.3	149	35.5	8.4	136	−1.39%	1.20%	−8.72%
Cycle # 21	35.9	8.3	149	35.5	8.4	136	−1.11%	1.20%	−8.72%
Cycle # 22	36.0	8.3	149	35.3	8.3	136	−1.94%	0.00%	−8.72%
Cycle # 23	36.0	8.3	149	35.6	8.4	136	−1.11%	1.20%	−8.72%
Cycle # 24	36.0	8.3	148	35.5	8.0	136	−1.39%	−3.61%	−8.11%
Cycle # 25	36.0	8.3	148	35.6	8.3	136	−1.11%	0.00%	−8.11%
Average	36.2	8.3	151	35.9	8.4	138	−0.94%	1.11%	−8.51%
Maximum	38.4	8.4	163	38.0	8.5	145	0.55%	6.25%	−7.95%
Minimum	35.8	8.0	148	35.3	8.0	136	−3.53%	−3.61%	−11.04%
Change	−6.77%	−4.76%	−9.20%	−7.11%	−5.88%	−6.21%			

*Axial Load = 474 kips, Shear Displacement = 2.48 inches.

The deformed isolators demonstrated less difference between the minimum and maximum EDC than the undeformed isolators. The rate of stabilization of energy dissipation over the imposed 25 cycles of test was similar between the deformed and undeformed bearings (see Figure 6). The lower amount of energy dissipated by the deformed bearings exceeds the minimum amount required by the original contract documents for a 5-cycle shear test.

Each isolation bearing was inspected for signs of damage or distress during and after the sustained load tests and combined compression and shear tests (Tests A, B, C, E, F, G, and H.) Upon completion of Test J, the isolation bearings were removed from the testing apparatus. The masonry and bottom load plates remained deformed in both bearings. One bearing was cut in half to permit inspection of the load plates, cap and lead core. Inspection of the cut bearing showed bent shim plates within the bearings; the amount of deflection in the shim plates ranged from a maximum at the bottom of the isolator, adjacent to the deformed load and masonry plates, to negligible deformation in the shims at the top of the isolator. The load plates and caps were intact, and the lead core did not look unusual. No signs of damage, cracking or delamination were observed inside the cut bearing.

Lack of either direct prior experience or information in the published literature describing a field condition such as this prevent prediction of future bearing behavior. Furthermore, these tests do not predict future performance of the installed bearings.

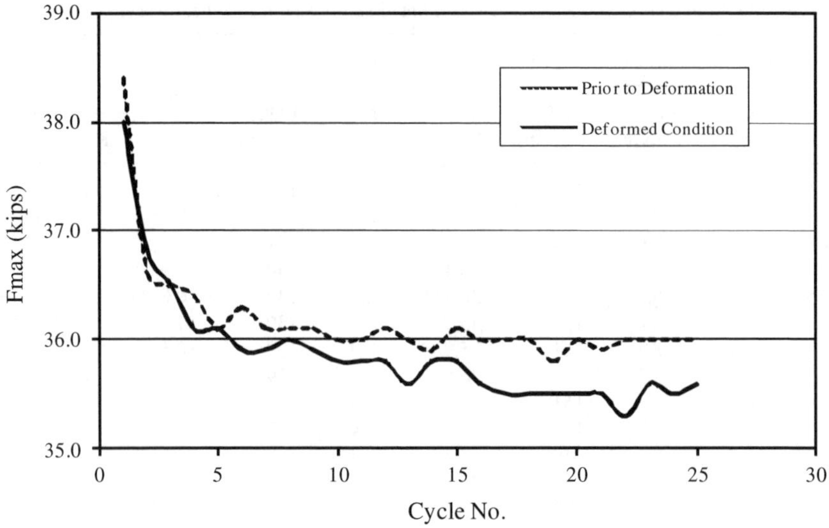

Figure 5. Variation in peak force (Fmax) over 25 cycles of test.

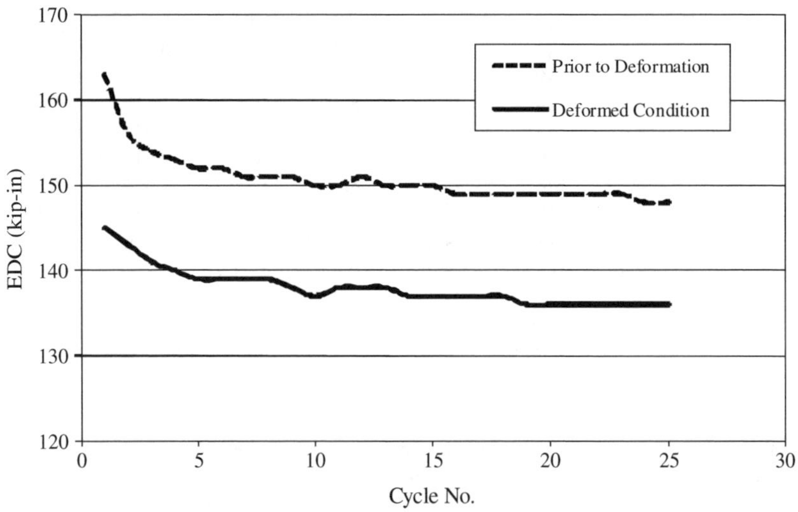

Figure 6. Variation in energy dissipation (EDC) over 25 cycles of test.

6 CONCLUSIONS

Data collected during this testing program was evaluated in accordance with the original contract requirements. Comparisons were also made between data collected from isolator tests in the deformed and undeformed conditions. Although they would have been rejected for out-of-flatness of the masonry plate, the deformation of the masonry plate by approximately one-half inch did not produce stiffness or energy dissipation values outside the ranges required by the specification.

Comparison of stiffness and energy dissipation between the deformed and undeformed conditions indicate that the effect of the deformed masonry plate on stiffness was less pronounced than the measured effect on energy dissipation. The data presented shows that during the five cycle tests the stiffness of the deformed isolators was slightly less than the undeformed, but that the energy dissipation was significantly less in the deformed condition. A similar effect was seen during the 25-cycle tests, in that the measured value of energy dissipation showed more impact from the deformation. The stability of all properties over 25 cycles of test in the deformed condition did not appear, however, to be changed significantly by the deformation (a change of 9.2% in energy dissipation over 25 cycles for the deformed bearings compared to 6.2% for the undeformed).

Visual inspection of the deformed isolators did not indicate any cause for rejection when evaluated in accordance with the project requirements (with the exception of the intentionally deformed plate). Furthermore, internal inspection of one isolator cut in half vertically did not suggest an unusual level of distress.

Although the authors are not familiar with tests of this nature performed in the past, the robustness of the isolators and their measured properties was not surprising in light of testing performed on damaged or rejected isolators. It has been seen that lead rubber elastomeric isolation bearings demonstrate stable and predictable performance even when subjected to a variety of conditions including excess axial load, excessive imposed rotation, bond failure, excess welding preheat, and now a bearing surface significantly out of flatness. Further research would be required to determine whether standard design or tolerance requirements result in unnecessary conservatism and additional costs to the bridge Owner.

REFERENCES

American Association of State Highway Transportation Officials (AASHTO) (1996 with Interims), *Standard Specifications for Highway Bridges*, Sixteenth Edition, Washington, D.C.

American Association of State Highway Transportation Officials (AASHTO) (1999 & 2000), *Guide Specifications for Seismic Isolation Design*, Washington, D.C.

Buckle, I., Constantinou, M., Dicleli, M., Ghasemi, H., 2006, *Seismic Isolation of Highway Bridges*, Multidisciplinary Center for Earthquake Engineering Research, Buffalo, New York.

Chapter 10

Humboldt Bay Middle Channel Bridge: 3D bridge-foundations-ground system

T. Trombetti & G. Gasparini
DISTART – Department of Civil Engineering, University of Bologna, Bologna, Italy

A. Elgamal & D. Forcellini
UCSD, Department of Structural Engineering, University of San Diego, California, USA

ABSTRACT: Soil-Structure interaction may play a major role in the seismic response of a bridge structure. Specifically, a significant reduction in soil stiffness and strength may result in permanent displacement of the abutments and foundations, thus imposing important kinematic conditions to bridge structure. This paper is aimed at showing the effects of this behavior referring to the Humboldt Bay Middle Channel Bridge, in California. The Finite Element model and nonlinear solution strategy are built in the open-source software platform OpenSees. The 3D nature of bridge response imposes significant computational challenges. The soil is modeled as a nonlinear material with a Von Mises multi-surface kinematic plasticity model so as to reproduce elasto-plastic shear response. The results obtained using 1978 Tabas earthquake record shows that changes in properties of the superficial soil layers dictate significantly different time histories of dynamic excitation at the various support points of the bridge (piers and abutments).

1 INTRODUCTION

PEER (Pacific Earthquake Engineering Research Center) initiated a number of testbed projects to synthesize the research products into a coherent methodology: the Humboldt Bay Middle Channel Bridge, near Eureka in northern California was selected as one of these testbed projects (Figure 1).

On September 16, 1978 Tabas earthquake record was elected as a potential site-specific rock outcrop motion at a hazard level of 10% probability of exceedance in 50 years. The Tabas Earthquake record is employed in this study to derive an incident earthquake motion along the FE mesh base with a process of deconvolution.

The spatial extent of the bridge-foundation-ground system is large, typically in the hundreds or thousands of meters, necessitating an appropriate finite element (FE) mesh to provide adequate modeling resolution. In order to model the geometric damping at the mesh base, the soil strata below is represented by transmitting boundary so as to handle spurious reflections of waves at the soil mesh base. Secondly, in view of the highly non linear properties of the foundation ground,

Figure 1. Sky view of Humboldt Bay Middle Channel Bridge (courtesy of Caltrans).

elaborate hysteretic constitutive models of soil materials are needed. The soil is modeled as a non-linear material with a Von Mises multi-surface kinematic plasticity model so as to reproduce elasto-plastic shear response. Based on downhole measurements of shear wave velocity, the soil profile is idealized into a surface crust layer and five underlying sub-layers. In particular, results of three different analyses are compared. For each case, the superficial layers are modeled with materials having different stiffness.

2 FE MODEL

System modeling and response computations are performed using OpenSees, an object-oriented, open-source FE analysis framework. The current version of OpenSees includes an extensive library of structural and soil material models, as well as a number of structural (e.g., beam-column, shell) and continuum elements. In the bridge-foundation-ground model, different types of elements are employed to represent the foundation and superstructure (Table 1)

2.1 *Soil model*

The foundation soil is composed mainly of dense fine-to-medium sand, organic silt, and stiff clay layers, with thin layer of loose and soft clay located near the ground surface. The average slope of the river channel from the banks to its center is about 7% (4 degrees).

Based on downhole measurements of shear wave velocity, about 0.25 miles north-west of the bridge, it was possible to define the soil profile. This profile is idealized into a surface crust layer, and five underlying sub-layers (see Figure 2).

This soil is modeled as a non linear hysteretic material with a Von Mises multi-surface kinematic plasticity model. In this regard, focus is on reproduction of the soil hysteretic elasto-plastic

Table 1. FE used in the model.

Structural element	Finite element
Longitudinal I-girders	3 D linear elastic beam–column
Transversal I-beam	3 D linear elastic beam–column
Deck	3D linear shell element
Piers	3D fiber–section
Piles	3D fiber–section
Expansion joint (3rd–6th Piers)	EQUALDOF (translation)
Deck-pier connection	EQUALDOF (translation)
Abutments-bridge connections	EQUALDOF (translation)

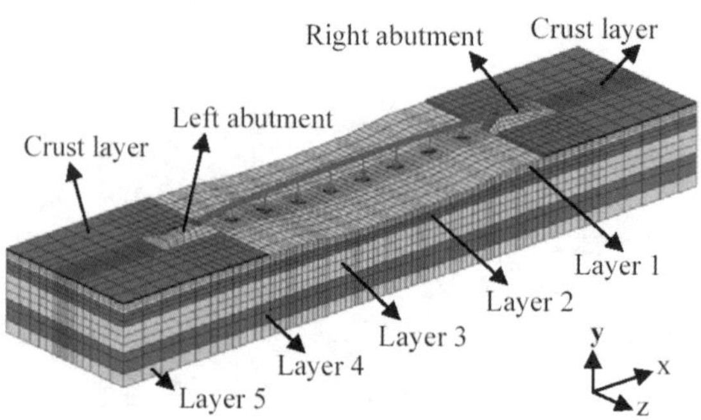

Figure 2. 3D FE model of bridge and idealized soil profile.

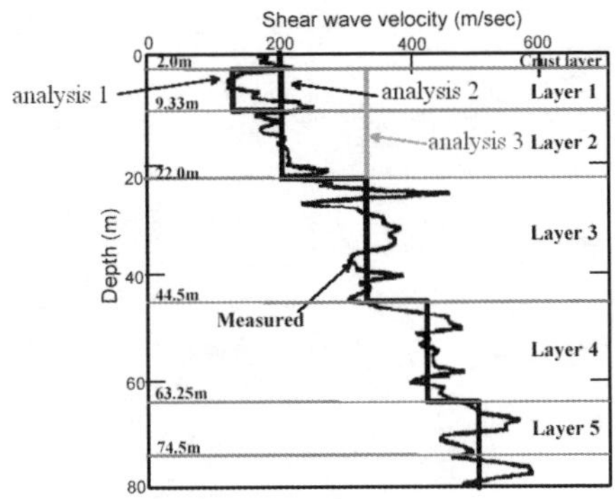

Figure 3. Idealized soil shear wave velocity profiles for the three analyses.

shear response (including permanent deformation; for more details see Lu et al. 2004, Elgamal et al. 2002, 2003, Yang et al. 2002, 2003).

2.2 *The analysis performed*

In this paper results of three different analyses are compared. For each case, the superficial layers are modeled with materials having different stiffness. Figure 3 shows the profiles of shear velocity with that the soil is modeled.

In the first analysis the superficial layer is modeled as a relatively soft soil material in order to study the effects of lateral soil spreading on the bridge foundation and superstructure (typical situation of a river deposited soft stratum). Shear strength of this soft layer is defined to be 10 kPa. The profile is idealized into a surface crust layer and five underlying sub-layers.

In the second analysis, layer 1 is modeled as a material with a middle stiffness (shear strength as 40 kPa). Evidently, this second case represents only a theoretical case: it is practically impossible to have in a river channel center.

In the third analysis, the first 44,5 m of soil were modeled as layer 3 (shear strength as 75 kPa) in order to enforce the stiffness of the layers: this profile is based on a unique layer of 74,5 m characterized with a variable shear modulus G (increasing with depth).

3 SPATIAL GROUND VARIABILITY

Computed ground surface motion at 3 representative locations (location 1 is the free field, location 2 is the abutments and location 3 is the river channel centre) are compared. Figure 4 shows the acceleration time history comparison in the three representative locations.

On the horizontal plan, in correspondence of locations 1 and 2, there are not big differences between the three analyses considered: maxima values are reached in correspondence of analysis 2. In location 3 accelerations show that increasing the stiffness of the layers (analysis 1–3) accelerations decrease and the damping of the entire soil increases. For vertical direction, instead, the hardening of the soil involves a substantial reduction of the accelerations.

4 RESPONSE OF THE BRIDGE STRUCTURE

In this paragraph we describe the response of the structure focusing on base and deck displacements for the three analyses.

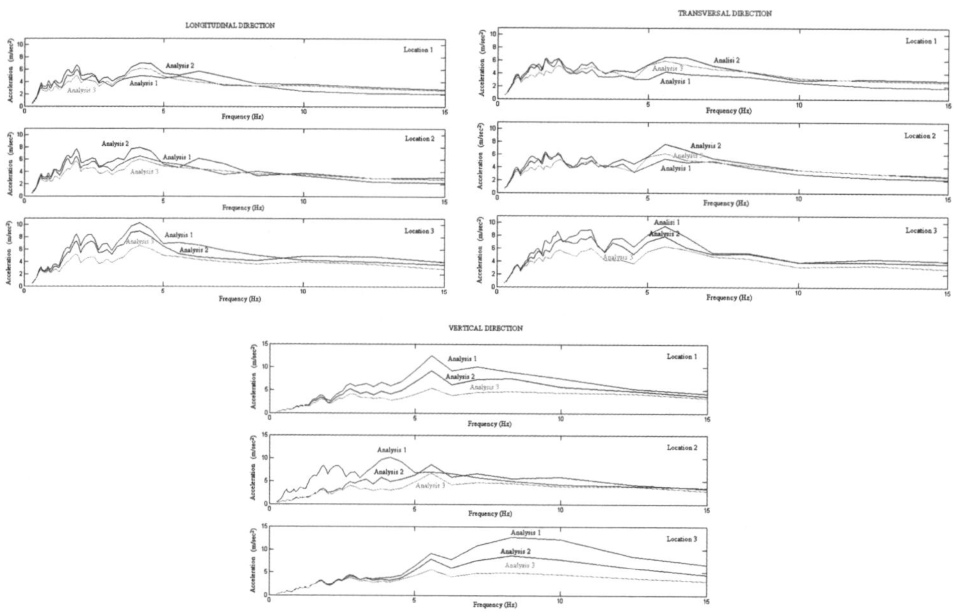

Figure 4. Acceleration time history comparison in the three representative locations: free field (location 1), abutments (location 2) and river channel centre (location 3) in the longitudinal (a), transversal (b), and vertical (c) directions, respectively for the three analyses.

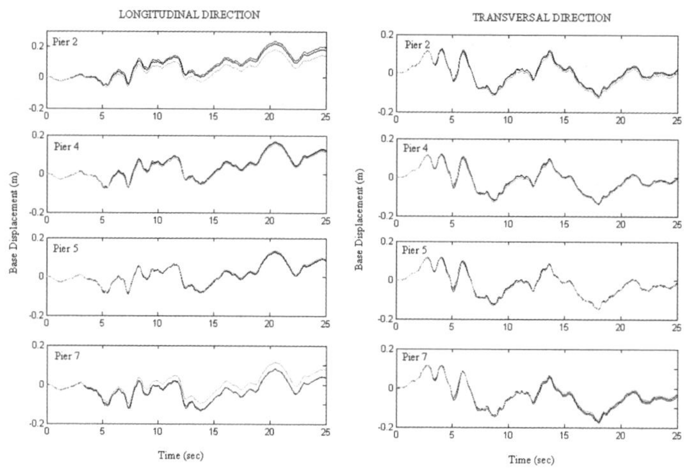

Figure 5. Base displacement time histories for the three analyses.

4.1 *Base displacements*

Figures 5 shows, in the same diagrams, base displacements (longitudinal and transversal directions) for the three analyses. The difference in displacement between any two pile caps increases continuously in time resulting in permanent pier top-to-bottom relative displacements.

As might be expected, much lower levels of permanent deformation are observed in the transversal direction. However, during earthquake excitation, the top-to-bottom relative displacement

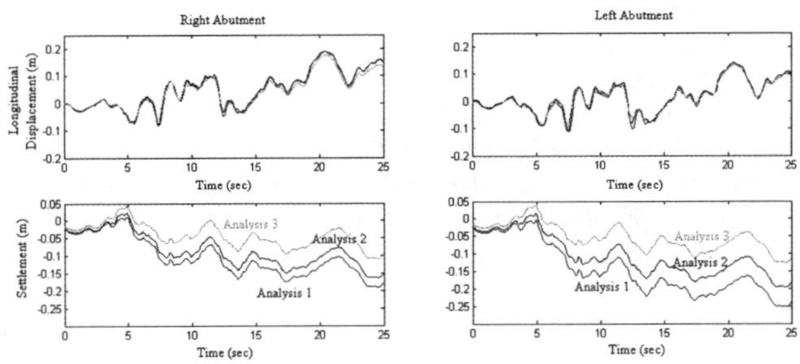

Figure 6. Longitudinal and vertical displacement (settlement) time histories for the three analyses.

of the piers is still quite large (as much as that in the longitudinal direction). This is a consequence of the relatively flexible configuration of the bridge superstructure in the transversal direction, in spite of the much higher moment of inertia of the piers cross-section in this direction as compared to the longitudinal direction.

Comparing analysis 1 and 2 is possible to observe that hardening the stiffness of the superficial layer does not cause a reduction in longitudinal displacements as instead occurs increasing the stiffness of layer 2. The maxima differences occur for external piers (2 and 7).

4.2 *Deck displacements*

Figure 6 shows longitudinal displacements and vertical displacements (settlements) in correspondence of the deck caused by the permanent deformations of the soil for the three analyses.

It is possible to see that the soil stiffness does not strongly influence the longitudinal displacements: three analyses give close results. Displacements reach mean permanent values of 0,15 m and 0,10 m respectively for left and right abutment.

Settlements, instead, depend strongly upon the stiffness of the soil: the difference in settlements between the three analyses increases continuously in time and the final settlements are very far each other.

5 CONCLUSIONS

The study conducted in this article shows that permanent ground deformations are related to the presence of superficial layers. In this case study, soil – structure interaction play a major role in the seismic response because reduction in soil stiffness and strength results in permanent displacement of the abutments and foundations. These important kinematic conditions to the bridge structure affect the bridge operability after the earthquake.

The study explores the influence of the soil stiffness on the lateral spreading of the Bridge-Foundation-Ground system. The parameters shown are the accelerations and the displacements of the foundations and of the bridge structure.

In this regard, the settlement of the bridge was found to be the most significant parameter.

REFERENCES

Elgamal A., Yang Z., Parra E., 2002. Computational modeling of cyclic mobility and post-liquefaction site response. *Soil Dynamics and Earthquake Engineering*, Vol. 22, pagg. 259–271, Elsevier, Amsterdam, Netherlands.

Elgamal A., Yang Z., Parra E., Ragheb A., 2003. Modeling of cyclic mobility in satured cohesionless soils. *International Journal of Plasticity*, Vol. 19, pagg. 883–905 Elsevier, Amsterdam, Netherlands.

Lu J., Yang Z., He L., Peng J., Elgamal A., Law K.H., 2004. Computational modeling of nonlinear soil-structure interaction on parallel computers. *Proceedings of the 13WCEE 13th World Conference on Earthquake Engineering*, paper n. 530, Vancouver, Canada.

Yang Z., Elgamal A., 2002. Influence of permeability on liquefaction-induced shear deformation. *Journal of Engineering Mechanics*, July 2002, pagg. 720–729, ASCE Publications, Reston, Virginia, USA.

Yang Z., Elgamal A., 2003. Application of unconstrained optimization and sensitivity analysis to calibration of a soil constitutive model. *International Journal for Numerical and Analytical Methods in Geomechanics*, Vol. 27, pagg. 1277–1297, Interscience, Wiley Press, New York, USA.

Yang Z., Elgamal A., Parra E., 2003. Computational model for Cyclic Mobility and associated Shear Deformation. *Journal of Geotechnical and Geoenvironmental Engineering*, December number, page 1119–1120, ASCE Publications, Reston, Virginia, USA.

4 *Bridge design, fabrication & testing*

Chapter 11

Design of Florida Avenue Bridge over the Inner Harbor Canal

E.T. Nelson
DMJM Harris, Richmond, VA

ABSTRACT: The $210 million Florida Avenue Bridge project is being designed to provide reliable access between St. Bernard and Orleans parishes over the Inner Harbor Navigational Canal (IHNC) in New Orleans, Louisiana. The project includes a five-span high-level bridge over the IHNC with a 470-foot center span. Bridge type studies were completed to determine the most viable structure type. Both cast-in-place segmental concrete box girder and steel plate girder alternates were selected for final design, with this paper focusing on the design of the segmental concrete alternate. The superstructure consists of a variable depth twin-cell box girder that is supported by voided box column piers and steel HP piles. The bridge will be built with form travelers using the balanced cantilever method of construction.

1 INTRODUCTION

The Florida Avenue Bridge project is part of the Louisiana Department of Transportation and Development TIMED (Transportation Infrastructure for Economic Development) program. This $4 billion improvement program is designed to enhance economic development through an investment in transportation projects, with the Florida Avenue Bridge project being one of three major bridge components of the TIMED Program.

The $210 million Florida Avenue Bridge project is being designed to provide reliable access between St. Bernard and Orleans parishes over the Inner Harbor Navigational Canal (IHNC) in New Orleans, Louisiana. The project includes over 10,000 feet of elevated viaduct and ramps and includes a 1,516-foot long high-level main span unit over the IHNC. The 5-span main unit with a 470-foot long center span is being designed for both steel plate girder and cast-in-place (CIP) segmental concrete box girder alternates. The approach structures and ramps include prestressed concrete Bulb-T girders and curved steel plate girders.

The project will be advertised for construction as two separate contracts. The first contract will include construction of the 5-span main unit over the IHNC and the second contract will include construction of the mainline approaches and ramps. This presentation focuses on the design of the 5-span main unit for the CIP segmental concrete box girder alternate.

2 PROJECT GEOMETRY AND DESCRIPTION

The alignment of the Florida Avenue Bridge is in an extremely congested area of New Orleans and places the 5-span main unit directly south of a new railroad lift bridge recently constructed by the Port of New Orleans. A photo of the current project site is shown in Figure 1.

In order to accommodate existing site features of the adjacent railroad lift bridge, the main span length crossing the IHNC is set at 470 feet. In addition to avoiding conflicts with the lift bridge, other challenges in developing the span layout were locating piers to avoid conflict with numerous underground utilities, flood walls, underground canals, pump houses, roadways, etc. Accommodating all

Figure 1. Current view of main span project site.

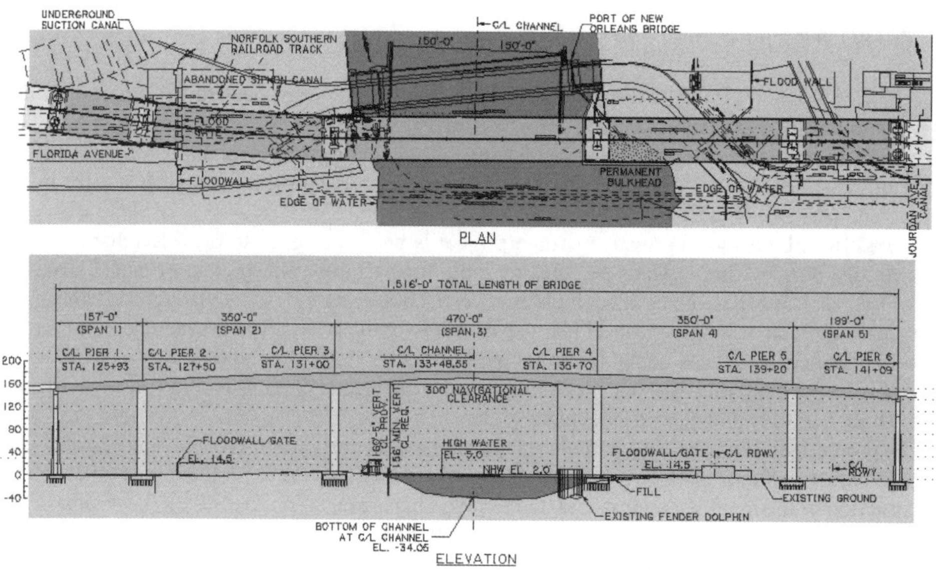

Figure 2. General plan & elevation CIP segmental concrete alternate.

of these obstacles and site features essentially dictated the span arrangement of 157′ – 350′ – 470′ – 350′ – 189′. The general plan and elevation for the CIP segmental concrete alternate is shown in Figure 2. The same span arrangement is used for both the CIP segmental concrete and steel plate girder alternates being developed in final design because of the limited space available to place footings.

The 470-foot long main span provides a 300-foot wide navigational channel and a minimum vertical clearance of 156 feet over the IHNC. The alignment is tangent for approximately 70% of the 5-span unit with the remaining portion horizontally curved in a radius of 3,500 feet. The typical section consists of four 12-foot travel lanes, two 8-foot outside shoulders, two 4-foot inside shoulders, and a 2-foot median barrier, for a total roadway width of 74 feet.

3 BRIDGE TYPE STUDIES

Bridge type studies were completed to determine the most constructible and economical structure types for this crossing. Conceptual studies included evaluations of the following types of bridges: steel plate girders, steel box girder, precast segmental concrete box girder, CIP segmental concrete box girder, and extradosed cable-stay. The CIP segmental concrete box girder and the steel plate girder alternates were determined to be the most viable structure types from these conceptual studies. Subsequently, preliminary designs and cost estimates were completed for these two alternates and the CIP segmental concrete box girder proved to be the least cost structure type. Although the concrete alternate was estimated to be approximately 30% less expensive than the steel alternate, the Louisiana DOTD opted to pursue both structure types through final design. The primary reason for this decision was to ensure that local Contractors, who are more familiar with steel plate girder structures, have the ability to bid competitively for this project.

4 CAST-IN-PLACE SEGMENTAL CONCRETE ALTERNATE

4.1 *Superstructure*

The superstructure is comprised of a variable depth, twin-cell trapezoidal box girder with a maximum depth of 26 feet at the main span piers and a minimum depth of 12 feet at mid-span. The box depth at the side span piers is 15 feet. The section height varies as a function of a circular curve with constant radius. Figure 3 illustrates the typical box girder cross section.

Segments are to be cast in balanced cantilever fashion with no more than one-half segment length out of balance at any time. The main span cantilever at Piers 3 & 4 has 13 segments extending out on

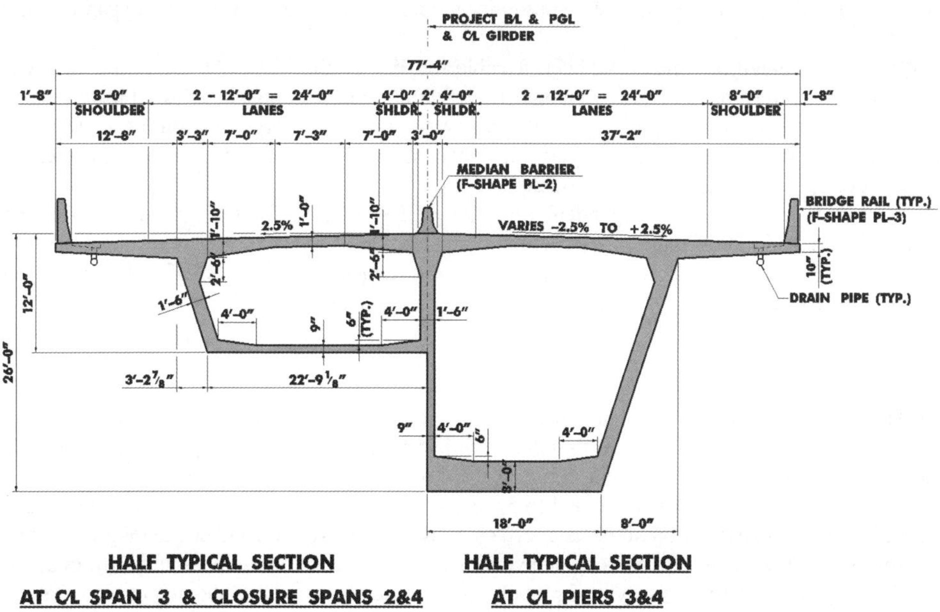

Figure 3. Typical box girder cross section.

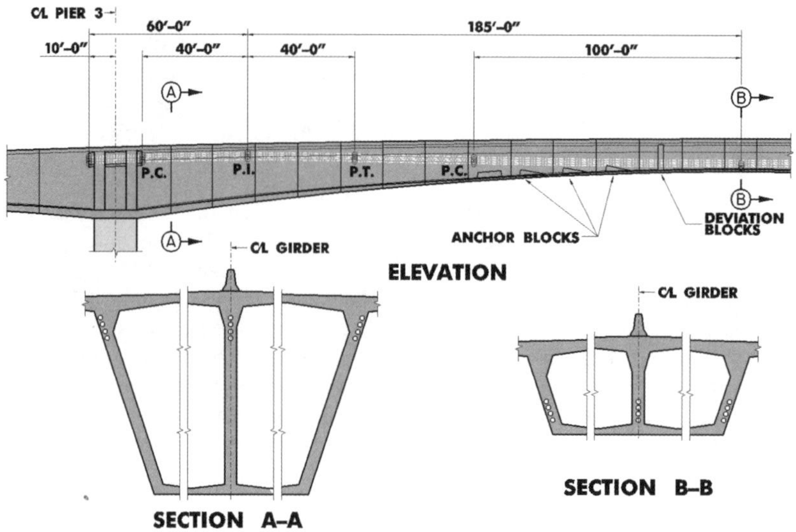

Figure 4. Continuity tendons in main span.

each side of the pier table, with all segment lengths being 16 feet. Closure segments are all 12 feet in length. The pier tables are fixed to the piers while the end span diaphragm segments bear on pot bearings.

All post-tensioning tendons are internal to the concrete superstructure. A combination of cantilever and continuity tendons is used longitudinally. Three top slab longitudinal cantilever tendons are needed for each cantilever segment cast. Twelve continuity tendons are provided in each of the spans, with additional bottom slab tendons in the main span over the channel and the adjacent side spans. The top slab deck is transversely post-tensioned with five tendons required for every 16-foot long segment. Vertical post-tensioning bars are also used in the webs of the pier tables to reduce shear stresses. Figure 4 illustrates the continuity tendons in the webs of the main span crossing the IHNC.

While a single-cell box may have been feasible for this width of box, the use of the third web helped optimize the amount of transverse post-tensioning required and also allowed for the placement of longitudinal post-tensioning in all three webs. The shear force distribution between the three webs due to dead load and live load was confirmed with a three-dimensional finite element model developed using the program SAP2000.

The two dimensional time-dependent program BD2 was used for the primary longitudinal analyses, with a SAP2000 staged construction model also developed to compare results with BD2. The superstructure design is being completed in accordance with the AASHTO LRFD Bridge Design Specifications.

4.2 *Substructure*

The substructure is comprised of cast-in-place voided box piers for the interior piers to minimize the concrete volume and reduce the foundation piling required. The voided piers are fixed to the superstructure creating frame action between the superstructure and substructure. The concrete strength used for the piers was increased to 5,000 psi to limit cracking due to stresses imposed on these piers during construction of the balanced cantilever superstructure and due to long-term creep and shrinkage movement. The piers are conventionally reinforced and range in height from 151 to 155 feet. Figure 5 illustrates the typical pier elevation and cross section of the main span piers.

The shape of the expansion joint piers is designed to closely match the shapes presented by the Louisiana DOTD to the public during the environmental assessment phase of the project.

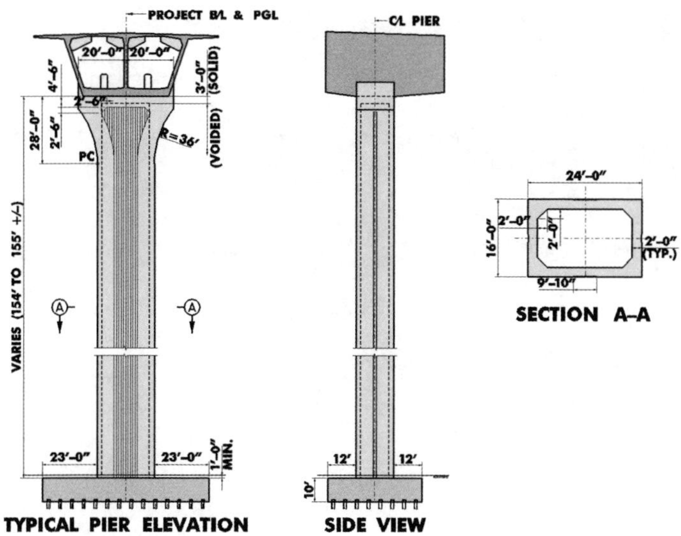

Figure 5. Main span pier elevation & cross section.

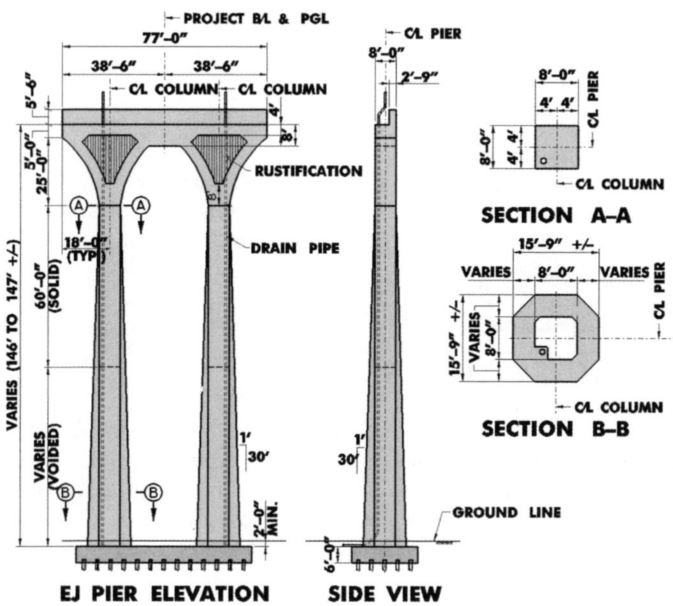

Figure 6. Expansion joint pier elevation & cross section.

These piers transition from the CIP segmental concrete box girder to prestressed concrete Bulb-T girders. Similar to the main span piers, the columns are conventionally reinforced and voided near the base to help reduce concrete weight and the size of foundations. Figure 6 illustrates the typical expansion joint pier elevation and cross section.

A major design concern was the overall stability of the soil in the project area. Not only are the banks of the IHNC subject to instability, but the area surrounding the canal has also experienced significant settlement. Compounding this instability is the fact that the new Florida Avenue Bridge foundations are in direct proximity to the railroad lift bridge foundations. As a result, vibration

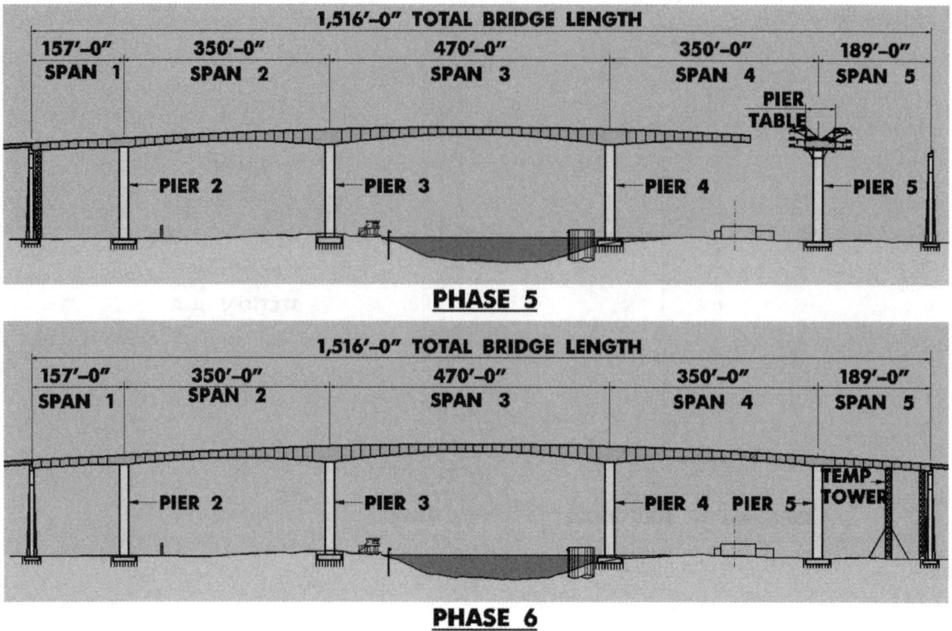

Figure 7. Erection sequence.

control and the use of non-displacement piles were key aspects to the foundation design. All pier foundations consist of concrete pile caps and steel HP piles to minimize vibration and soil displacement impacts on the adjacent structures. Pre-construction condition surveys of the approach roads leading up to the railroad lift bridge and the railroad bridge operations building are also stipulated in the Project Specifications.

5 CONSTRUCTION SEQUENCE

The 5-span main bridge will be built with form travelers using the balanced cantilever method of construction. The 16-foot typical segment, 12-foot closure segment and 50-foot pier table length were optimized based on discussions with contractors and form traveler suppliers. Erection by progressive cantilever is required at both end spans. Figure 7 provides a schematic representation of the erection sequence envisioned for the progressive cantilever phases.

6 CONCLUSION

Existing site constraints and navigational restrictions dictated the long span arrangement for the main span unit of the Florida Avenue Bridge. However, both steel plate girder and cast-in-place segmental box girder alternates proved to be viable structure types and both alternates were carried forward through final design.

Final design of the main span for the Florida Avenue Bridge was completed earlier this year and design of the approaches is currently ongoing. The construction contract for the five-span main unit over the IHNC is planned for advertisement in March 2008, with the scheduled completion to open the bridge for traffic in 2011.

REFERENCES

American Association of State Highway and Transportation Officials (AASHTO), *LRFD Bridge Design Specifications*, Third Edition, 2004 and subsequent interim specifications through 2006 (AASHTO: Washington, DC).

Bridge Software Institute, *FB-MultiPier*, 457 Weil Hall, Gainesville, FL 32611.

Computer & Structures Inc., *SAP2000*, 1995 University Avenue, Berkeley, CA 94704.

Interactive Design Systems, *Bridge Designer BD2*, 16885 Via Del Campo Court, San Diego, CA 92127.

Louisiana Department of Transportation and Development (DOTD), *Bridge Design Manual*, Fourth English Edition, May 2003.

Louisiana Department of Transportation and Development (DOTD), *Standard Specifications for Roads and Bridges*, 2006 Edition.

NY Associates Inc., *Final Environmental Assessment for New Florida Avenue Bridge over the Inner Harbor Navigational Canal*, August 2004.

Chapter 12

Heat curving HPS 485W bridge I-girders

A.N. Gergess
University of Balamand, El-Koura, Lebanon

R. Sen
University of South Florida, Tampa, Florida, USA

ABSTRACT: Heat curving is widely used for fabricating curved steel bridge I-girders. Curving is accomplished by asymmetric heating of the flanges of the straight girder. Heat is applied along the girder length continuously or intermittently with the heated width varying from 1/12 to 1/4 of the flange width depending on the curvature. Curvature develops after the girder cools to ambient conditions. Current practice limits the maximum temperature to 620°C for conventional Grades 250 and 345 steels. The "Guide for Highway Bridge Fabrication with HPS 485W Steel" recommends investigating heat curving of HPS 485W at 705°C. This paper evaluates the validity of the 705°C temperature using non-linear finite element analysis. Other fabrication issues relating to heat curving stiffened and hybrid girders are also addressed. Results show that the maximum temperature can be somewhat lower. Stiffeners may reduce the curvature by up to 10% while hybrid girders with top and bottom flanges made of different steel grades require different heating profiles.

1 INTRODUCTION

Heat curving is usually employed for fabricating conventional Grade 250 ($F_y = 36$ ksi) and 345 ($F_y = 50$ ksi) steel girders for curved bridges. In this method, the top and bottom flanges of a straight fabricated girder, with or without intermediate transverse stiffeners, are simultaneously heated along *one* edge (Fig. 1) at temperatures above the re-crystallization or work-hardening range (Wick 1960). Figure 1 shows two common heating methods: continuous (flange tips are continuously heated along their length, Fig. 1a) and intermittent V-heating (flange tips are heated in truncated triangular wedge shaped areas spaced at regular intervals along the girder's length, Fig. 1b).

(a) Continuous Heat (b) Intermittent V-Heat

Figure 1. Heat curving.

The asymmetric heat application induces unequal expansion and contraction thereby curving the girder. Curvature develops in a concave shape along the heated edge during heating and reverses after cooling (Fig. 1). Several heat-cool cycles may be required before a girder attains its desired curvature.

The 345 MPa (50 ksi) AASHTO limit (AASHTO 1996) on yield stress was not an issue until the recent success of the newly developed High Performance Steel, HPS 485W (*W stands for corrosion-resistant weathering steel*), which yield strength of 485 MPa (70 ksi) exceeds the limit. The "*Guide for Highway Bridge Fabrication with HPS 485W Steel*", states that temperatures as high as 677°C (1250°F) do not adversely affect the mechanical properties of the HPS 485W base metal and recommends investigating heat curving of HPS 485W steel at 705°C (1300°F) (AASHTO 2000). The aim of this paper is to evaluate the validity of these higher limits for HPS 485W girders on the basis of previous numerical analysis and experimental results (Brockenbrough 1970a, b). The paper also addresses important fabrication issues, mainly heat-curving stiffened and hybrid girders.

2 COMMON FABRICATION PRACTICE IN THE STEEL SHOP

Continuous heating and V-heating are most economically and commonly used for fabricating curved steel I-girders. Continuous heat (Brockenbrough 1972) is generally used for radii of curvature R smaller than 300 m (985 ft). Intermittent V-heating (Fig. 1b) is used for longer radii ($R > 300$ m (985 ft)) (Brockenbrough 1973). The top and bottom flanges should be heated simultaneously at the same rate of heating. The main parameters affecting heat curving are the heated flange width and temperature that both depend on the radius of curvature (R) to be induced in the girder (CALTRANS 2002).

For continuous heat (Fig. 2a), the heated flange width varies from 1/12 to 1/4 flange width (depending on the desired radius of curvature, Brockenbrough 1970a, b). For V-heat, heating can proceed up to the web/flange juncture; the heating width may be extended beyond the web/flange juncture to a distance equal to 1/8 flange width or 75 mm (3 in.), whichever is smaller in case V-heat is used for $R \leqslant 300$ m (985 ft) (Davidson et al. 2004, Fig. 2b). For V-heating, the angle at the wedge should be limited to 30 degrees and the base of the triangle should not exceed 254 mm (10 in.) (Fig. 2b, Brockenbrough 1973).

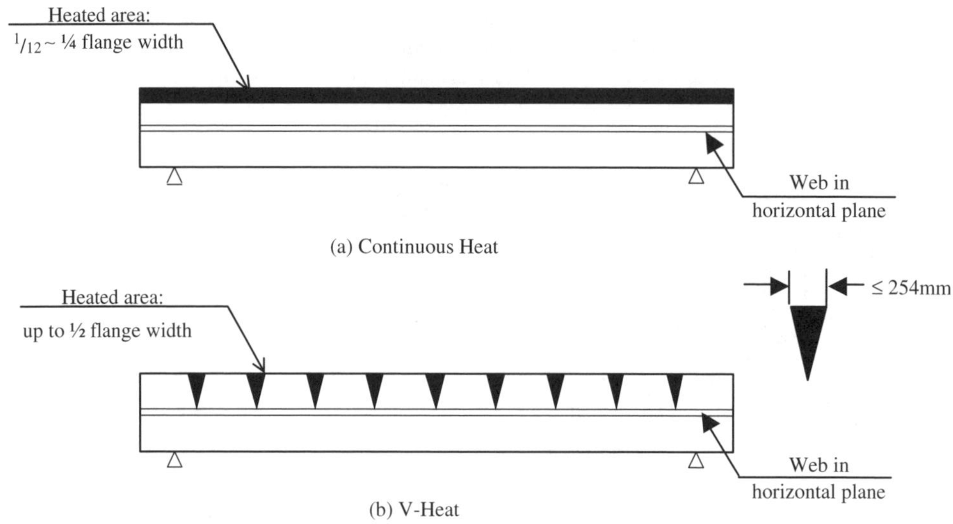

Figure 2. Heated width.

The heating temperature should not exceed 620°C (1150°F) for conventional steel grades (Grade 250 and Grade 345) and 705°C (1300°F) for HPS 485W steel. In the steel shop, these limits are complied with using temperature indicating crayons. After heating, the girder should be allowed to cool naturally. Artificial cooling methods may be employed only after the girder has cooled to 315°C (600°F).

The heat curving operation can be carried out with the girder placed in a horizontal (Fig. 3a) or vertical position (Fig. 3b). If placed vertically, the girder should be braced laterally or attached to a rigid platform at the middle (Fig. 4) in order to ensure that it will not overturn during heating. When placed in a horizontal position, supports should be provided at the ends of the girder and at intermediate positions in order to obtain a uniform curvature. The distance between the intermediate supports should be such that the self-weight bending stresses in the flanges are less than 186.2 MPa (27,000 psi) (Fig. 3b, Davidson et al. 2004).

In stiffened girders, intermediate transverse stiffeners can be placed before or after heat curving (Fig. 1). If placed before, they should be attached only to the web (welding to the flanges is carried out after heat curving). Bearing stiffeners should be attached after heat curving. Longitudinal stiffeners should be heat curved separately and then welded to the girder.

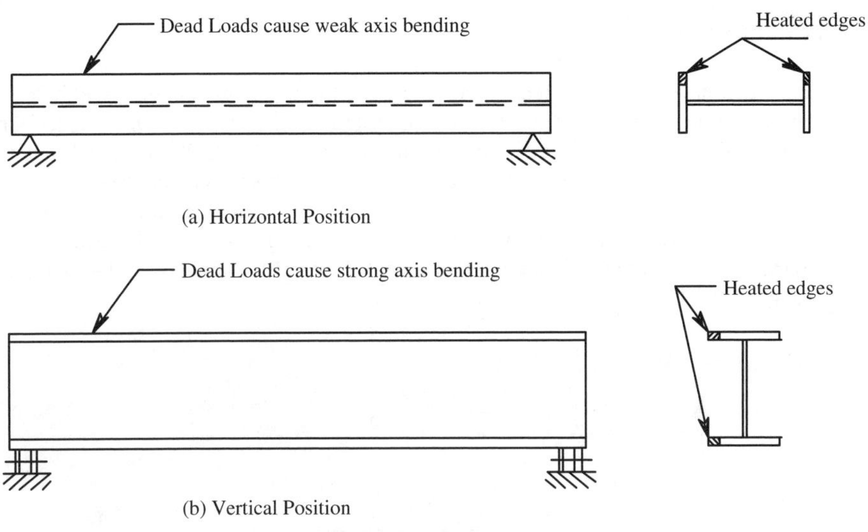

Figure 3. Girder's position during heat-curving.

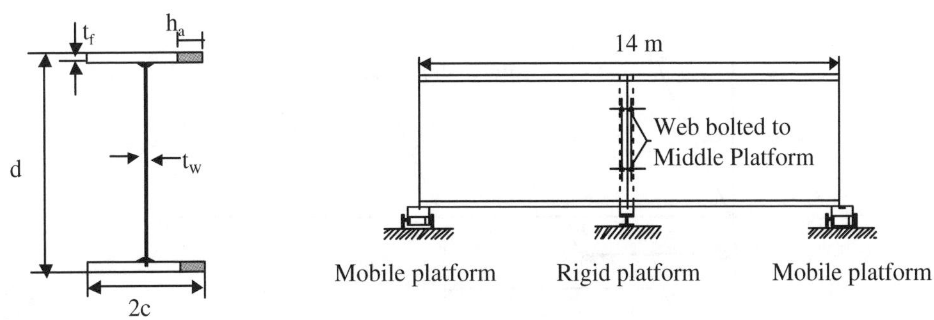

Figure 4. Test girder (Brockenbrough 1970a, b).

Cambering is also required before heat curving, taking into consideration the loss in camber that may occurs due to heat curving (Hilton 1984). In general, the girder's camber should be checked after completion of the heat curving operation.

3 APPROACH

A three-dimensional, non-linear finite element model was developed for this study using MSC/NASTRAN software (MSC/NASTRAN 2000). The model incorporating material and geometric non-linearity was first calibrated using test data (detailed information on calibration may be found in Gergess 2001). The calibrated model was then used for predicting the response of an HPS 485W girder that was geometrically identical to the one tested (Brockenbrough 1970a, b) and subjected to the same experimental heat/cool regime. Appropriate adjustments were made to the heat/cool cycles to obtain curvatures comparable to those for the test girder. Analysis in this paper is based on continuous heat (Fig. 2a) and girder positioned vertically (e.g. selfweight neglected, Fig. 3b) with its web bolted at mid-length to a fixed platform and the bottom flange placed on mobile platforms at the ends to permit lateral movement (Fig. 4).

4 PARAMETERS

The parameters that have the most effect on curvature are the flange thickness t_f, width 2c, heating temperature T, heated width h_a (Fig. 4), material yield point F_y (Fig. 5, an idealized stress-strain curve is used, e.g. elastic-perfectly plastic) and the initial residual stresses (that develop from fabrication of the straight girder). The dimensions used in this paper were selected from the US Steel test (Brockenbrough 1970a, b). It was shown previously that the radius of curvature to flange width ratio, R/2c relates directly to heating temperature e.g. for a specific radius of curvature R and flange width 2c, the heating temperature could be easily obtained from fabrication aids (Brockenbrough 1972, 1973).

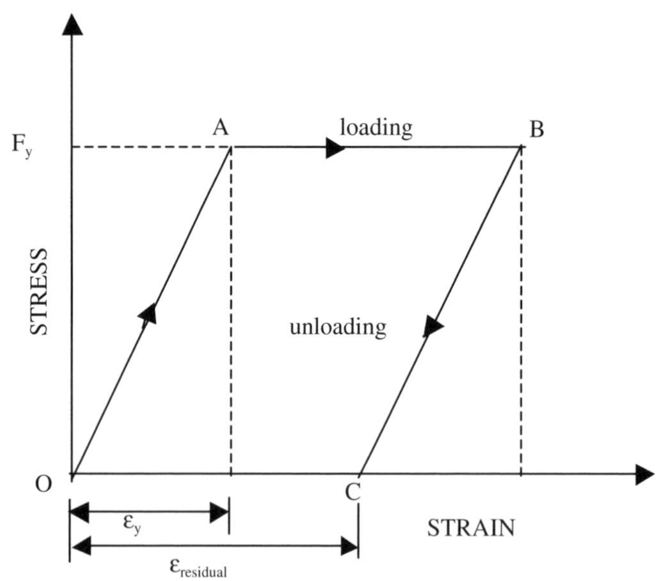

Figure 5. Idealized stress-strain curve.

The US Steel test girder was 46 ft (14 m) long, with the following properties: flange thickness $t_f = 61$ cm (24 in.), flange width $2c = 5.1$ cm (2 in.), web depth $d = 116.8$ cm (46 in.) and thickness $t_w = 1.27$ cm (1/2 in.) (Fig. 4). The heating conditions included heating the through thickness girder flange along edge strips of two widths, $h_a = 8.9$ cm (3.5 in.), 1/6 flange width heated and 13.3 cm (5.25 in.), 1/4 flange width heated to six values of maximum temperature (409°C (768°F) to 544°C (1011°F)). The heat/cool regimes are illustrated in Table 1. The radii of curvatures (R) that developed after each heating operation are also shown in Table 1 (Sen et al. 2003).

The ambient temperature material properties for HPS 485W steel are yield stress $F_y = 485$ MPa (70 ksi), the modulus of elasticity E = 200 MPa (29,000 ksi) and coefficient of thermal expansion $\alpha = 0.000011$/°C (0.00000629/°F). The variation in steel properties with temperature was also incorporated. Reductions in yield stress and modulus of elasticity are shown in Fig. 6 (the ratio corresponds to the modified material property at temperature T divided by the ambient temperature property). Increase in the coefficient of thermal expansion α is given function of the designated temperature T (°C) and the coefficient of thermal expansion at ambient temperature $\alpha' = 0.000011$/°C as $\alpha = \alpha'(0.98 + 0.000544T)$ (Brockenbrough 1970a).

Note that the Grade 250 steel temperature-dependence based on short-time elevated-temperature tensile tests (Brockenbrough 1970a, Brockenbrough and Merrit 1999) was also used for HPS 485W because of its similar chemical composition though lower carbon content (refer to Table 2 for chemical composition).

Table 1. Heating cycles for US steel test girder (Brockenbrough 1970a, b).

Heaing cycle	Heating temperature °C (°F)	Heated flange width cm (in.)	Radius of curvature m (ft)
1	295 (563)	8.9 (3.5)	539 (1770)
2	207 (405)	8.9 (3.5)	539 (1770)
3	544 (1011)	8.9 (3.5)	200 (655)
4	510 (950)	8.9 (3.5)	178 (583)
5	365 (689)	13.3 (5.25)	147 (481)
6	409 (768)	13.3 (5.25)	121 (397)

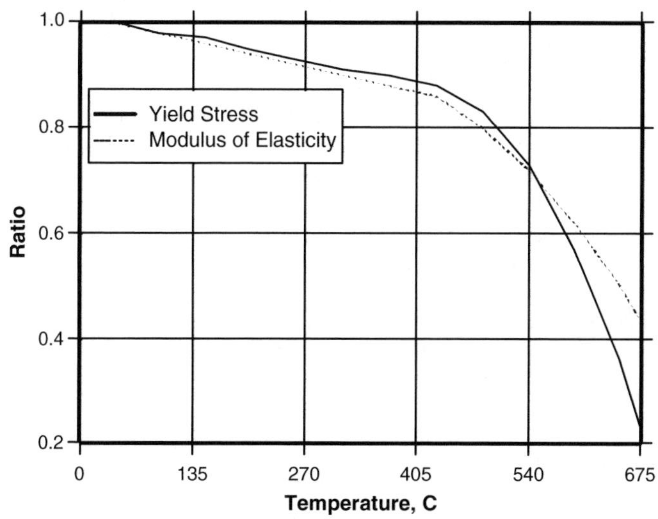

Figure 6. Temperature-Dependent steel properties.

Table 2. chemical composition of structural steel (AASHTO 2000).

Element	ASTM A709 Grade 485W Composition, %	ASTM A709 Grade 250 Composition, %	ASTM A709 Grade 345 Composition, %
Carbon	0.11 max	0.28 max	0.23 max
Manganese	1.15–1.3	0.8–1.2	1.35
Phosphorus	0.02 max	0.04 max	0.04 max
Sulfur	0.006 max (calcium treated)	0.05 max	0.05 max
Silicon	0.35–0.45	0.15–0.4	0.15–0.4
Copper	0.28–0.38	0.2 min (if specified)	0.2 min (if specified)
Nickel	0.28–0.38	–	–
Chromium	0.5–0.6	–	–
Vanadium	0.05–0.07	–	0.01–0.15
Molybdenum	0.04–0.08	–	–
Aluminum	0.01–0.04	–	–
Nitrogen	0.015 max	–	0.015 max

The distribution of initial residual stresses in the HPS steel was assumed to be the same for Grade 250 steel (it was determined from US Steel's data for gas-cut flanges (Brockenbrough 1972)). Although they are static stresses, they can influence the mechanical integrity of the heat curved section. It was previously shown and confirmed by the finite element calibration model (Sen et al. 2003) that they could increase the girder's curvature by up to 15% (Brockenbrough 1970a).

In general, plate girder fabrication introduces residual stresses and camber loss. This is particularly true for heat curved girders (Shin and Walter 1981). In case of curved girders, residual stresses consist of two parts: one due to manufacturing of the plate girders as straight and the other induced from the curving process. The residual stress distribution pattern induced during the curving process depends on the fabrication procedure. In the U.S. Steel study (U.S. Steel 1973), it was reported that the residual stress pattern is a function of dimensions and material properties of the straight girder and the heat curving procedure.

5 FINITE ELEMENT MODEL

The finite element analysis was conducted using NASTRAN computer software (MSC/NASTRAN 2000) in which material and geometric non-linearity were considered. The model can accurately idealize the girder geometry, stiffness, support conditions, initial residual stresses, and temperature loading. The flanges and webs were modeled using four-noded iso-parametric plate elements with in-plane bending stiffness. The finite element mesh had a total of 1739 nodes, 1656 elements and a maximum aspect ratio of three. Details of the model may be found elsewhere (Sen et al. 2003).

The analysis was carried out in steps to determine initial residual stresses and the application of each heat/cool cycle. The program automatically combines results from the individual steps to provide both intermediate and final results.

6 RESULTS OF THE INVESTIGATION

Results of the investigations for HPS 485W steel, hybrid and stiffened girders are summarized in this section.

6.1 *HPS 485W steel girder*

For HPS 485W steel, the primary goal was to determine curvatures resulting from different temperatures increases. Calibration of the finite element model for the grade 250 steel test girder showed that heating cycles 3 and 6 were the most critical (Table 1). Therefore, only these two cycles are considered in the analysis of the HPS 485W steel girder.

The analysis was first performed for the same heat/cool cycles used in the US Steel tests (h_a = 8.9 cm (3.5 in.) and T = 544°C (1011°F), h_a = 13.3 cm (5.25 in.) and T = 409°C (768°F)). For T = 544°C (1011°F) and ha = 8.9 cm (3.5 in.), the radius of curvature for HPS 485W was much larger – 341.7 m (1121 ft) compared to the measured value of 199.6 m (655 ft) for grade 250 US steel test girder. For T = 409°C (768°F), h_a = 13.3 cm (5.25 in.), the radius for HPS 485W was also much greater – 292.3 m (959 ft) compared to 121 m (397 ft) measured value for grade 250 US steel test girder. Consequently, it was concluded that heating temperatures for HPS 485W girders were higher than that required for the test girder to obtain comparable curvature. Results of subsequent investigations are summarized in Table 3.

In the initial trials, temperatures were kept below the current limit of 621°C (1150°F), (AASHTO 1996). Subsequently, higher temperatures of up to 704°C (1300°F) (upper limit suggested by AASHTO for investigation for heat curving HPS485W sections), were considered (AASHTO 2000).

1. For heating cycle 3 (h_a = 8.9 cm (3.5 in.), Table 1), the temperature was increased from 544°C (1011°F) to the current AASHTO limiting temperature of 621°C (1150°F) for conventional steel (AASHTO 1996). The corresponding radius of curvature in the HPS 485W girder was 252.4 m (828 ft), 26% less than the 341.7 m (1121 ft) value determined for T = 544°C (1011°F) but still 26% larger than the measured value of 199.6 m (655 ft) for grade 250 US steel test girder.
2. The limiting temperature of 621°C (1150°F) was then increased in heating cycle 3 (Table 1) and the radius of curvature was found to reduce substantially. For 649°C (1200°F) applied to h_a = 8.9 cm (3.5 in), the radius was obtained as 158.9 m (521 ft), 20% *less* than, 199.6 m (655 ft) the measured radius for grade 250 test girder.
3. The effects of heating cycle 6 were then examined (h_a = 13.3 cm (5.25 in.)). Heating cycle 6 is applied after heating cycle 3, e.g. to the deformed girder shape that developed after cooling in heating cycle 3. The initial trial consisted of a temperature of 621°C (1150°F) applied over h_a = 8.9 cm (3.5 in.) (heating cycle 3) followed by a temperature of 579°C (1075°F) applied over a heated width h_a = 13.3 cm (5.25 in.) (heating cycle 6, Table 1). The radius of curvature was determined to be 166.7 m (547 ft), 37% more than the measured radius of 121 m (397 ft) for grade 250.
4. In the next trial, a temperature of 649°C (1200°F) was applied over h_a = 8.9 cm (3.5 in.) (heating cycle 3) followed by a temperature of 579°C (1075°F) applied over a heated width h_a = 13.3 cm (5.25 in.) (heating cycle 6, Table 1). The radius of curvature that developed was 125.1 m (411 ft), only 4% more than the measured radius of 121 m (397 ft) for grade 250.

Table 3. Results of the investigation of the HPS 485W steel girder.

Heated width h_a cm (in.)	Heating temperature °C (°F)	Radius of curvature R m (ft)	Ratio to test girder's Radius (R_m)*
8.9 (3.5)	544 (1011)	341.7 (1121)	1.71
13.3 (5.25)	409 (768)	292.3 (959)	2.42
8.9 (3.5)	621 (1150)	252.4 (828)	1.26
13.3 (5.25)	579 (1075)	166.7 (547)	1.38
8.9 (3.5)	649 (1200)	158.9 (521)	0.80
13.3 (5.25)	579 (1075)	125.1 (411)	1.04

* R_m = 199.6 m (655 ft) for h_a = 8.9 cm (3.5 in.), 121 m (397 ft) for h_a = 3.3 cm (5.25 in.).

It may be concluded that a temperature of 649°C (1200°F) for HPS 485W steel (compared to 544°C (1011°F) for grade 250 steel) applied over 1/6 flange width would induce comparable curvatures between grade 250 and HPS 485W steel. A temperature of 579°C (1075°F) for HPS 485W steel (compared to 409°C (768°F) for grade 250 steel) applied over 1/4 flange width would induce comparable curvatures between grade 250 and HPS 485W steel. It should be noted that those temperatures are below the 705°C (1300°F) recommended temperature (AASHTO 2000).

6.2 *Hybrid girders*

For hybrid girders in which typically the top flange (or bottom in continuous beams) and the web are made of the same steel grade while the other flange is made of different steel, the main effect examined was the relative movement between the two flanges if the same heating cycle were used for both flanges.

The hybrid girder investigated was the test girder in which both the top flange and the web were Grade 250 steel while the bottom flange was HPS 485W. The relative movement in the finite element model was obtained from the unequal lateral offset in the top and bottom flanges. At 544°C (1011°F), the relative movement was 3.56 cm (1.4 in.). The main concern was whether such a large differential movement could distort the section. In this event, it was necessary to apply different heating temperatures to the top and bottom flanges. It should be noted that the girder behavior

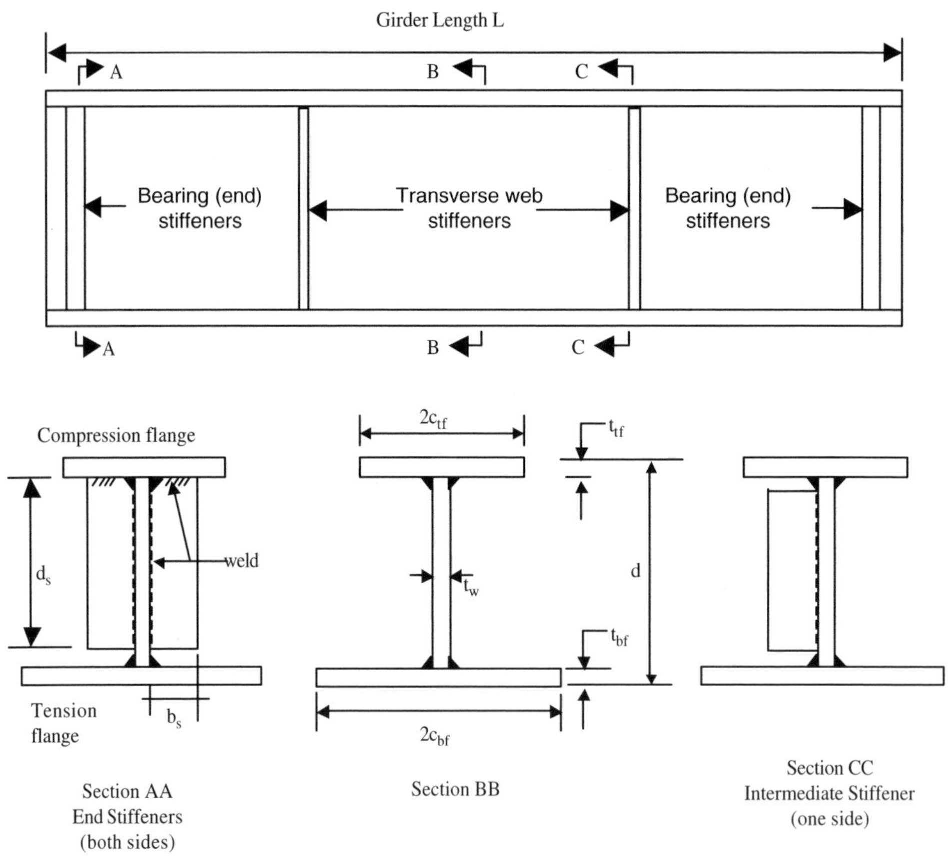

Figure 7. Stiffened girder.

during heating and cooling is not influenced by the vertical web element for the support conditions shown in Fig. 4.

6.3 *Stiffened girders*

For stiffened girders (Fig. 7), investigations were made for the effect of intermediate stiffeners placed along the girder's length. The stiffeners were modeled using four-noded, isoparametric plate elements with in-plane bending stiffness (as for the flanges and web), and were only attached to the web. The effect of stiffeners on induced curvature depends on (1) size of the stiffener (mainly thickness t_s) and (2) web depth d (Fig. 7). The effect of stiffeners was noticeable when five intermediate stiffeners equally spaced at 2.5 m were used. The thickness of the stiffener was the same as the web thickness 1.27 cm (1/2 in.) (Fig. 4).

For Grade 250 steel (girder and stiffeners) and at a heating temperature of 544°C (1011°F), heated width h_a = 8.9 cm (3.5 in.), the induced radius of curvature was 218 m (715 ft), 9.2% more than the induced radius of curvature of 200 m (655 ft) for the un-stiffened girder. If the number of stiffeners is reduced to four, the increase in the radius of curvature is only 5%. If two stiffeners are used, the effect is minimal (only 1%).

7 CONCLUSIONS AND RECOMMENDATIONS

Results from this investigation indicate the need for using higher temperatures for heat curving HPS 485W sections. The analysis suggests that the optimal maximum temperature is 649°C (1200°F), higher than the current AASHTO limit (*8*) of 621°C (1150°F) but lower than the 705°C (1300°F) temperature recommended by AASHTO 2000. As this temperature does not affect the base strength (AASHTO 2000), consideration should be given to its future adoption for use with HPS 485W sections.

For hybrid girders, different heating temperatures are required to prevent distortion. It is recommended that heating temperatures for the top and bottom flanges be based on homogeneous girders, with support conditions in which the ends (placed vertically) are free to move but with the web bolted at mid-length to a center platform.

For stiffened girders, it was found that the effect of stiffeners on curvature is minimal mainly because the girder's support does not restrict lateral movements during heating.

Overall, the finite element analysis suggests that curving HPS steel requires relatively minor modifications to current practice.

ACKNOWLEDGEMENTS

The author greatly acknowledges the financial support of the Flom Fellowship at the University of South Florida, Tampa, Florida and the assistance and cooperation of Tampa Steel Erecting Co, Tampa, Florida, USA.

REFERENCES

AASHTO, 1996, Standard Specifications for Highway Bridges, 16th Edition, Washington, DC.

AASHTO, 2000, Guide for Highway Bridge Fabrication with HPS70W Steel, *published by the American Association of State Highway & Transportation Officials*, Washington, DC.

Brockenbrough, R.L., 1970a, Theoretical Stresses and Strains from Heat-Curving, *ASCE, Journal of Structural Division,* July, vol. 96, no. ST7, pp. 1421–1444.

Brockenbrough, R.L., 1970b, Experimental Stresses & Strains from Heat-Curving, *ASCE, Journal of Structural Division,* July, vol. 96, no. ST7, pp. 1305–1331.

Brockenbrough, R.L., 1972, Fabrication Aids for Continuously Heat-Curved Girders, United States Steel Corporation, Pittsburgh, PA, April.

Brockenbrough, R.L., 1973, Fabrication Aids for Girders Curved with V-Heats, United States Steel Corporation, Pittsburgh, PA, January.

Brockenbrough, R.L., Merrit, F.S., 1999, *Structural Steel Designer's Handbook*, 3rd Edition, McGraw-Hill, NY.

California Department of Transportation, 2002: http://www.dot.ca.gov/hq/esc/oe/specifications/SSPs/2002-SSPs/Sec_10/49-59/55-510_B07-30-99_DU.doc. Accessed April 16, 2004.

Davidson, J. Abdalla, R., Madhavan, M., 2004, Stability of Curved Bridges during Construction. Prepared by UTCA (University Transportation Center for Alabama), UTCA Report Number 03228, December. pp. 29–31.

Gergess, A., 2001, *Cold Bending and Heat Curving of Structural Steel I-Girders*, PhD Dissertation, Department of Civil and Environmental Engineering, University of South Florida, Tampa, FL, August.

Hilton, M.H., 1984, Deflections and camber loss in heat-curved girders, *Transportation Research Record* 950(2), National Research Council, Washington, D.C.

MSC/NASTRAN for Windows, The MacNeal-Schwendler Corporation, 2000, Finite Element Modeling and Postprocessing System, Los Angeles, California.

Sen, R., Gergess, A.N., and Issa, C., 2003, Finite element modeling of heat-curved I-girders, *Journal of Bridge Engineering*, ASCE, 8(3), pp. 153–161.

Shin, S.W., and Walter, G.H., 1981, Case histories of residual stress related component failures, *Residual Stress for Designers and Metallurgists*, American Society for Metals, Chicago.

United States Steel Corporation (USS), 1973, Fabrication aids for continuously heat-curved girders, AISC Marketing, Inc., Chicago, IL.

Wick, C.H., 1960, *Chipless Machining*, Industrial Press, Inc., New York.

Chapter 13

Testing of a novel flexible concrete arch system

S.E. Taylor, D. Robinson & A.E. Long
Queen's University of Belfast, Northern Ireland

A. Gupta
Macrete Ltd, Toomebridge, Northern Ireland

ABSTRACT: This paper describes the testing of a novel flexible masonry concrete arch system which requires no centering in the construction phase or steel reinforcement in the long-term. The arch is constructed from a 'flat pack' system by use of a polymer reinforcement for supporting the self-weight of the concrete voussoirs and behaves as a masonry arch once in the arch form. The paper outlines the construction of a prototype arch and load testing of the backfilled arch ring. Some comparisons to the results from analysis software have been made. The arch had a load carrying capacity far in excess of the current Highways Agency design wheel loads.

1 INTRODUCTION

1.1 *Background*

Masonry arch bridges have been used for 4000 years and today they still play an important role in the road network of the UK and other areas of the world. There are currently ~70,000 masonry arch bridges in the UK and more in other countries in Europe (Harvey, 2007) . However, the rapid rise in labor costs associated with the construction of masonry arch bridges had made them less cost effective than their reinforced and pre-stressed counterparts. Never the less, many of these more recent steel reinforced concrete bridges have had to be repaired due to corrosion or replaced due to lack of carrying ability to meet new European loading standards (Highways Agency, 2001 and 1995). The repair or replacement of bridges, environmental and aesthetic consideration must receive priority and account taken of the whole life cost of a bridge structure. This means that a masonry arch bridge which can be transported flat, lifted and erected rapidly is an attractive option for small bridges which make up the majority of the bridge stock in the UK and Ireland.

The arch system developed under a Knowledge Transfer Partnership (KTP) between Queen's University Belfast and Macrete Ltd. Uses the arch form, plain structural elements and eliminates of corrodible reinforcement there by meeting the requirement of a more sustainable and reduced whole life cost bridge form. A set of concrete voussoirs have been laid contiguously in a horizontal form with the tapered gap at the bottom. A grid of polymeric reinforcement is placed on top and a thin layer of concrete is added. When lifted into position, cracks form in the top concrete which allows rotation to take place and the arch profile is formed from the 'flat-pack'. Polymeric reinforcement is very effective in this application as the loading on this element is short term and only occurs when the arch is being lifted, it also has the advantage of being non-corrosive. A 5 m span full size prototype arch has been constructed from flat pack, monitored during backfilling operations (Taylor et al, 2006) and is due to be tested to up to six time the current wheel loading. This paper will describe the testing of the novel arch system and compare to load prediction from analysis such as ARCHIE.

2 ARCH DETAIL

2.1 *Construction of the voussoirs*

There are two options for the construction of the arch unit. The voussoirs can be pre-cast individually, laid contiguously horizontally with a layer of polymer grid material placed on top. The individual voussoirs are then interconnected by an in-situ layer of concrete which is placed on top (Figure 1). Alternatively, the arch unit can be made in a single casting operation by using a shutter with wedge formers spaced to simulate the tapered voussoirs. The arch unit can be cast in convenient widths to suit the design requirement, site restrictions and available lifting capacity. When lifted, the wedge shaped gaps close, concrete hinges form in the top layer of concrete and the unit is supported by tension in the polymer grid. The arch shaped units are then placed on a pre-cast footings or anchor blocks. When in the final arch position, the self-weight is carried by compression in the arch ring and the arch behaves as an un-reinforced masonry arch.

2.2 *Polymeric reinforcement detail*

The advancements in composite material technology and the ability of the polymer in this system to be sufficiently strong yet flexible has provided the key to the success of the arch. Material tests were carried out on samples of the polymeric reinforcement. The average tensile strength is summarized in Table 1 and a typical failed sample shown in Figure 2.

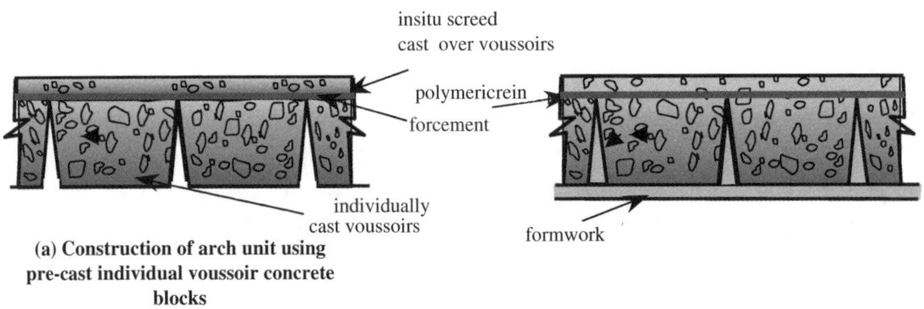

Figure 1. Flexible concrete arch construction.

Table 1. Tensile strength of polymeric reinforcement.

Sample no.	Load at failure (kN)	Tensile strength (kN/m width)	Maximum creep at ultimate load
1	2.01	57.4	
2	2.35	67.1	
3	1.36	38.9	
4	1.81	51.7	
Average		**53.8**	0.2

Figure 2. Typical failed specimen of polymer reinforcement.

2.3 *Test arch construction details*

The arch details were as given in Table 2.

Details of the lifting procedure fort he test arch is illustrated in Figure 3. The last photograph shows the arch sitting in its final position on the correctly sloped seating unit.

Table 2. Test arch details.

Voussoir dimensions	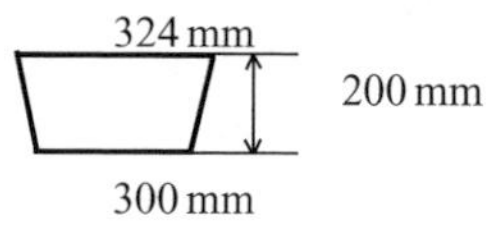
Clear span:	5.00 m
Effective span:	5.24 m
Internal height:	2.00 m
Depth of arch ring:	0.240 m (40 mm top screed)
Width of arch ring:	1.00 m
Polymer Reinforcement 150/100	
Tensile strength	kN/m
(2 layers over middle 17 blocks and 1 layer for remaining outer 3 blocks)	
Arch ring concrete compressive strength*	~55 N/mm^2
Backfill	lean mix concrete to 0.4 m above arch extrados

* is based upon average 28 days cube compressive tests and concrete in the arch ring in excess of 28 days.

Figure 3. Arch lifting process and final position in seating units.

3 TESTING OF 5 M SPAN ARCH RING WITH CONCRETE BACKFILL

3.1 *Summary of test*

The 5 m span prototype flexible concrete arch system, as shown in Figure 4, was monitored during concrete backfill operation and this has been presented in a previous paper (Taylor et al, 2006). It was then tested in accordance with the requirements of the relevant bridge loading category and following the guidelines in BS8110 (1985) for the testing procedure. A simulated static wheel load was applied at the mid span and the third span of the arch ring. It is recognized that, in practice, there will be a dynamic amplitude factor above the static load. The single wheel load and factors of safety used to establish the test loads were based upon the requirements in BD91/04 (Highways Agency, 2004). Bridges are designed under static load conditions with factors of safety applied to these loads. An impact factor of 1.8 is recommended in BD91/04 and this takes into account dynamic amplitude effects on an arch bridge form. The single wheel load, for this category of bridge, is 5.75 t.

This equates to an ultimate design load (ULS) of : $5.75 \times 1.65 \times 1.8 = 17.1$t and incorporating a contingency factors of safety of 1.1 for the test gives: $17.1\text{t} \times 1.1 = 18.8$ t

However, in both tests the ULS design load was exceeded and the maximum applied load was 35 t. That is, the full test loads were six times the single wheel load and nearly twice the ULS design load. Digital photos and observations were made throughout the duration of the test.

3.2 *Instrumentation and testing procedure*

The instrumentation set-up for the mid span and third span load tests is shown in Figure 4. Deflection transducers were used to monitor both horizontal and vertical deflections and vibrating wire strain gauges were used to measure crack openings at the joints between voussoirs. A typical test arrangement is depicted in Fig. 4. A circular concentrated load was applied at the loading position via a 300 mm diameter steel plate bedded on soft board (Fig. 4). The application of load was

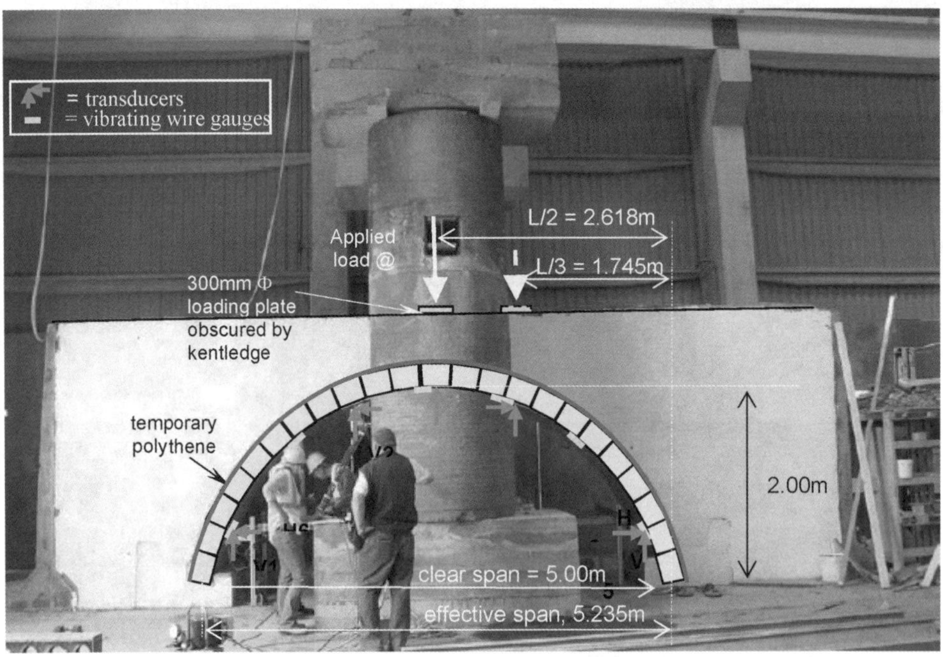

Figure 4. Instrumentation and test-set-up for arch ring and arch with backfill.

from an accurately calibrated 500 kN hydraulic jack system and the test rig was assembled with the top beam horizontal about both axes thus minimizing eccentricity effects. The simulated static wheel load was applied to mid span and third span position. The load was applied incrementally and deflection and strain measurements were recorded at each incremental loading. A service test load of 10 t was applied twice at each load position prior to the application of the full test load. As the loading was increases, the load was held at each of the load stages stated in section 3.1. The behavior of the arch and the backfill was observed throughout.

4 TEST RESULTS

4.1 *Observed behavior*

No cracking was observed under the two test loads of 10 t under both the mid span and third span load conditions. The first cracking appeared at an applied load of 20 t in the concrete backfill under the mid span load (the first test position). For both loading positions, the crack started on a line adjacent to the perimeter the loaded area and propagated towards the arch ring voussoirs. Under mid span loading, the crack followed the vertical line of the 18 mm plywood stop end at the mid span which had been used between the two sides of the backfill to facilitate demolition. Under third span loading the crack propagated diagonally towards the intrados.

The plywood stop end had been placed across the whole width of the arch (i.e. 1 m length) and this probably caused additional cracking in the concrete backfill. There was no visible opening in the joints between the voussoirs and no further cracks developed under the full test load. Under third span loading and at the full test load of 352 kN, the joint directly below the plywood stop end had opened by 2 mm. This opening was partly due to the differential settlement in the backfill due to presence of the plywood across the full width of the arch unit which provided a shear plane. Cracking also occurred horizontally adjacent to the anchor block. After unloading, there was visible recovery in the deflections and cracking in the arch system. The openings in the voussoirs also closed.

4.2 *Deflection measurements*

Figure 5 shows typical load vs. deflection results and the overall deflections at maximum applied load (that is, the vector sum of the maximum horizontal and vertical deflections) have been summarized in Figure 6 which shows the exaggerated deflected shape. It can be seen that the maximum deflection under the mid span load was 2.3 mm inwards at an applied load of 340 kN (~34 t). The maximum deflection under the third span loading was 10.3 mm outwards at the third span at an applied load of 352 kN. A deflection of 10.3 mm is equivalent to the (effective span/508) and within acceptable limits for deflection. It should be noted that a plywood stop end had been used between the two sides of the backfill to ease demolition. This caused a higher degree of cracking at the mid span compared to a continuous backfill and, coupled with the polythene liner, probably gave a conservative prediction of the deflections compared to an arch ring without a liner or plywood at mid span.

It can be seen from load versus deflection results the reading that at an applied mid span load of 200 kN the rate of deflection increases for similar load increments. This was due to cracking in the backfill at the position of the plywood (Figure 7). The rate of change in deflection also changed at an applied third span load of 200 kN. This was also due to cracking in the concrete backfill. However, the maximum deflections, for both loading conditions, were less than span/500 at an applied load of ~34 t which is nearly six times the single wheel load for this category of bridge. The recovery of the arch after loading was good. For example, for the full test load at mid span, the maximum deflection at V3 was 2.0 mm at an applied load of 340 kN. After unloading, the permanent deflection was 0.3 mm. This equates to an 85% recovery in deflection which is within the acceptable limits given in BS 8110: Pt 2: Section 9 (British Standards, 1985).

The recovery rate for each of the tests based upon the maximum deflections is Table 3 summarises. From the load versus deflection, it was noted that there is a shift in the reading at an applied

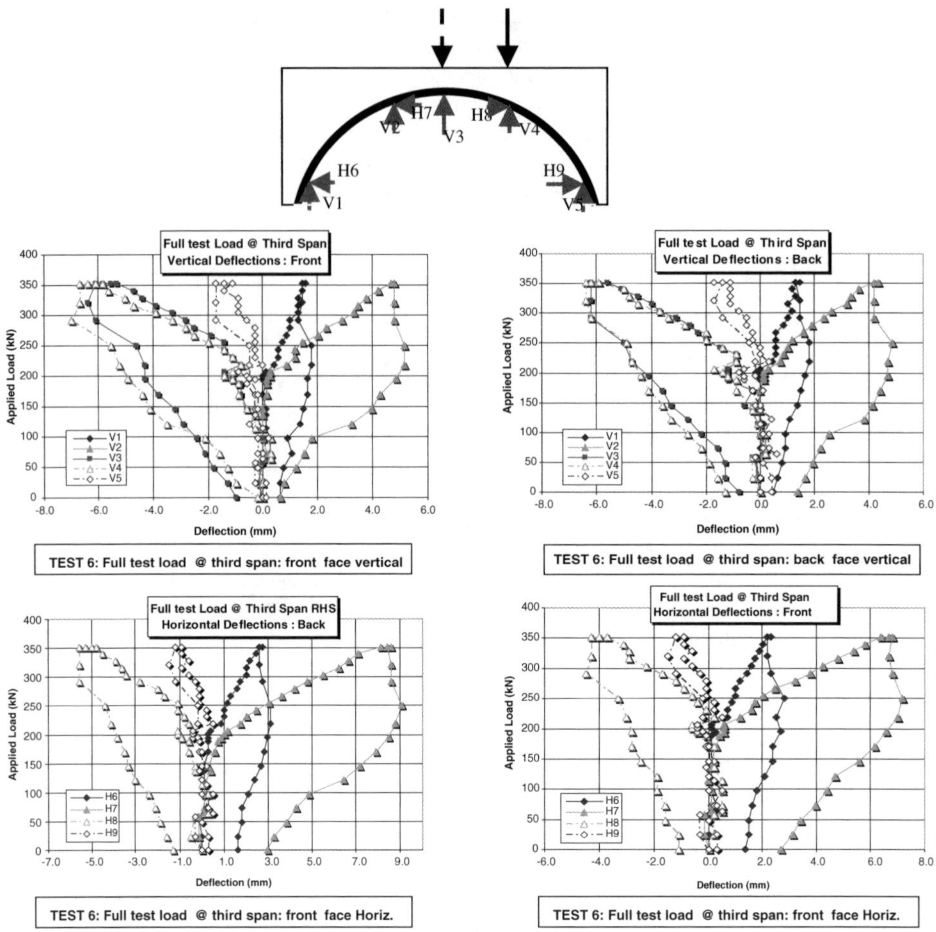

Figure 5. Applied load vs. deflection for Test 6: full test load at third span.

mid span load of 200 kN. This was due to cracking in the backfill as discussed in Section 3. The rate of change in deflection also changed at an applied third span load of 200 kN. This was also due to cracking in the concrete backfill. However, the maximum deflections, for both loading conditions, were less than span/500 at an applied load of ~34 t (six times the single wheel load).

4.3 *Strain measurements*

The strain measurements were very low with the mid span load indicating low levels of stress in the arch ring. Under the third span loading, the strain measurements were also low. However, the largest opening occurred at the mid span voussoir right hand side joint which did not have a gauge. The joint opening was approximately 2 mm under the maximum applied load at third span.

5 ANALYSIS OF THE ARCH SYSTEM

5.1 *NLFEA analysis*

Other researchers, such as Fanning and Boothby (2001), have shown the benfits of modeling arch behavior using Non linear finite element analysis (NLFEA) . Abaqus NLFEA was used to analyse the

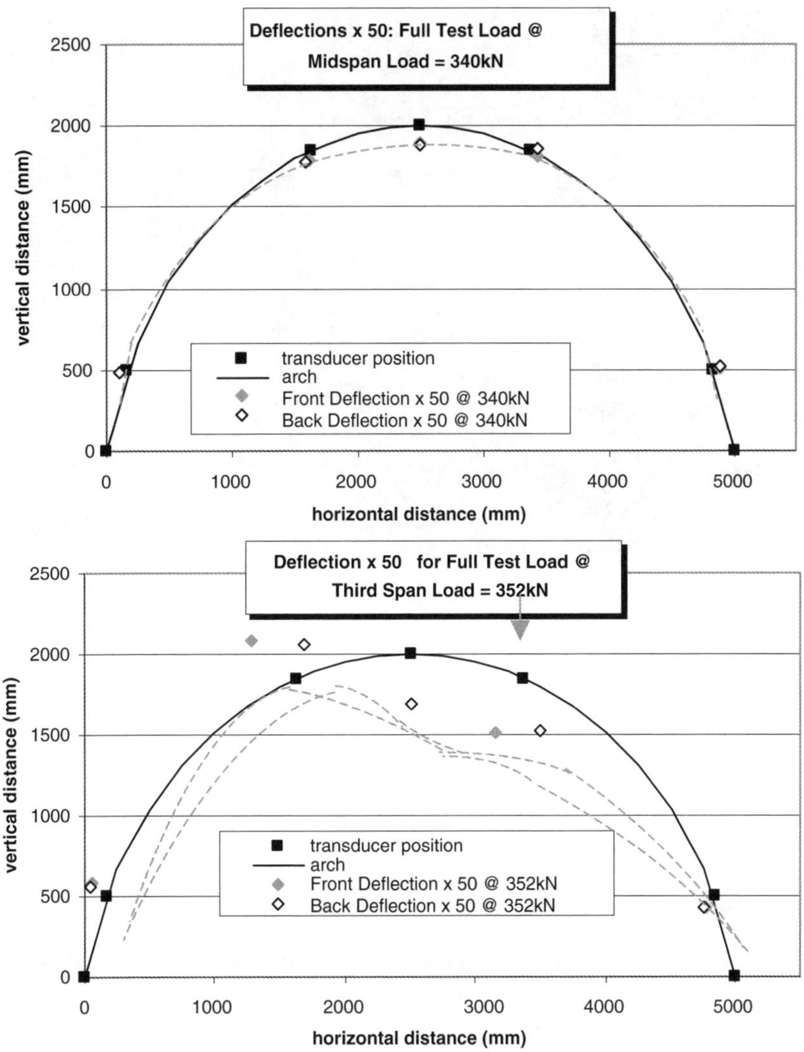

Figure 6. Applied load vs. deflection for Test 6: full test load at third span.

arch and two contrasting approaches for predicting the collapse were investigated. In the first, the material was modeled using a plastic material model for the concrete in the arch ring. The ring was assumed to be homogeneous. The use of the plasticity model allowed the formation of cracks in the concrete to be simulated. As a consequence the approach will be able to predict the formation of the hinges within the arch. The collapse of the arch occurred when sufficient hinges formed and a mechanism occurred. Figure highlights the formation of plastic hinges in the arch ring under the action of a concentrated load in the centre of the arch. Figure represents the stage in the loading where the formation of hinges at the quarter and three-quarter position along the arch has just started. The addition of these hinges results in a mechanism being formed and the collapse of the arch.

The second approach modelled the individual blocks using a contact analysis to simulate the interaction between the blocks. The contact face was established using a Coulomb friction law. To enable this analysis, the explicit finite element method was required. The explicit method is generally used to simulate transient dynamic response but by applying the load at a suitable rate a pseudo-static analysis was achieved. Figure 9 shows the collapse of the arch under the action of a

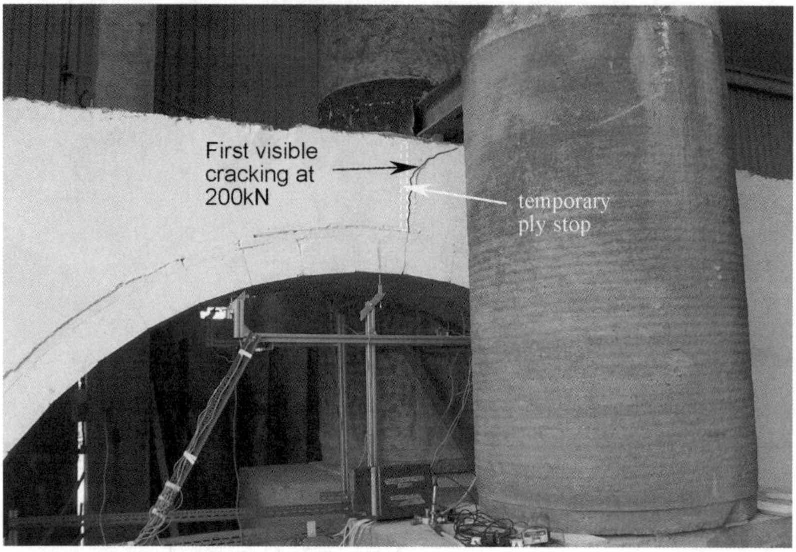

Figure 7. Full test load at third span at an applied load of 200 kN.

Table 3. Summary of recovery in deflections.

Test No.	Position	Maximum vertical deflection (mm)	Recovery (%)
3: 340 kN @ mid span	B V2/H7	−1.5	80
	B V3	−2.3	74
6: 250 kN @ third span	B V2/H7	5.0	70
	B V4/H8	−6.5	79

load at mid span, indicating that slipping or the opening of the joints occurred. Analysis of the arch using NLFEA is on-going and models requires further development to include the effects of the polymer reinforcement and interaction with the backfill before they can be compared effectively with the test models.

5.2 *ARCHIE analysis*

The arch was analysed using ARCHIE (Obvis, 2006), a numerical analysis package which takes into account the arch backfill. It is important to note that this software is also used by the DRD Road Service in Northern Ireland for load assessment analysis of their arch bridges. Therefore, assessment of the load carrying capacity of the flexible arch system using ARCHIE was a critical task in the development of the arch system. The arch unit was analysed under different wheel loading conditions. A typical case of arch unit analysis is shown in Figure 10.

An arch unit of the required geometry can be created and loaded with the standard wheel loads. A line of thrust is indicated in Figure 8. Under design loading, the position of the thrust line in the arch unit gave information about the stability of the unit. Furthermore, ARCHIE was able to demonstrate the change in the thrust line by changing the height of the Backing Material (BM) at the springing level and the effect of changing the Passive Pressure (PP). The passive pressure factor is equivalent to the 'At Rest' pressure coefficient $(1-\sin\phi)$.

Therefore, for a particular loading condition, arch ring depth and using a the appropriate for BM the passive pressure required to resist the arching thrust is given. Alternatively, the passive pressure factor can be fixed and the minimum arch ring depth established for the given loading conditions.

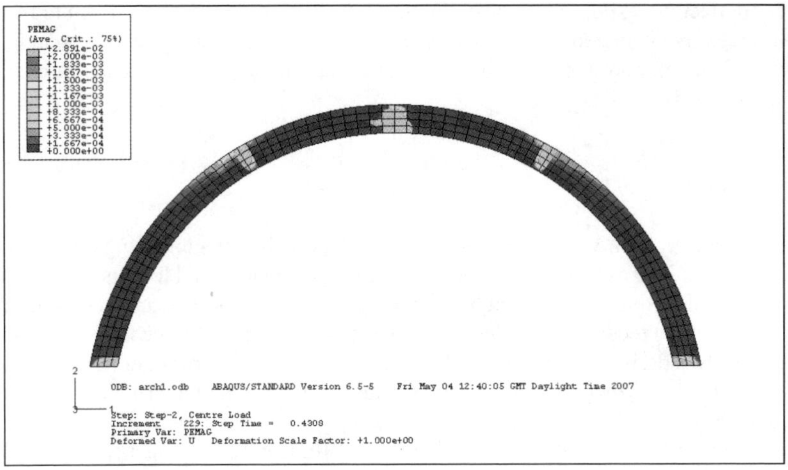

Figure 8. NLFEA results using the homogenous model.

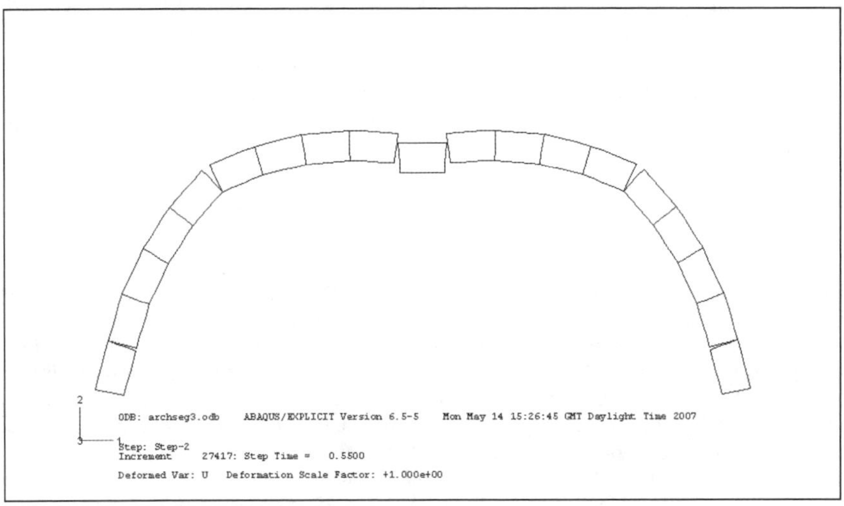

Figure 9. NLFEA results using the explicit model with contact analysis.

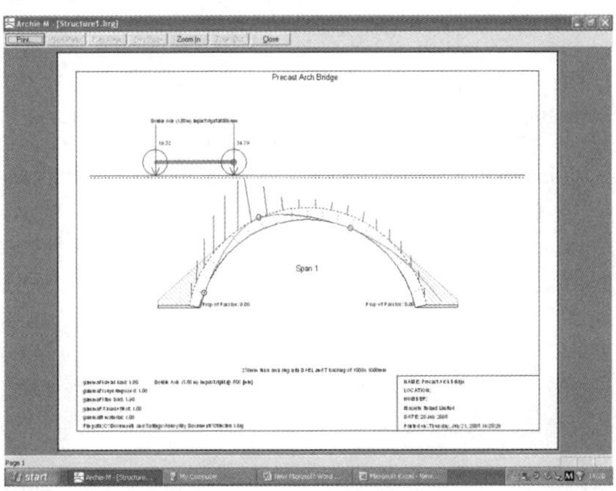

Figure 10. ARCHIE analysis.

A similar deflection response was given in the ARCHIE analysis in comparison to the test load results. However, the predicted ultimate capacity was conservative based on the actual load carrying capacity of the arch system (which was greater than the applied test loads in excess of the design ultimate loads).

6 CONCLUSIONS

The 1 m prototype arch ring with concrete backfill was capable of supporting a midspan load of 34 t and a third span load of 35 t which is nearly six times the wheel load for this category of bridge and nearly twice the ultimate load including ULS, dynamic and contingency factors of safety. The arch ring showed good recovery in deflections and recovery in the joint openings after the removal of all load. This was despite the presence of the plywood stop end across the width of the backfill at mid span and the polythene layer between the arch ring and the backfill. These were used to facilitate demolition.

The maximum deflection, with third span load, was 10 mm and is equivalent to (span/508) which is within acceptable limits for deflection. The maximum deflection occurred at the third span and there was a 69% recovery in the maximum deflection after the removal of the all load. The amount of recovery, for an applied load which was twice the ULS design load (which included a 1.8 dynamic impact load factor), suggests that the maximum loading was not ultimate capacity of the arch system. The strain values were very low at maximum applied loads indicating low levels of stress in the arch ring. The results from ARCHIE gave a similar deflected shape to the measured results although the predicted load capacities were conservative. An analysis of the arch using NLFEA gave good prediction for the behavior of the arch system and this work is on going.

REFERENCES

ARCHIE-M: Masonry Arch Bridges and Viaduct Assessment Software, Version 2.0.8, OBVIS Ltd. UK.

British Standards Institute, 1985, BS 8110: Part 2: Structural use of concrete: Code of practice for special circumstances Section 9: Appraisal and testing of structures and components for construction, London.

British Standards Institute, 1985, BS 8110: Part 2: Structural use of concrete: Code of practice for special circumstances Section 9: Appraisal and testing of structures and components for construction, London.

Fanning P J and Boothby T E, 2001, Three-dimensional modelling and full-scale testing of stone arch bridges, Computers and Structures, vol. 79, no. 29–30, pp. 2645–2662.

Highways Agency (UK), BD37/01, 2001, Departmental Standard, Loads for Highway Bridges (used with BS5400: Pt2) Design Manual for Roads and Bridges, Volume 1, Section 3, Part 14, Department of Transport, Highway and Traffic.

Highways Agency (UK), BD44/95, 1995, Departmental Standard, The assessment of concrete highway bridges, Design Manual for Roads and Bridges, Volume 1, Section 3, Department of Transport, Highway and Traffic.

Highways Agency (UK), BD 91/04, 2004, Departmental Standard, Unreinforced masonry arch bridges, Design Manual for Roads and Bridges, Volume 2, Section 2 Special Structures, Part 14.

Taylor S E, Gupta A, Kirkpatrick J, Long A E, Rankin G I B and Hogg I, 2006, *Development of a novel flexible concrete arch system*, 11th International Conference on Structural Faults and Repairs, Edinburgh.

5 *Bridge construction & rehabilitation*

Chapter 14

Westfield Great River Bridge

M. Ennis
STV Incorporated, Boston, MA., USA

ABSTRACT: The Great River Bridge, built in 1939 is located in downtown Westfield, a City in Western Massachusetts. The through truss bridge is a landmark for the City, forming perhaps the most distinctive structure in the downtown area. The project scope initially involved the rehabilitation of the 368 foot long, two-span structure. However the project has grown to include the design of two other bridge structures, four landscaped parks, several thousand feet of urban roadway, and two miles of railroad track.

1 DESCRIPTION OF PROJECT

Westfield, founded in 1669, is located 100 miles west of Boston, and 85 miles east of Albany. The city now has a population of 40,000. Historically, local industry involved the production of bricks, cigars and whips, later bicycles, textile machinery and precision tools production. The City has been dubbed 'Whip City' for its most famous product, the buggy whip.

The Great River cuts through the downtown center of Westfield. The Great River Bridge carries Elm Street over the river. Elm Street constitutes the main thoroughfare through the downtown, and connects the city center to the Massachusetts Turnpike, which is located 2 miles north of the river. The Great River Bridge provides the only vehicle crossing of the river for several miles.

Built in 1939, the bridge superstructure consists of a twin span, continuous Warren through truss. The truss chords are formed by riveted, built up steel sections. The total bridge length is 368 feet. Although the bridge is not registered as a 'Historic Structure', the bridge does have historic significance since it is believed to be the earliest continuous Warren through truss bridge in the Commonwealth of Massachusetts.

Figure 1. Aerial view of Great River Bridge.

Figure 2. 1939 Photograph of original construction.

Figure 3. Westfield Great River Bridge.

In 1994, STV Incorporated and the BSC Group were contracted to perform an evaluation of the bridge. The evaluation found that the chord elements were structurally sound, but that the concrete deck, steel stringers, and steel floorbeams were all in need of replacement. However, at the functional level, the evaluation found that the bridge was generating a choking point for vehicular traffic. Elm Street is sufficiently wide to accommodate four lanes of traffic, but the Bridge itself only accommodates three lanes of traffic. Adding to the traffic congestion is the eleven span, CSX railroad viaduct which crosses over North Elm Street immediately north of the Bridge. Despite a

Figure 4. Substandard roadway clearance at Viaduct.

severe dip in the Elm Street roadway profile at the viaduct, the vertical clearance is only 12 foot, 9 inches for vehicles.

The combination of restricted traffic capacity on the bridge, and the sub-standard vertical clearance immediately north of the bridge serve to generate chronic traffic congestion in the downtown area.

Through an iterative process with the City, the Designer, and the Massachusetts Highway Department, a bridge reconstruction plan was developed. The plan aimed to provide relief for the worsening traffic congestion, while preserving the original Great River Bridge Structure. The plan involved:

- Rehabilitate the existing Great River Bridge,
- Construct a 'Sister Bridge', matching the truss configuration of the Great River Bridge, immediately down river, and
- Reconstruct the CSX railroad viaduct at a higher elevation.

In order to implement this work, several other construction tasks were required, namely:

- Reconstruct 5,000 feet of urban roadway to accommodate the Sister Bridge.
- Demolish several buildings to facilitate the bridge/roadway construction.
- Reconstruct 30,000 linear feet of rail road track, including crossovers and spur lines to accommodate the 5 foot rise in track profile.

Finally, the City saw an opportunity to enhance the downtown area with several aesthetic features, namely:

- Four landscaped parks in the vicinity of the bridges,
- A free standing clock tower,
- Brick paved, and tree lined sidewalks, and
- Stamped and colored concrete in the traveled way at roadway intersections.

STV/BSC received notice to proceed with final design from the City in 2001 for the scope of work outlined above. As of April 2007, the design is complete, and the low bidder has been selected. Construction is anticipated to begin in late spring. The BSC Group, our Joint-Venture Partners, was responsible for Permitting and Traffic Signalization design. Our sub-consultants included Landworks Collaborative, VHB and Vine Associates.

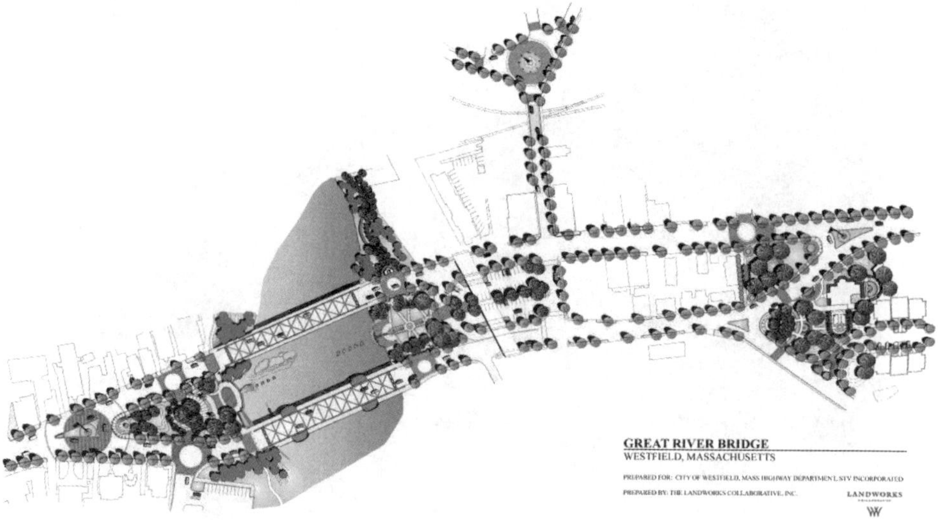

Figure 5. Rendering of proposed work (Landworks collaborative).

2 REHABILITATION OF THE GREAT RIVER BRIDGE

A critical design step in the implementation of this plan involved assuring that the original truss structure could meet the design requirements of the current AASHTO Code. Of particular concern was the reinforced concrete center pier. The wall type pier is lightly reinforced and is clad in masonry block. The original bridge has a fixed bearing for each Warren truss frame located on this pier, with expansion bearings at each abutment. Consequently, all horizontal longitudinal loads acting on the bridge superstructure were being resisted solely by the pier structure, through the two fixed bearings. Analysis of the structure found that the pier structure could not safely carry the horizontal loads specified by standard AASHTO design.

Another concern for the bridge related to the existing rocker type bearings, which employ a large diameter steel pin to connect the bearing assembly to the bottom chord of the truss frames. These steel pins are fracture critical, and are prone to sudden failure. In addition the rocker type bearings present a potential for instability during a large seismic event.

A two step approach was initiated to address the problem of the lightly reinforced concrete pier, namely to more accurately determine the horizontal design loading, and to redistribute the design horizontal loadings through redesigned bearing assemblies, to both abutments in addition to the pier.

To more accurately define the lateral seismic loading, a Site Specific Seismic Analysis was commissioned to be performed to more accurately determine the seismic forces. This analysis was performed by Prototype Engineering. A response spectrum specific to the Great River Bridge was developed. Using this site specific response spectrum, seismic design loads for the bridge were significantly reduced compared to employing the standard response spectrum supplied by AASHTO.

In order to redistribute the horizontal loads into all six bearings, dynamic isolation type bearings were considered. But the use of large elastomeric pads was first investigated, as a 'low tech', and cheaper alternative. With only six bearings supporting the entire superstructure, the loads acting on each bearing is significant, and consequently the size of the neoprene pads that would be required presented a concern. A maximum dimension of 48" was established to assure that the neoprene pads could be readily fabricated.

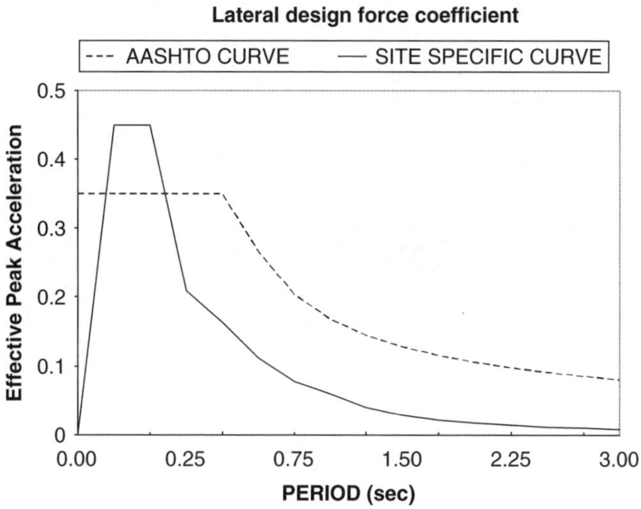

Figure 6. Response spectrum.

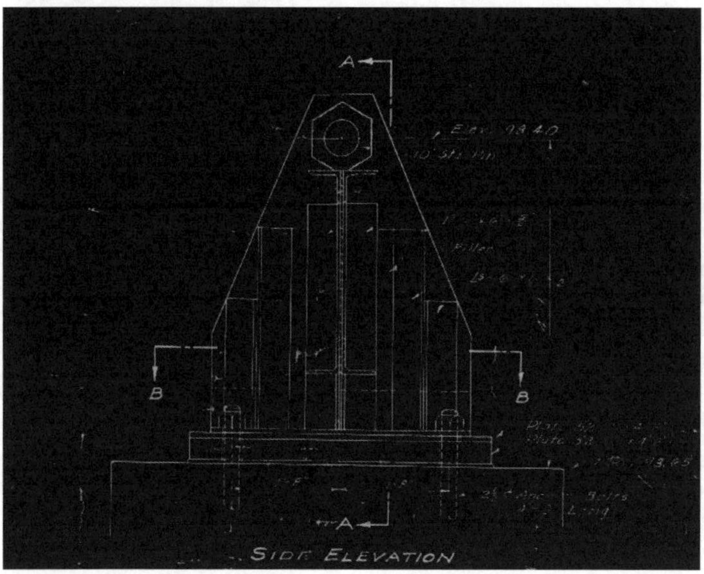

Figure 7. Elevation of existing bearing at Pier.

Several iterations of design were performed, incorporating multiple different neoprene pad dimension configurations. Changes in the pad's height, length and thickness all impacted the lateral stiffness value of the pad. By modifying the lateral stiffness values of the individual bearings, the lateral load acting on the bridge can be distributed into each of the substructure elements as desired. Through this iterative process neoprene pads were sized for both the pier and the abutments so that the bridge could safely resist the design seismic event without the need to replace the existing pier.

Another challenge for the design team was the detailing of the replacement bearing assemblies, which incorporate the neoprene pads. The truss frames will remain in place during construction.

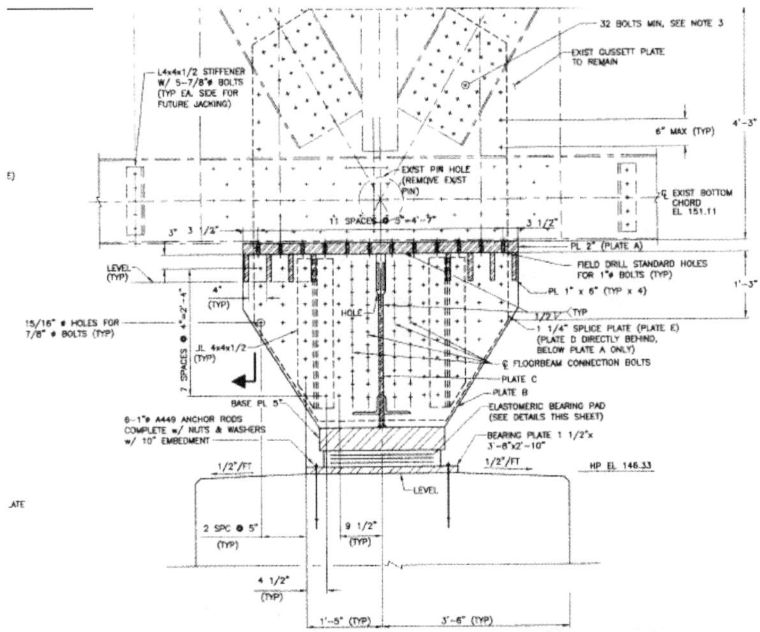

Figure 8. Elevation of proposed bearing.

Consequently the steel truss frames will need to be temporarily jacked up and supported, to allow for the removal of the existing rocker type bearings and steel pins. The new bearing assemblies can then be installed. These assemblies will need to be bolted to the lower chord and gusset plates of the existing truss. The challenges for the designers in detailing these assemblies include:

- Load on the neoprene pad to be evenly distributed,
- Height of proposed assembly to match existing,
- Adequate load transfer from both gusset plates and lower chords
- Floorbeam moment connection to inside face of assembly,
- Sidewalk bracket moment connection to outside face of bearing.
- Assure constructability.

Figure 8 shows the final configuration of the proposed bearing assembly. Through both the analysis and the design stages, the proposed bearing assemblies for the Great River Bridge constituted the largest challenge in the design of all three bridges on this project.

3 SUMMARY

Through the efforts of the design team and in partnership with both MassHighway and the City, a landmark bridge structure for the City is not only preserved, but becomes the corner stone of an urban revitalization project.

REFERENCES

Bridge Type Study – Great River Bridge, 1995, STV Incorporated/BSC Group.

Existing Conditions Report, 1995, STV Incorporated/BSC Group.

Historic Resources Survey – Great River Bridge Improvement Project, 1994, McGinley Hart & Associates.

Site Specific Seismic Response Spectrum, 2002, Great River Bridge – Prototype Engineering, Inc.

Chapter 15

Renewing the Crooked Fork Creek Bridge

G.S. Wilson
Palmer Engineering, Nashville, Tennessee, USA

ABSTRACT: The State Route 62 Bridge over Crooked Fork Creek in Morgan County, Tennessee was originally designed in 1940. After more than six decades of service to the citizens of this rural community, the bridge had become structurally deficient and functionally obsolete. The Tennessee Department of Transportation (TDOT) decided an extensive rehabilitation was needed to address the structural problems and improve its functionality. The project entailed a complete replacement of the original superstructure as well as repair and modification of the existing substructure units. All work was accomplished without construction within the channel of the creek. To avoid the need for a lengthy detour, construction activities were phased and traffic control designed so that one lane of traffic could be maintained across the bridge throughout the duration of the project.

1 INTRODUCTION

Built in 1940, the State Route 62 Bridge over Crooked Fork Creek, in Morgan County, Tennessee had become structurally deficient from deterioration and functionally obsolete due to its substandard width and safety features. The Tennessee Department of Transportation assigned Palmer Engineering the task of bringing the structure up to modern day standards in the most cost effective manner while being sensitive to the environment and minimizing construction impacts to the traveling public.

2 ORIGINAL BRIDGE

The bridge was originally built as three simple spans with lengths of about 12.8 m (42′) each for a total bridge length of 38.9 m (127′–9¼″). The entire bridge was in a five degree horizontal curve and was superelevated at 8.34%. The roadway profile within the bridge and approach area placed the structure in a 0% grade. State Route 62 crosses Crooked Fork Creek on a skew, which adds to the geometric complication of the bridge layout. Each substructure unit was skewed 20 degrees to the long chord of the bridge.

In cross-section, a 203 mm (8″) cast-in-place concrete deck was supported by four W30 × 116 steel beams. The diaphragms were C10 × 15.3 shapes. The deck had 152 mm (6″) by 229 mm (9″) tall concrete curbs and had a total travel width between curbs of 7.3 m (24′–0″). The bridge railings were fabricated steel sections supported by the curbs and the fascia beams.

The abutments were the spill-through type with spread footings bearing on rock. The piers were wall-type with spread footings bearing on rock.

3 CONDITION ASSESSMENT

The first step of the project involved a field inspection to assess the bridge's condition so that a plan could be devised that would best meet TDOT's goals. The guardrails approaching the bridge

Figure 1. Above bridge before rehabilitation.

were substandard as were the original barrier rails which also had areas of collision damage and missing sections. The original concrete deck had been overlaid with multiple layers of asphalt throughout the years. Prior to rehabilitation the total asphalt thickness on the bridge was 203 mm (8″), which made the riding surface near the top of the original curbs (Fig. 1).

Visual inspection from below revealed that the concrete deck was deteriorated in multiple locations with heavy spalling, cracking, leaching, and exposed rebar. There was damage to the concrete curb with areas of severe scaling and exposed rebar. The deck drains were not functioning as some drains had been paved over and others were clogged with debris. The joints, located above each substructure unit, were leaking and the steel beams exhibited areas of heavy corrosion and section loss. In some areas, previous repairs for section loss and holes in the beam webs had already been performed. Bearings for the beams were in need of repair or replacement and exhibited heavy corrosion, pitting, and missing bolts.

The piers and abutments exhibited areas of delamination, minor scaling, and spalling but were in relatively good condition (Fig. 2). Embankments at the spill-through type abutments had areas of missing rip-rap that had been placed to prevent washout and loss of backfill material.

4 SUPERSTRUCTURE UPGRADE

After the bridge's condition was evaluated, upgrade options were investigated. Typical options range from merely repairing the structurally deficient items to a complete bridge replacement. In the end, it was decided that the superstructure would be completely replaced, thereby eliminating the structural deficiencies as well as providing a wider bridge with today's safety standards.

The new cross-section is comprised of a 210 mm ($8^1/4$″) cast-in-place concrete deck supported by four AASHTO PCI Standard Type 1 beams. The new deck has a total width of 11 m (36′–0″) and

Figure 2. Below bridge before rehabilitation.

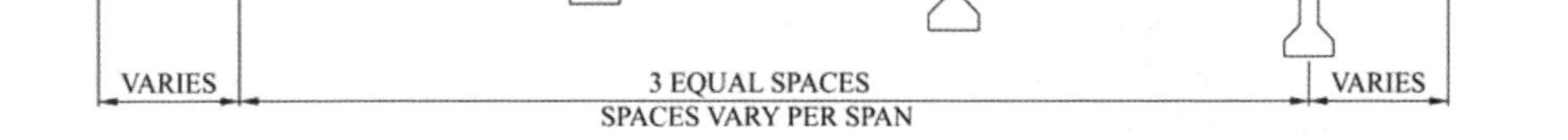

Figure 3. New cross section.

accommodates two 3.6 m (12′–0″) travel lanes with 1.5 m (4′–10″) shoulders (Fig. 3). An 80.5 km per hour (50 mile per hour) design speed required a superelevation of 7.2% for the new bridge deck.

The original simple span configuration with its open joints was eliminated. The new bridge superstructure was made continuous for live loads and integral with the abutments. The entire bridge is now jointless which should enhance long-term durability.

During field inspection of the original bridge, signs of high water were noticed by the engineers. Subsequent interviews with local residents confirmed that the bridge had been overtopped during periods of high water. As a result, a TDOT standard bridge rail that utilizes a post and rail system rather than a solid parapet was used on the new bridge to allow flow through the barrier during high

Figure 4. Above bridge after rehabilitation.

water (Fig. 4). The bridge approaches were made safer by way of new guardrail with safety end terminals placed on each approach.

5 REHABILITATED SUBSTRUCTURES

As previously mentioned, the original substructure units were in good condition and exhibited only small areas of delamination and spalling. Rather than replace the abutments and piers, they were repaired, modified, and incorporated into the rehabilitated bridge (Fig. 5).

The existing abutment backwall and wingwalls were removed and the abutments were widened and made taller to accommodate the new, wider superstructure. The different superelevation of the new bridge was also easily accommodated during the rehab. To create a jointless structure, the abutments were made integral with the superstructure and concrete approach slabs were added to the bridge ends.

To avoid the necessity of working within the channel of the creek and the complication of widening the solid wall-type piers all the way to their foundations, only the pier caps were modified. The caps were made taller to accommodate the new superstructure and a 1.4 m (4′–7″) cantilever was added to each side of each pier to carry the fascia beams.

6 PHASED CONSTRUCTION

State Route 62 is an important thoroughfare to this area of Tennessee. While closing the bridge to traffic during the rehabilitation project would have sped up construction, it would have put a burden

Figure 5. Below bridge after rehabilitation.

Figure 6. Phased construction and traffic control.

on the people who travel across it daily. A detour route would have taken drivers at least 37 km (23 miles) out of their way. So it was of critical importance that traffic control be designed to maintain some level of operation on the bridge for drivers even while the renovation process was underway. Therefore, construction was completed in phases and a traffic light system was employed enabling one lane of traffic to travel across the bridge throughout the entire rehabilitation process (Fig. 6).

7 CONCLUSIONS

Though the bridge underwent an extensive rehabilitation, construction activities were phased and traffic control designed so that one lane of traffic could be maintained on the bridge throughout the duration of construction utilizing a traffic light system. The safety and functionality of the bridge were improved with a much wider travel-way, updated bridge rails, new approach guardrail, and improved deck drainage. Although once structurally deficient, the bridge now has a state of the art superstructure utilizing prestressed beams and a jointless bridge deck. The substructure units were repaired, modified, and incorporated into the rehabilitated bridge. The original bridge provided over six decades of service to the citizens of Tennessee and the rehabilitated bridge should serve them well for many years to come.

ACKNOWLEDGEMENTS

Jamison Construction was the prime contractor for the bridge rehabilitation. The Tennessee Department of Transportation is the owner of the SR 62 Bridge over Crooked Fork Creek. Ray Henson served as the TDOT's on-site construction inspector and Rocky Christy was the TDOT manager for this project. Wayne Seger oversees the Bridge Inspection and Repair Division for the TDOT and served as a technical reviewer for this paper. The author wishes to express thanks to each of them for their cooperation and support.

REFERENCES

American Association of State Highway and Transportation Officials 17th Edition. 2002. *Standard Specifications for Highway Bridges.*

Tennessee Department of Transportation. 1995. *Standard Specifications for Road and Bridge Construction.*

Chapter 16

Rapid delivery!
New Jersey overnights bridge rehabilitation for Trenton bridges

Harry A. Capers, Jr.
Arora and Associates, Lawrenceville, NJ, USA

Xiaohua 'Hannah' Cheng
New Jersey Department of Transportation, Trenton, New Jersey, USA

ABSTRACT: The aging highway bridge continuously renewed while accommodating traffic flow. The traveling public is demanding that this rehabilitation and replacement be done more quickly to reduce congestion and improve safety. Conventional bridge reconstruction is typically on the critical path because of the sequential, labor-intensive processes of completing the foundation, the substructure, the superstructure infrastructure in the United States is being subjected to increasing traffic volumes and must be components, railings, and other accessories. Bridge systems can allow components to be fabricated off site and moved into place quickly while maintaining traffic flow. Depending on the specific site conditions, the use of prefabricated bridge systems can minimize traffic disruption, improve work-zone safety, minimize impact to the environment, improve constructability, increase quality, and lower life-cycle costs. New Jersey Department of Transportation clearly demonstrated the truth of these statements with their approach to the replacement of the superstructures of two structurally deficient bridges carrying a freeway section of Route US 1 through the capitol city of Trenton.

The project involved completely replacing the superstructures of the two bridges with new ones designed for a 75 year life made off site and installed over three weekend shutdowns of 57 hours each. A project of this magnitude would typically take the Department approximately two years to design as a traditional deck replacement and bridge rehabilitation project. The project approach saved more than 22 months and an estimated design and construction savings, including delay-related user costs in excess of $2M.

The results? An extremely happy motoring public.

1 INTRODUCTION

The aging highway bridge infrastructure in the United States is being subjected to increasing traffic volumes and must be continuously renewed while accommodating traffic flow. The traveling public is demanding that this rehabilitation and replacement be done more quickly to reduce congestion and improve safety. Conventional bridge reconstruction is typically on the critical path because of the sequential, labor-intensive processes of completing the foundation, the substructure, the superstructure components, railings, and other accessories. Bridge systems can allow components to be fabricated off site and moved into place quickly while maintaining traffic flow. Depending on the specific site conditions, the use of prefabricated bridge systems can minimize traffic disruption, improve work-zone safety, minimize impact to the environment, improve constructability, increase quality, and lower life-cycle costs. New Jersey Department of Transportation clearly demonstrated the truth of these statements with their approach to the replacement of the superstructures of two structurally deficient bridges carrying a freeway section of Route US 1 through the capitol city of Trenton.

The project involved completely replacing the superstructures of the two bridges with new ones designed for a 75 year life made off site and installed over three weekend shutdowns of 57 hours each. A project of this magnitude would typically take the Department approximately 2 years to design as a traditional deck replacement and bridge rehabilitation project. The project approach saved more than 22 months and an estimated design and construction savings, including delay-related user costs in excess of $2M. The results? An extremely happy motoring public.

This paper will outline the application of prefab technology on a project where traffic control issues demanded rapid bridge construction techniques. The paper will also demonstrate a project where decision makers took the risk of total facility shutdown to allow for rapid construction.

2 NEED FOR ACCELERATED BRIDGE CONSTRUCTION

Bridge engineers have successfully used accelerated bridge construction practices for many years around the globe. Our good fortune in this country to experience continued growth has also had the result of us developing a greater dependence on our transportation infrastructure and less tolerance to interruptions caused by taking lanes out of service for routine maintenance. With the advent of high performance materials, emerging advanced technologies, the FHWA is attempting to provide leadership in meeting the public's expectations as illustrated in their Vision and Mission Statements. Their vision speaks to 'Vital Few' important items to be focused on by the agency, those being Safety, Congestion, Environment Streamlining and Stewardship.

In focusing on their vision and goals their work has led to the recommendation for modular prefabricated construction, among other things. The concept of prefabricated elements and systems is being researched as well as applied and put to use in building bridges.

To obtain information about technologies being used in other industrialized countries, a scanning tour of five countries was made in April 2004. The overall objectives of the scanning tour were to identify international uses of prefabricated bridge elements and systems and to identify decision processes, design methodologies, construction techniques, costs, and maintenance and inspection issues associated with use of the technology. The scanning team was, therefore, interested in all aspects of design, construction, and maintenance of bridge systems composed of multiple elements that are fabricated and assembled off-site. The elements consisted of foundations, piers or columns, abutments, pier caps, beams or girders, and decks. Bridges with span lengths in the range of 20 to 140 feet were the major focus, although longer spans were of interest if a large amount of innovative prefabrication was used.

The focus areas of the study were, therefore, prefabricated bridge systems that

1. Minimize traffic disruption,
2. Improve work zone safety,
3. Minimize environmental impact,
4. Improve constructability,
5. Increase quality, and
6. Lower life-cycle costs.

One of the findings of the team was the use of prefabricated superstructure systems either in part or in total by several other countries as a solution to project requiring minimum down time of the facility and for which construction access was difficult.

The typical sequence of constructing bridge superstructures in the United States is to erect the concrete or steel beams; place temporary formwork or stay-in-place steel or concrete panels, place deck reinforcement, cast deck concrete and remove formwork if necessary. Elimination of the need to place and remove formwork for the deck can accelerate construction. Two systems to accomplish this were identified during the tour.

The use of full depth prefabricated concrete decks reduces construction time by eliminating the need to provide cast-in-place concrete. During the tour, it was observed that precast panels were used on steel beams to produce both composite and non-composite members. Composite action was developed through the use of studs located in pockets in the concrete deck slab. The use of full depth prefabricated concrete decks provides a means to accelerate bridge construction using a factory produced product.

3 RAPID DECK REPLACEMENT

3.1 *Route 295 over Creek Road emergency repairs*

On May 30, 2002, a tractor trailer traveling along I-295 South struck the overpass for Creek Road over I-295 near milepost #26 in Bellmawr, Camden County, New Jersey near the border of Gloucester County. The impact of the oversized and over height piece of earth-moving equipment that the tractor-trailer was carrying shattered the facia beam and the fifth interior beam causing serious damage to and potential catastrophic failure of the overpass. (Figure 1)

After initial inspection by NJDOT bridge staff, the overpass was found to be structurally unsound requiring closure of both I-295 mainline and Creek Road until temporary repairs could be made. The bridge was strengthened with a new, reinforced beam that was installed from the top of the bridge deck. Options to repair the bridge from under the deck were evaluated, but ultimately rejected, because they would have further restricted the bridge's maximum height safety requirements.

The mainline of Route I-295 was reopened after the installation of a temporary support beam but replacement of a portion of the deck and structural members down to the abutments was required before Creek Road could be reopened.

Creek Road is a major route supporting commercial and emergency services for the City of Bellmawr, and any traffic diversions on Creek Road or from I-295 onto surrounding network roads would cause unacceptable abnormally heavy traffic in those areas. Such conditions constituted an emergency in the region and it was determined by the Governor's office that damage to the overpass required immediate repairs on an emergency basis.

Figure 1. Exposed prestressing strands of damaged interior girder of Creek Road Bridge.

Figure 2. Night time erection of deck panels over I-295.

The Governor's declaration of an emergency allowed the department's structural engineers the ability to look to various proprietary prefabricated bridge systems to rapidly put the bridge back into service. The chosen strategy was to advertise for a superstructure system consists of a prefabricated composite concrete deck and stringer unit usually produced for simple spans manufactured by a supplier.

Several stipulations were that the systems would be designed in accordance with current American Association of State Highway and Transportation Official's Bridge Design Standards, would utilize weathering steel, concrete with corrosion inhibitors to increase service life and would provide a shallower superstructure depth than existed prior to the accident.

The permanent repair was made utilizing Inverset deck panels as supplied by Fort Miller Company, of Schuylerville NY. (Figure 3) The panels provided are cast inverted in forms suspended from the support girders and when turned upright the decks are precompressed which should provide for increased durability. Panels come in lengths up to about 100 feet with widths usually in the 8 to 12-foot range. Stringer spacing of the panels was arranged to permit the construction of new bearing seats under the existing bridge between existing stringers prior to demolition of the damaged superstructure. It should be noted that these bearing seats were also prefabricated to minimize construction time on site. Spacing the stringers to allow this advance construction of the bridge bearing seats also resulted in closer spacing allowing for a nine-inch increase in vertical underclearance.

Upon completion of fabrication and delivery on site, required modifications to the bearings, staged demolition of existing structure and erection of the new panels was performed on successive weekend shutdowns tremendously minimizing traffic impacts to the interstate and Creek road.

3.2 *The Olden Avenue Project*

Each day the New Jersey Department of Transportation's Route 1 carries more than 50,000 vehicles through the City of Trenton, Mercer County, the capitol of the State of New Jersey. Route 1 serves

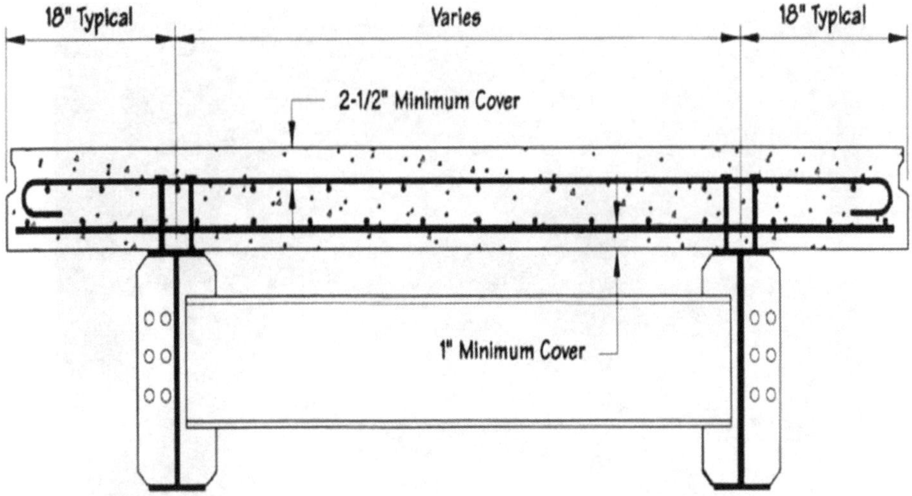

Figure 3. Inverset Deck Panel.

as a vital link to adjacent Pennsylvania and is a heavily traveled land service route for those communities along the Northeast Corridor between New York City and Philadelphia. Just north of the City of Trenton in Lawrence Township, Route 1 divides into two Routes, Route 1 business which carries local traffic to points in the city and Route US 1 Freeway which was intended to provide faster assecs to downtown Trenton, the State Capitol Offices and the crossing of the Delaware River between Trenton and Morrisville, Pennsylvania.

The freeway section configuration of the highway was constructed on embankment and provides for two lanes of traffic in each direction divided by a Jersey Barrier along its entire length creating significant challenges for maintenance of traffic for any work performed within its limits. However, three bridge decks on the Route 1 Freeway mainline, one at the Olden Avenue Connector and two at Mulberry Street, had deteriorated to the point where they were in need of constant maintenance.

In 2005 the replacement of these three bridge superstructures was undertaken in a project that was to become the NJDOT's first Hyperbuild project. The term 'Hyperbuild' was coined in 2004 by NJDOT Commissioner Jack Lettiere. Hyperbuild projects acknowledged the tremendous need to minimize traffic impacts for all of the previously mentioned reasons and recognized the potential savings millions of dollars in design, construction, and road user costs that could be realized. The Commissioner's vision was to reduce the time from initiation of design to the opening of the finished project to traffic by utilizing innovative methods of design and procurement for certain types of projects. Such projects had to have a well-defined scope and, if possible, require limited right-of-way acquisition, utility relocations and environmental impacts.

The three bridges in the project were actually located at two points along the mainline of Route US 1 separated by approximately a half mile. The Route 1 Bridge over the Olden Avenue Connector is a highly-skewed steel girder bridge with concrete deck. (Figure 4) The 35.0-ft wide single-span bridge has a span of 86.8 feet and carries two lanes of traffic. The Route 1 Bridge over Mulberry Street consists of two parallel bridge superstructures on a common abutment with a median barrier separating each direction of traffic. The 82.2 foot wide bridge has a 60.0 foot long single-span and carries four lanes of traffic over Mulberry Street.

Preliminary review of the sites of these deficient bridges indicated that they met these criteria. Using the lessons learned during the emergency repairs of the Creek Road Bridge over Route I-295, NJDOT's structural engineers also concluded that a similar approach of using prefabricated superstructure panels for the rapid replacement of these bad decks would be an effective strategy due to the very limiting project site and need for rapid project development and construction.

Figure 4. Deteriorated decks of the Route 1 Bridge over the Olden Avenue Connector.

Each superstructure was designed using 5 full-length segments of varying width, each with two Grade 50W steel girders and a 9-inch thick composite concrete deck (Inverset) system. The 86.8-ft long bridge span over Olden Avenue utilized W36x182 girders, and the 60-ft long bridge spans over Mulberry Street utilized W30x99 girders. Segment sizes considered the transportability and erection restrictions associated with urban nature of the project site.

The 15 segments were designed and fabricated at The Fort Miller plant in Schuylerville, New York, assembled at the plant to verify field tolerances, and trucked to an airport parking lot near the bridge. The segments were required to be onsite 24 hours prior to the start of demolition of the existing bridge. The contract specified high performance concrete to be used for all concrete on the job.

Construction was planned to occur using only weekends to shut downs mainline traffic to minimize disruption along the corridor. Each of the three bridges was allowed a 57-hour window commencing Friday evening with all activity off of the roadway and both lanes re-opened before the morning rush on Monday. If this window was exceeded, a Lane Occupancy Charge would be assessed, up to $10,000 per day.

As is typical on NJDOT projects, incentives were also included on this project to encourage the contractor to minimize onsite construction time even further than 57 hours per bridge. For the bridge over the Olden Avenue Connector, an incentive of $1,500 per hour was specified if the work was completed in less than 57 hours, not to exceed a maximum of $27,000. For each bridge over Mulberry Street, an incentive of $2,000 per hour was specified if the work was completed early, not to exceed $36,000.

Liquidated damages were also specified. The contractor would be charged $1,500 per hour if he took longer than 57 hours to open the bridge over the Olden Avenue Connector to traffic, and $2,000 per hour if he took longer than 57 hours to open either of the bridges over Mulberry Street. Also, the contractor would be charged $4,200 per day if the bridges weren't substantially completed by the specified completion date in the contract, and an additional $900 per day if all work was not completed within three months following that.

Figure 5. Initiation of erection of Route 1 Southbound Bridge over Mulberry Street.

The engineer's estimate for this project was $3.8M. The low bid of $3.5M from Neshaminy Constructors, Inc. was 8% or $297,000 less than the engineer's estimate. There were five bidders on this project. The second lowest bid was 10% higher than the low bid.

The Route 1 Bridge over the Olden Avenue Connector was replaced during a weekend closure in August 2005. The Route 1 Southbound Bridge over Mulberry Street was replaced during a weekend closure in September 2005, followed by the Route 1 Northbound Bridge over Mulberry Street during a weekend closure in October 2005. Design and construction would have taken 22 months using conventional methods.

The Route 1 Bridge over the Olden Avenue Connector was closed at 7 p.m. on a Friday in August 2005, and traffic was rerouted onto a five mile detour. The bridge was demolished in place using conventional methods. The existing abutments were repaired and new bearing seats constructed. The prefabricated superstructure was then erected. The longitudinal joints between superstructure segments were then sealed, and the expansion joints at the ends of the span were completed. The cast-in-place parapets were connected to the outside segments with bars in threaded inserts.

The Route 1 Southbound and Northbound bridges over Mulberry Street were closed at 7 p.m. on a Friday in September and October 2005, respectively, and traffic was rerouted onto a five mile detour for Southbound Mulberry, while on and off ramps were used for Northbound Mulberry. (Figures 5 and 6) The construction methods and time required to replace these bridges were similar to the bridge over the Olden Avenue Connector. Parapets and median barriers were cast-in-place concrete.

All three bridges were opened in less than the required 57 hours. The bridge over the Olden Avenue Connector was opened in 56 hours, the bridge over Southbound Mulberry was opened in 51 hours, and the bridge over Northbound Mulberry was opened in 54.5 hours. With all three bridges opened well before Monday morning rush hour, the contractor earned an $18,500 incentive.

Figure 6. Erection of final deck panel of Route 1 Southbound Bridge over Mulberry Street.

4 CONCLUSIONS

Each of the three bridges in the New Jersey DOT's first Hyperbuild project was replaced in a weekend, during a total of six days over three consecutive months. The replacements were completed in significantly less than the 22 months required for conventional design and construction, and they were completed under budget. The design and construction savings, including delay-related user costs, are in excess of $2M.

Each bridge is expected to see a 75–100 year service life due to the quality of its prefabricated superstructure, the use of high performance concrete, and the attention given to connection details. Conventionally constructed bridges have an average minimum 50-year life in New Jersey.

And the project clearly demonstrated the benefits that can be reaped by applying accelerated construction strategies; Less Construction Time, Increase Work Zone Safety, Less Maintenance – More Durability – Higher Quality, Reduce User and Life-Cycle Cost and Minimization of Traffic and Environmental Disruption. And of course, an extremely happy motoring public!

REFERENCES

Capers, Lopez, Lewis, Nassif, August, 2003, *Applying The Principals Of Get In, Get Out, And Stay Out Designing And Detailing Of Highway Structures For Rapid Construction In New Jersey*, 2003 NYC Bridge Conference, New York, NY.

Capers, Harry A., July, 2005, Hyperbuild! Rapid Bridge Construction Techniques in New Jersey, Transportation Research Board 6th International Bridge Engineering Conference, Boston. MA.

McGreevey, Governor James E., *Executive Order #28*, State Of New Jersey, Trenton, NJ.

New Jersey Department of Transportation, May 30, 2002 News Release, *Creek Road Bridge over 295 damaged in afternoon accident*, Trenton, NJ.

New Jersey Department of Transportation, June 5, 2002 News Release, *Creek Road Bridge repairs move forward to alow reopening on Monday Morning*, Trenton, NJ.

Ralls,Tang, Bhidé, Brecto, Calvert, Capers, Dorgan, Matsumoto, Napier, Nickas, Russell, March 2005, *Prefabricated Bridge Elements and Systems in Japan and Europe,* Washington, D.C.

Roads and Bridges Magazine, November, 2005, *NJDOT completes first Hyperbuild project,* Scranton Gillette Communications, Inc. Arlington Heights, Il.

Chapter 17

Accelerated construction of precast concrete piers on the Route 70 over Manasquan River Bridge Replacement Project

Eric Yermack
Arora and Associates, P.C., Lawrenceville, New Jersey, USA

ABSTRACT: The Route 70 over Manasquan River Bridge Replacement Project utilized an innovative precast substructure solution on a project requiring difficult coordination of highway and marine traffic, environmental constraints and community involvement. The 25-foot high, 724-foot long bridge, which is supported on five architecturally treated precast High Performance Concrete (HPC) in-water piers, crosses a navigable waterway in the coastal region of the State of New Jersey. The precast pier column and cap components were fabricated offsite, delivered via barges and trucks and assembled using post-tensioning. Pier foundations were constructed at the waterline within precast concrete cofferdam shells, which provided pile driving templates, served as architecturally treated formwork for the footings and eliminated construction of traditional cofferdams. The use of precast concrete substructures allowed the New Jersey Department of Transportation (NJDOT) to successfully address the project needs and resulted in the accelerated construction of a high quality signature bridge with a savings of over 15 months in construction over conventional methods.

1 INTRODUCTION

It has long been a goal of the Federal Highway Administration (FHWA) and NJDOT to implement a "Get in, get out, stay out" approach to bridge and highway construction projects (Capers et al. 2003). One of the primary methods of implementing this strategy is to employ prefabricated bridge elements and systems. In recent years, the use of precast concrete bridge components has increased dramatically. Precast components offer significant advantages over cast-in-place concrete including rapid construction, improved durability, reduced environmental impacts and a reduction of onsite labor resulting in improved work zone safety. While precast superstructure components have seen widespread use, precast substructure components have had limited application. The use of precast substructures can have tremendous impacts on construction schedules through time saved in establishing a work zone, forming, placing reinforcement, pouring concrete, stripping formwork and curing, all of which can be accomplished offsite and in parallel with other construction operations (Capers 2005).

2 BACKGROUND

The existing Route 70 Bridge over the Manasquan River (Bridge No. 1511-150) crosses a navigable waterway and is considered a gateway to both Monmouth and Ocean Counties in the coastal region of the State of New Jersey, and it serves vehicular, pedestrian and marine traffic. Constructed in 1936, the bridge is 625-ft long with a single leaf bascule span over the navigation channel. The 17 approach spans are supported on reinforced concrete pile bents. The existing bridge was in poor condition. The pile bents had been repaired but continued to deteriorate at the waterline, the abutments had experienced movement and the deck exhibited cracks, spalling and leakage of efflorescence. The movable span had been retrofitted with a sprinkler system to provide cooling during the summer months and prevent the movable span from becoming stuck in the open position. The bridge had been given an overall sufficiency rating of 20.6. The bridge also did not meet current geometric requirements. It only provided 11-ft travel lanes, a 50-ft navigation channel and 15-ft vertical underclearance. Due to its low underclearance, the bridge was required to be opened on demand for the passage of marine traffic, which caused disruptions and impeded the flow of vehicular traffic on the Route 70 corridor.

3 THE PROJECT

Since the bridge was structurally deficient and functionally obsolete, the NJDOT recommended it for replacement. NJDOT challenged its design consultant, Arora and Associates, P.C., to provide a design that would allow for the accelerated construction of the project by using precast concrete substructure components that could be fabricated off-site, shipped to the project site and then quickly assembled. To satisfy environmental in-water work restrictions, it was critical to develop a structural system for the bridge piers that would allow the contractor to complete the pier construction as quickly as possible and with a minimum of environmental disturbances.

The proposed replacement bridge is the centerpiece of a $52 million project that will carry the dualized section of Route 70 across the Manasquan River and provide the missing link along the Route 70 corridor (see Figure 1). The project will also provide for the long-term regional vehicular and marine transportation needs along the Route 70 corridor and the Manasquan River. In addition to these considerations, the NJDOT and FHWA have a goal of eliminating movable bridges, where possible, to reduce annual operating and maintenance costs and to provide for a more reliable, passive transportation infrastructure.

The proposed bridge section will have a 12-ft median, two 12-ft lanes and one 10-ft shoulder in each direction. Sidewalks and parapets are included on each side of the bridge resulting in an overall bridge width of 94′-8″. The 724-ft long, high-level, fixed bridge will consist of twin structures, with each having 2 three-span continuous superstructure units (119′-120.25′-120.25′) comprised of PCEF Bulb Tee Girders spaced at 8′-0″ on center. The superstructure will be supported on two abutments and five architecturally treated in-water piers with pile foundations. With this span arrangement, the proposed bridge foundations could be constructed with minimal conflicts with

Figure 1. Architectural rendering of the proposed replacement bridge.

the existing bridge foundations. Marine traffic needs would also be accommodated with the bridge underclearance being increased to 25-ft, the navigation channel widened to 75-ft and the centerline of channel shifted 15-ft towards the centerline of the river.

In addition to the bridge replacement and roadway widening, the project included many additional elements. A new bridge fender system and a public fishing pier were both designed using Fiber Reinforced Polymer (FRP) composite piles and lumber. There were also retaining walls, bulkheads, ramps, traffic signals, water quality stormwater management retention basins and manufactured treatment devices, highway lighting, ITS improvements and utility relocations.

4 CONTEXT SENSITIVE DESIGN

Early in the project, NJDOT requested that bridge aesthetic recommendations be developed to enhance this gateway structure and provide a well-balanced, aesthetically pleasing structure that would fit into this unique river environment. The needs of the surrounding communities of Brick Township and the Borough of Brielle were also taken into consideration through a series of public meetings, which resulted in a revision to the bridge height, changes to the project footprint and the addition of noise walls.

Arora studied the project aesthetic issues with its subconsultant H2L2 Architects/Planners, LLP and the NJDOT Bureau of Landscape and Urban Design. Traditional solid-shaft and multi-column pier types constructed on plinths at the waterline (NJDOT 2002) were compared against more creative architectural concepts. The process resulted in a pier architectural design with each pier being supported on a masonry faced plinth and having a pair of prismatic vertical columns at the centerline of the bridge and inclined tapered columns sloping outward towards the bridge fascias. A cap beam would then connect the tops of the columns. In the final condition the piers would appear uplifting and have two symmetrical trapezoidal openings. In addition to the distinctive pier treatment, the parapets, sidewalks, retaining walls and noise walls associated with the project also received architectural treatments.

During the Final Design phase of the project, the bridge was designated as the "September 11 Memorial Bridge" by the State of New Jersey in remembrance of the 166 people from Monmouth and Ocean Counties who died in the September 11, 2001 terrorist attacks. The project scope was then enhanced to include additional architectural treatments, pylon monuments, plaques and signage with landscaping.

5 ENVIRONMENTAL

In-water work restrictions were stipulated in the environmental permit conditions issued by the United States Army Corps of Engineers in their Nationwide Permit 23 and the New Jersey Department of Environmental Protection in their CAFRA and Waterfront Development Permits. Specifically to protect the winter flounder and anadromous (alewife) fish runs during migration and spawning a timing restriction of January 1st to April 30th was imposed to prohibit in-water construction activities and to reduce the possibility of increased turbidity.

Additional environmental considerations were to minimize impacts to coastal wetlands, subtidal/intertidal shallows and State open waters. Our investigation of the project site also uncovered the presence of salt laden soils, arsenic and beryllium in the riverbed sediments. To address these conditions, we sought to reduce the project footprint in the riverbed and minimize the amount of riverbed sediments that would need to be removed to construct the pier foundations.

6 TRAFFIC CONTROL

Route 70 is a heavily traveled regional corridor with a 2 Way A.D.T. (2005) of 32,300 vehicles. Since Route 70 is also a coastal evacuation route, NJDOT required that two lanes of traffic be maintained in each direction during construction. Rather than building a new structure on a parallel alignment, it was decided to construct the project in stages. This had the added benefit of minimizing the amount of right of way purchased to construct the project.

After construction of the eastbound bridge structure, approximately 3-ft from the south fascia of the existing bridge, all traffic would be transferred onto the newly constructed first half of the bridge. Traffic would be maintained in four 10′-11″ wide temporary lanes, which would utilize the entire bridge deck surface including the sidewalk and shoulder areas. Demolition of the existing structure could then be performed, followed by the construction of the second, westbound half of the bridge next to the north fascia of the eastbound structure. A final stage would then be required to shift traffic into its final lane configuration. The project will widen the bridge from 56′-10″ to 94′-8″ and shift the centerline of Route 70 by 28′-10″.

7 PRECAST SUBSTRUCTURE SOLUTION

Using the proposed construction sequence, there would be a total of ten pier halves constructed in the river. The demolition of the existing bridge and construction of bulkheads at the proposed abutments would also be subject to the same in-water work restrictions. If a rapid method of pier construction was not utilized, the project schedule could become susceptible to impacts from winter concrete construction restrictions. Therefore, it was critical to design the piers so that they could be completed quickly and the superstructure construction advanced before cold weather could delay construction.

To meet these challenges, Arora and Associates, P.C. selected a structural system for the piers consisting of architecturally treated precast concrete cofferdam shells, columns and caps connected through post-tensioning. 8,000 psi HPC was used for all of the precast bridge elements for added strength and durability.

8 FOOTINGS

The foundations utilized 24-inch diameter concrete-filled pipe piles driven to an estimated tip elevation of −110. For both the eastbound and westbound structures, groups of 37 piles were used at the fixed piers, 26 piles were used at the continuity piers and 32 piles were used at the expansion pier. The typical footing size for each half of each pier was 30-ft wide by 49.5-ft long. Rather than constructing the footings below the riverbed, which is 16-ft deep in some locations, and within traditional braced steel sheeting cofferdams, the pier foundations were constructed at the waterline within precast concrete cofferdam shells. The cofferdam shells provided driving templates for the piles, served as architecturally treated formwork for the footings and minimized disturbances to the riverbed.

The contract plans detailed the architectural and dimensional requirements for the cofferdams and provided nominal reinforcement for shipping and handling of the units. The cofferdams were faced with a #1104 random cut stone pattern and coated with a clear epoxy waterproofing seal coat. This gave the appearance of the pier footings being faced with wet granite masonry at the waterline. The contract documents allowed for the contractor to introduce joints and fabricate smaller sections, which could facilitate casting, shipping and erection of the individual components. The contractor was also given the responsibility of selecting his own method of temporary support for the cofferdam shells.

In December 2005, the construction contract was awarded to George Harms Construction Co., Inc. of Howell, New Jersey. To construct the foundations, the contractor chose to drive pilot piles with a template around the perimeter of each pile group and used these piles to support a temporary frame from which the cofferdam sections could be hung (see Figure 2).

The cofferdams were fabricated in sections varying in length from 7.2-ft to 14.5-ft, trucked to the site and then loaded onto barge platforms. The individual sections were then hoisted into place

Figure 2. Perimeter piles driven with a template.

Figure 3. Precast cofferdam section hoisted into place.

and bolted together using couplers consisting of 1 ¼″ anchor bolts, 4″ structural tubing and 1″ threaded rods (see Figure 3). The remaining piles for each footing were then driven through openings in the floor of the cofferdam shell. To facilitate the pile driving, a vibratory hammer was used to advance the piles 60-ft through the upper riverbed muck layer. An impact hammer was then

Figure 4. Work being performed inside a cofferdam. Piles are being filled with concrete. Structural support of the cofferdam has been transferred to the heads of the piles.

used to drive the piles another 50-ft to the estimated pile tip elevation. After a 7-day setup, each test pile was restruck to verify they had achieved a capacity greater than the 800 kip ultimate pile driving resistance.

The remaining work for each footing was to:

- Seal the annular spaces around the pile heads,
- Place tremie concrete,
- Dewater the cofferdam,
- Cut the piles off at 6-inches above the floor slab,
- Transfer support of the cofferdam to the pile heads through a series of smaller support beams and hanger rods,
- Concrete the piles, and
- Make a mass pour of footing structural concrete (see Figure 4).

9 PIER COLUMNS AND CAPS

Arora designed the pier columns to be constructed from hollow segmental units connected by post-tensioning strands extending from anchorages cast in the footings to tie points in the cap beams. Precast manufacturers were consulted during design to determine a preferred segment height for fabrication. 4-ft high segment sections with a 9-inch wall thickness were selected for design. However, the contract plans allowed for the contractor to modify the segment heights for his convenience and method of construction. 7-ft deep by 5-ft wide hollow prestressed concrete box beams were selected for the cap beams. The cap beams and the columns have rounded ends and sides,

Figure 5. Precast pier columns and a cap beam have been loaded on a barge and are being prepared for erection.

respectively. The post-tensioning design was based using ASTM A416 seven wire, grade 270, low relaxation strands.

During the shop drawing development process, the contractor again exercised the provisions in the contract documents to modify the precast concrete columns. He elected to fabricate the columns as complete units of approximately 16-ft in length rather than the 4-ft segments shown on the plans. He had mobilized a number of large barge-mounted cranes, which gave him the ability to easily handle the larger concrete members. Using complete column units cut the column erection sequence from four steps to a single step. The contractor also proposed substituting 1¾″ diameter, ASTM A775, grade 150, threadbar for the strands. This was preferred due to the sloping outer columns. Since the proposed post-tensioning system performed the desired function, it was determined to be equivalent, and the requested substitution was allowed. The precast pier column and cap components were then fabricated offsite, delivered via trucks, loaded onto barges and assembled in place using the post-tensioning threadbar (see Figure 5).

10 CONSTRUCTION SCHEDULE

During the pier construction the contractor operated on a six-day workweek to advance the pier construction as quickly as possible. He also employed several crews, which moved from one pier location to the next, performing the same tasks for each pier. Once an element of a pier was constructed the crew performed the same series of tasks for the construction of the same element on the next pier, and likewise for each step in the pier construction process. In this way the pile foundations, cofferdams, columns and cap beams were rapidly constructed for all of the piers (see Figure 6).

The in-water work, which initially consisted of pile driving, began on July 1, 2006. The contractor's crews worked steadily moving from one pier to the next until the last pier cap on the eastbound half of the bridge was completed on October 23, 2006. The schedule for the construction of the eastbound piers 1 through 5 is illustrated in Figure 7. Also shown are schedules for typical precast and cast-in-place piers.

The construction of all the precast eastbound piers, including their pile foundations, was accomplished in 96 working days. A typical individual precast pier was constructed in 63 working days. The overall duration required to construct five precast piers was only 33 working days longer than the time it took to construct a single pier. This was due to the process involved in constructing all

Figure 6. Crews work on multiple piers at the same time. Piers are shown in various stages of completion.

Activity ID	Activity Description	Orig Dur	Early Start	Early Finish
Rt. 70 / Manasquan - Precast Pier Construction				
Construct all Eastbound Precast Piers (P1 - P5)				
A1000	PILES	54	01JUL06A	21SEP06A
A2000	PRECAST COFFERDAMS	45	12JUL06A	19SEP06A
A3000	CAST-IN-PLACE PIER FOOTINGS	38	01AUG06A	25SEP06A
A4000	PRECAST PIER COLUMNS	31	23AUG06A	05OCT06A
A5000	PRECAST PIER CAPS	37	30AUG06A	23OCT06A
Construct Single Precast Pier (Typical)				
B1000	DRIVE TEST PILES	7	11JUL06A	19JUL06A
B1100	TEST PILE ORDER LENGTHS	5	18JUL06A	26JUL06A
B1200	FURNISH PILES	7	25JUL06A	04AUG06A
B1300	DRIVE PRODUCTION PILES	14	04AUG06A	23AUG06A
B2000	INSTALL PRECAST COFFERDAM	10	07AUG06A	18AUG06A
B3000	POUR FOOTING CONCRETE	3	24AUG06A	28AUG06A
B4000	ERECT PIER SEGMENTS/PT	8	19SEP06A	27SEP06A
B5000	ERECT PIER CAP	10	30SEP06A	13OCT06A
Conventional Pier Construction				
Construct Single Cast-In-Place Pier (Typical)				
D0500	CONSTRUCT COFFERDAM	15	11JUL06*	31JUL06
D0600	EXCAVATE AND DEWATER	3	01AUG06	03AUG06
D1000	DRIVE TEST PILES	7	04AUG06	14AUG06
D1100	TEST PILE ORDER LENGTHS	5	15AUG06	21AUG06
D1200	FURNISH PILES	7	22AUG06	30AUG06
D1300	DRIVE PRODUCTION PILES	14	31AUG06	20SEP06
D1350	SETUP - PLACE TREMIE CONC	2	21SEP06	22SEP06
D1360	TREMIE CURE	5	23SEP06	27SEP06
D1400	CONSTRUCT FOOTING	3	28SEP06	02OCT06
D1500	CONSTRUCT PIER COLUMNS	5	03OCT06	09OCT06
D1600	CONSTRUCT PIER CAP	11	10OCT06	02NOV06

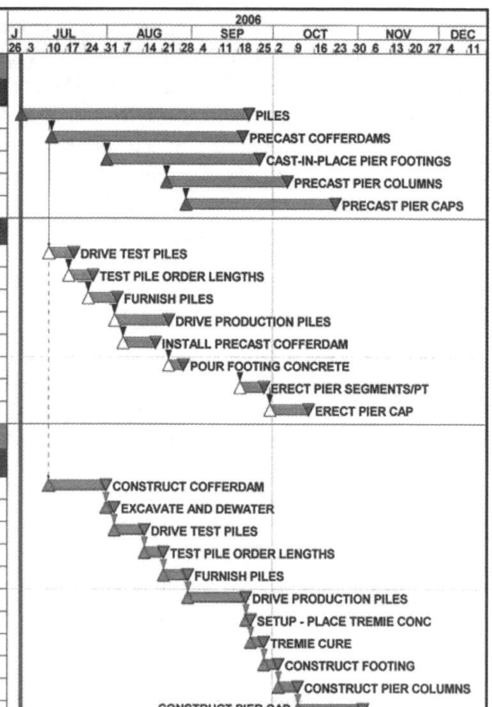

Figure 7. Construction schedules are shown for all of the precast eastbound piers 1 through 5, a typical precast pier and a cast-in-place pier.

five piers at the same time with multiple crews. Since there was slack in each of the pier construction activities, the contractor could have achieved a far shorter construction duration for a single pier. However, he was working towards the greater goal of constructing all of the piers in the shortest amount of time. When all five piers are looked at together, all pier construction activities are on the project critical path. A truer representation of the duration of the precast pier construction can be obtained by simply dividing the overall duration of 96 working days by 5 piers, which results in a duration of approximately 19 working days per pier.

The effect of using precast components can also be quantified by comparing the construction duration of a conventional pier to that of a precast pier. If a conventional cofferdam and cast-in-place construction were used, it is estimated that a pier could be constructed in 82 working days. Using a six-day workweek would improve the overall duration of constructing a cast-in-place pier from 115 calendar days to 99, but this would not decrease the total number of working days required. During the actual construction of the eastbound half of the bridge, the contractor ordered the production piles ahead of time and started driving production piles before all the piles had been furnished to the site. This yielded a savings of 4 working days. Since the pile foundations would be similar for each type of pier, this savings should also be considered for the cast-in-place pier, and the duration should be reduced to 78 working days.

Based on this comparison, the cast-in place construction duration would have been 59 days longer or more than 4 times the precast construction duration. Doubtless efficiency for multiple pier construction would also be realized for cast-in-place piers, and the multiple cast-in-place pier duration should be used for a better comparison. However, it is not expected that the duration would approach the 19 working day efficiency that was achieved for the precast piers.

With the implementation of the precast pier system, the contractor was able to advance the superstructure construction and the first girders were erected on September 29, 2006 (see Figure 8). This early start on the bridge superstructure coupled with a mild winter allowed the contractor to complete pouring the bridge decks in early January 2007. If conventional pier construction had been employed, valuable schedule time would have been lost to the construction of temporary cofferdams, forming of the pier footings, columns and cap beams, curing and finishing the substructure concrete. The pier construction would have extended into December 2006, and there would have been no chance of completing the eastbound bridge deck before Spring 2007.

Because of the accelerated bridge construction using precast components, the contractor was able to shift traffic over to the newly constructed first half of the bridge by April 13, 2007. The existing

Figure 8. Eastbound bridge piers 1 through 5 have been completed.

bridge is currently being demolished down to the waterline during the remaining in-water work restriction period. After July 1, 2007, the contractor expects to complete the in-water demolition work and begin driving piles for the construction of the pier foundations for the westbound half of the bridge. The project is approximately 500 calendar days ahead of schedule and is anticipated to be substantially complete in August 2009.

11 CONCLUSION

The Route 70 over Manasquan River Bridge Replacement Project utilized a precast concrete substructure solution to meet the project needs and facilitate the construction of the bridge. The precast pier system was detailed on the contract plans to allow the contractor and his fabricators to modify the design so that there would be a maximum economy in materials used, reduced costs to the owner, NJDOT, and the most efficient method of construction could be employed by the contractor. The in-water piers were constructed with an average duration of approximately 19 working days per pier when considering the overall duration of the pier construction activities. Once the construction of the project is completed, the use of precast substructures will have resulted in a high quality architecturally treated signature bridge constructed over 15 months ahead of schedule.

ACKNOWLEDGEMENTS

The author would like to acknowledge the contributions of the following team members, without whom this project would not have been possible:

- Pankesh Patel, P.E. – Project Manager, New Jersey Department of Transportation
- David Wallis – Resident Engineer, New Jersey Department of Transportation
- Alan Haring, P.E. – Project Engineer, Arora and Associates, P.C.
- Andrew Baran – Constructibility, Arora and Associates, P.C.
- H2L2 Architects/Planners, LLP – Architect
- Robert Harms – Vice President, George Harms Construction Co., Inc.
- Jeff Brantly – Sr. Project Engineer, George Harms Construction Co., Inc.

REFERENCES

Bureau of Structural Engineering, New Jersey Department of Transportation, New Jersey Department Of Transportation Bridges and Structures Design Manual, Fourth Edition, 2002, NJDOT, Trenton, NJ

Capers, Lopez, Lewis, Nassif, August 2003, Applying The Principals Of Get In, Get Out, And Stay Out Designing And Detailing Of Highway Structures For Rapid Construction In New Jersey, 2003 NYC Bridge Conference, New York, NY

Capers, Harry A., July 2005, Hyperbuild! Rapid Bridge Construction Techniques in New Jersey, Transportation Research Board 6th International Bridge Engineering Conference, Boston. MA

Chapter 18

Improving tomorrow's infrastructure: Extending the life of concrete structures with solid stainless steel reinforcing bar

R.E. Schnell
Talley Metals Technology, Inc., Hartsville, SC, USA

M.P. Bergmann
New York State Department of Transportation, Long Island City, NY, USA

ABSTRACT: Stainless steel reinforcing has been used in numerous structures throughout North America, including the Progreso Port Authority Bridge, Yucatan, Mexico, in 1937; the Haynes Inlet Slough Bridge, North Bend, OR, USA, in 2002; the Belt Parkway Bridge over the Ocean Parkway, Brooklyn, NY, USA, in 2004; and Woodrow Wilson Memorial Bridge on the Capitol Beltway, Washington, DC, USA in 2006. Recent advances in concrete technology have provided structural designers with materials which can easily last over 100 years, and the life of many concrete structures today is limited by the reinforcing. Improvements in the life of the reinforcing can translate directly into extended life of the structure. Current projections by several transportation agencies show that the use of solid stainless steel reinforcing bar in bridge decks will more than double the life of the bridge deck. While solid stainless steel reinforcing bar can increase the cost of the bridge deck by as much as 12% (compared to carbon steel reinforcing), the economic value of the longer life outweighs the initial higher cost. In most cases, the additional cost of solid stainless steel reinforcing bar represents less than 1.5% of the total cost of the structure. Most concrete structures are designed with minimum concrete cover over the reinforcing bar which is required to protect the reinforcing bar from corrosion. Where the reinforcing bar is completely resistant to corrosion, the cover can be reduced, saving costs of concrete and reducing the total weight of the structure. In some structures, design savings made possible by the use of solid stainless steel reinforcing bar will offset as much as 100% of the initial cost increase from using the stainless reinforcing.

1 INTRODUCTION

Corrosion of carbon steel reinforcing bar has been a serious issue for highway agencies around the world for many years. In the United States, these problems appeared in southern coastal states as long as 75 years ago, and appeared in many northern states after the use of deicing salts became common about 50 years ago. It would have been impossible, in those early years of bridge design and construction, for bridge and civil engineers to have foreseen the number of vehicles, and the huge loads that are being transported on these bridges today. In addition to the load concerns, deterioration due to the chloride salt content, either from the deicing salts employed, or the salt spray in coastal regions, has severely impacted our bridge and roadway infrastructure. For the last 35 or 40 years, rebar corrosion has been one of the most important issues facing bridge engineers. Upon entering the 21st century, engineers are now being confronted with an enormous amount of deteriorating bridges, and new solutions are constantly being evaluated daily to address these rising concerns.

The Federal Highway Administration (FHWA) along with many of the various state Departments of Transportation (DOT's) began experimenting with methods to extend the life of concrete carbon steel reinforcing bar around 1970, as a result of these corrosion issues. The

FHWA has also been tasked with the problem of seismic retrofit, due in part to the seismic activity which can occur in various parts of the United States, so high strength and excellent ductility are paramount in preserving structural integrity, in addition to corrosion resistance. Other FHWA projects include innovative bridge research and construction, and value pricing projects based on full life cycle projections. Any or all of the above mentioned projects may require a re-evaluation of the types of reinforcing materials currently being used.

2 MATERIALS EMPLOYED FOR REDUCING REINFORCING BAR CORROSION

2.1 *Epoxy coated rebar*

One of the first methods developed is still the most common: coating carbon steel with an epoxy coating. The epoxy coating is intended to protect the carbon steel from moisture and from salts, and to electrically isolate a rebar mat, from other nearby mats which may be at different potentials.

Early bridge decks constructed with epoxy-coated reinforcement bar (ECR) did not exhibit the desired long life. Analysis of early failures showed that poor adherence, or the poor quality of the epoxy coating, allowed corrosive salts to penetrate. The concrete mixtures of that time had fairly high permeability, and the epoxy coatings provided only 5 to 10 years of additional life.

Subsequent testing showed that a principal cause of corrosion is the different potentials between the top and bottom mats in the deck. Many states began to use ECR in both the top and bottom mats for this reason (McDonald, et.al., 1998, and Samples, et.al., 1999). However, the presence of uncoated composite shear studs in many bridge decks will provide an anode to initiate corrosion at defects in the top ECR mat. For this reason, the benefits of ECR bottom mats are limited.

The Concrete Reinforcing Steel Institute established a producer certification program for ECR, and the life of bridge decks using ECR is now in the range of 35 to 50 years in northern states where deicing salts are used (Humphreys, 2004).

The principal advantage of ECR is to provide longer life than that of uncoated carbon steel. Disadvantages include poorer bond with cement paste, fragility of the coating, adherence of the coating, and the limited life of the coating. While CRSI's certification program has substantially improved the initial quality of epoxy coatings, some studies have shown that damage to coatings during handling and concrete placement can be ten times the defects from the coating process itself (Samples, et.al., 1999).

2.2 *High performance concrete (HPC)*

Many agencies around the world have developed varieties of "high performance concrete" (HPC) in the last 15 years. Most of these mixes use substantially lower amounts of Portland cement than previous mixes, while adding fly ash, ground granulated blast furnace slag, and/or silica fume in various proportions. These mixes show a reduced heat of hydration and a slower strength gain than many of the older mixes. They generally tend to have less shrinkage, less microcracking, and a much lower permeability than the more "conventional" mixes.

Many tests have shown that corrosion rates in bridge decks are related to the amount of cracking (Smith, et.al., 1996, & Fanous, et.al., 2000). HPC bridge decks are more durable than those constructed with older mixes, and many agencies believe they can consistently achieve 50 years life. Disadvantages are the slower curing times and, in general, the higher initial costs.

2.3 *Galvanized rebar*

Many agencies began using galvanized carbon steel reinforcing bar over 30 years ago. The galvanizing on carbon steel rebar has two functions: it protects the steel from corrosive chemicals, and it provides a sacrificial anode so that the steel itself will not corrode until the zinc coating is exhausted. Some agencies have had good results with galvanized reinforcing bar, but the overall record of galvanized reinforcing bar is similar to ECR (Burke, 1994, & McDonald, et.al., 1998).

An HPC deck with galvanized reinforcing bar is generally estimated to have a life of 35 to 50 years. The advantages of galvanizing include a better bond to the cement (compared to ECR), and a less fragile coating. Disadvantages include more price volatility, limited life of the coating, and the fact that galvanized rebar cannot be used in a placement with uncoated steel (because the coating will sacrifice itself to protect the uncoated steel nearby).

2.4 *"Zn-ECR" coatings*

A US producer has recently introduced reinforcing bar which is spray-coated with molten zinc and then epoxy-coated. Although it would appear that this product would have significantly longer life than ECR or uncoated galvanized rebar, further tests are needed. Some preliminary tests have shown that the life of bridge decks constructed with this product will be longer than any product except stainless steel (Clemena, et.al., 2004).

However, the tests were not done with uncoated steel in the same placement. Since many actual bridge decks have uncoated shear studs, defects in the epoxy coating could create a site for accelerated corrosion.

This product would appear to have all of the same limitations as ECR or galvanized rebar, such as poor bond, fragile coating, variations in coating thickness, etc.

2.5 *Microcomposite multistructural formable steel (MMFX-II™)*

This proprietary alloy is a low-carbon 9% chromium alloy with unusually high tensile mechanical properties. Tests have shown that it provides significantly longer life than uncoated carbon steel reinforcing bar, and will probably provide longer life than ECR or galvanized steel (Clemena, et.al., 2004). Some states now accept this material as a substitute for ECR, and some have discontinued the use of ECR entirely in favor of MMFX-II™ or other materials with longer life.

While data is incomplete, it appears that an HPC deck, in conjunction with the use of MMFX-II™ reinforcing bar, will have a life in the range of 30 to 50 years. Advantages of MMFX-II™ include a good bond to the cement paste (compared to ECR), no problems with handling a fragile coating, and a higher yield at 0.2% deformation. Disadvantages include a sole source, poor ductility, and higher initial costs than ECR or galvanizing.

2.6 *Fiber reinforced plastic (FRP) rebar*

This is the most recently developed material. It has been used in a few experimental structures. While the material itself will never corrode, it does have a limited life span. FRP does lose strength with age, and most experts in this field estimate a life of 65 to 90 years in service conditions before the loss of strength is unacceptable (GangaRao, 2007). The principal problems with FRP reinforcing bar are high initial cost, low elastic modulus (generally requiring FRP to be used at least one size larger in deck designs), impossibility of bending (requiring prefabricated bends spliced to straight bars), and poorer bond with cement paste (comparable to ECR).

Another unanswered question with FRP is the value of thermal conductivity. Most designers have assumed that reinforcing bar serves several purposes: structural strength, crack control, and equalizing temperature (to reduce thermal stress.) FRP reinforcing bar has much lower thermal conductivity than any metal, and will not equalize thermal stress as well as metal reinforcing. The authors have found no research on the probability of cracking from thermal stresses when non-conducting reinforcing bar is used.

2.7 *Stainless steel clad rebar*

Two companies, one in the United Kingdom and one in the United States, have produced carbon steel rebar with a stainless steel cladding in recent years. This material has the potential of providing comparable life to solid stainless steel at lower cost. Tests have shown that the only deterioration that occurs in this material is at the cut ends (Clemena, et. al., 2004), which must be capped, to avoid corrosion of the carbon steel base.

However, its principal disadvantage is its limited availability. The only US plant is not currently in production, and the UK-produced material may not be used on Federally funded highway projects in the United States. Since the clad material is not readily available at this time, it is not practical for designers to specify it, and it will not be considered further here.

2.8 *Solid stainless steel rebar*

Solid stainless steel reinforcing bar has been used successfully in very corrosive environments for over 70 years. In 1937, the Progreso Port Authority, in the Port of Progreso, Yucatan, Mexico, constructed a bridge using stainless reinforcing rebar, AISI Type 304, due to the aggressive chloride environment of the saltwater where this bridge was built. Almost 70 years later, this bridge is still standing, and being used daily. According to the local authorities, this bridge has not had to undergo any type of major repair work, throughout the life of this structure. A sister bridge, built to offset the heavy traffic flow in this area, was constructed in the 1960's, using standard carbon steel rebar. That bridge has been out of service for many years, because the deck and foundation have almost completely disintegrated, due to a complete loss of the carbon steel reinforcing bar.

Tests by the FHWA and various states show that solid stainless steel reinforcing bar will last at least 100 years in typical northern state conditions (McDonald, et. al., 1998). The most commonly used alloys today are Type 316LN and Type 2205, which have significantly better corrosion resistance than Type 304. Even though uncoated solid stainless steel rebar is exposed to potential differences between mats, the corrosion threshold is an order of magnitude higher than for carbon steel. Some tests with a stainless steel top mat and an uncoated carbon steel bottom mat showed that the top mat actually became slightly anodic, and the bottom mat corroded while the top mat was undamaged.

The obvious advantages of solid stainless steel reinforcing bar are extremely long life, excellent corrosion resistance, and high strength with good ductility, good bond to the cement, no fragile coating, and no need of end caps. The disadvantage is the expense of the higher initial cost. Typically, solid stainless steel costs 2.5 to 4.0 times the cost of carbon steel. However, new design life requirements, such as 100+ years, demand that bridge engineers evaluate both the overall construction costs and the total life cycle costs, as they decide what materials will give them their best option. With maintenance and replacement costs measured in billions of dollars, due to the corrosion of carbon steel reinforcing bar in the United States, the total life cycle cost of bridge and highway structures, should far outweigh the initial cost of materials.

Recently, Talley Metals, a Division of Carpenter Technology Inc. introduced a new lower cost stainless steel alloy, EnduraMet 32®, which is suitable for concrete reinforcing bar. Corrosion resistance and most structural properties are similar to AISI 316LN or 2205. However, the low nickel, and its metallurgically balanced alloy content, reduces its cost dramatically. Typical purchase costs for EnduraMet 32® are from 1.5 to 2.0 times the cost of carbon steel, or about one half the cost of AISI 316LN or 2205.

The standard specification that covers stainless steel reinforcing bar is ASTM A-955, and EnduraMet 32® meets all of the strength requirements of the various grade levels, and far exceeds the ductility requirements, making it easy to form, while maintaining its superior strength. Corrosion macrocell testing, which measures the corrosion rate of steel rebar, including stainless, in a simulated concrete pore solution, has demonstrated that EnduraMet 32® far exceeds the proposed ASTM requirement of 0.25 μm/yr. avg., by attaining 0.015 μm/yr. avg. in a 15 week test period.

The FHWA's slogan, "Get in, Get out, and Stay out", which is commonly used, in describing the need to minimize any disruptions to traffic flow, is intended to improve the public's perception regarding the rehabilitation of road and bridge structures. The use of solid stainless reinforcing bar, in critical bridge decks and components, will significantly improve the life of these structures, thus meeting the FHWA's intention.

2.9 *Comparison of alternatives*

Bridge designers have the choice of various types of reinforcing bar as outlined above. The choice of material will depend on life span, reliability, and economic issues such as initial capital cost and total life cycle cost.

New bridges in most states today are designed for a 75 year life span, and some major structures are designed for 100+ years. In the past, most bridge agencies have accepted the fact that a 75-year bridge will require at least one major rehabilitation during that period. However, especially in urban areas, major rehabilitations have proven to be very expensive and have caused substantial disruptions to normal traffic flow. Bridge owners have been looking for more durable materials, and some of the materials described above can provide substantially longer life at relatively low cost.

FRP reinforcing and the various solid stainless steel options all can provide bridge deck with a life span of 75 years or more. The "Zn-ECR" material may achieve this life span, but more testing will be needed. However, when a designer considers other structural properties such as bond to the cement paste, the FRP and Zn-ECR materials are no better than "conventional" ECR. The solid stainless steel reinforcing bar options alone, have the durability to last over 75 years (and most will last over 100+ years), and all have optimum structural properties.

As noted above, the stainless steel options may have the highest costs. Bridge designers cannot arbitrarily select a more expensive material because it will last longer. Most agencies use life-cycle cost comparisons when selecting different materials for bridges (and highways), and this practice is encouraged by FHWA. The section below is intended to illustrate the economic comparisons between selected rebar options and to give guidance to bridge designers when they are selecting materials for new bridges and for major bridge, or roadway, rehabilitations.

3 ECONOMIC COMPARISONS

Most decisions to use materials with more or less durability are based on cost. Since projected life of concrete bridge elements is always greater than 25 years, a simple cost comparison cannot be used. The FHWA and most state agencies use a life-cycle cost comparison, using an estimated discount rate based on interest minus inflation. Historically, this rate has always been near 4%, and that figure will be used throughout this paper.

As noted above, a well constructed HPC deck with ECR in top and bottom mats can reasonably be expected to last 35 to 50 years in most northern states. An identical deck with solid stainless reinforcing could last as much as 120 years, but no one has projected the life of the concrete itself that far.

Current costs for both carbon steel and stainless steel are rising rapidly. The best available figures today are that the purchase cost of stainless steel (AISI 316 or 2205) will be about 2.5 to 4.0 times the purchase cost of carbon steel. Placement costs are virtually identical. In the New York City area, rebar placement cost is generally equal to the purchase cost of the carbon steel. Thus, in the NYC area, in place costs for solid stainless steel are 1.75 to 2.25 times the cost for ECR.

The price of deck reinforcing (ECR) generally represents about 10% to 14% of the cost of the entire bridge deck. Assuming the average of 12% for ECR, solid stainless steel would represent an increase in cost of 9% to 15% of the entire deck, compared to ECR.

Assume that a bridge deck constructed with ECR will last 40 years, and will then be replaced at current costs. The present worth of the 40-year replacement is equal to 20.83% of the cost of the deck today. However, the cost of related construction items such as demolition, barriers, railing, joints, and maintenance & protection of traffic must be added to the deck costs. If the related elements add about 25% to the deck costs, the present worth of the 40-year replacement is 26.04% of the cost of today's construction. This compares favorably with the 9% to 15% increase in costs to use solid stainless steel instead of ECR.

Obviously, in highly congested areas such as central city arterials, maintenance & protection of traffic costs are unusually high. The high cost of detours, and the high cost of deck repairs which become necessary near the end of the life of the deck, make the comparison even more favorable to the stainless steel reinforcing.

Table 1. Comparison of initial cost and life cycle costs of bridge decks with various types of reinforcing.

Reinforcing type	ECR, galvanized	MMFX-II™	FRP	Solid stainless	EnduraMet 32®
Initial deck cost (compared to ECR)	100.00%	103.00%	106.00%	112.00%	106.00%
Estimated life (yrs.)	40	50	65	100	100
Present worth of deck replacement at end of life	26.04%	18.12%	10.35%	2.77%	2.10%
100-year life cycle cost as a percentage of initial cost of ECR deck	130.22%	121.12%	115.21%	114.77%	108.62%

Design Assumptions:

1. Present worth of deck replacement and 100-year life cycle costs assume 25% for related costs of replacement (M&PT, demolition, etc.)
2. 100-year life cycle cost assumes replacement with identical deck design at end of each life span. Remaining salvage value at 100 years is deducted
3. FRP values assume equivalent linear quantities, with all bars 1 size larger than steel bars
4. "Solid stainless" assumes AISI 316LN or 2205

The following table illustrates the relative cost of new bridge decks constructed with ECR (or galvanized rebar), MMFX-II™, FRP, Solid Stainless, and EnduraMet 32®. While the longer-lived options (FRP and stainless) have a higher initial cost, the life cycle costs of these decks are actually lower than the "conventional" ECR deck.

4 DESIGN IMPROVEMENTS AVAILABLE WITH NON-CORROSIVE REINFORCING

All of the comparisons above assume that all decks are designed identically, using the Standard Specifications for Highway Bridges or "empirical" methods. However, the use of non-corroding reinforcing will allow design savings in other areas.

4.1 *Reduced deck thickness*

Most bridge owners require a minimum cover over the top mat of reinforcing between 50 mm (2″) and 75 mm (3″). The common standard in many US states is 62 mm (2.5″) while New York requires 75 mm. New York also allows a designer to reduce the top mat cover by 25 mm (1″) if non-corroding reinforcing is used in the top mat. Since NYSDOT's "standard" bridge deck with ECR is 240 mm (9.5″) thick, the use of non-corroding reinforcing allows a reduction in deck concrete volume of 10.52%, with a corresponding reduction in dead load of the deck.

Concrete material and placing costs represent about 9% to 10% of the cost of a bridge deck. Thus, the 10.42% reduction in thickness will reduce the initial cost of the deck by approximately 1%. Since the cover over the top steel is not included in the flexural design of the deck, there is no loss in structural capacity from the reduced slab thickness.

Reduction in dead weight of the deck will reduce the total dead load of the structure. For a typical multi-span continuous steel plate girder structure with spans in the range of 60 m (200 ft), the deck dead load represents about 65% of the total dead load, and about 40 to 45% of the total dead plus live load. The demand on the girders will thus be reduced by about 4%. For the more common continuous structures, this analysis assumes that there will be very little savings of structural steel in the positive moment areas, because the reduction in deck thickness will effectively reduce the area of the composite girder flange. However, since composite action is not assumed in negative moment areas, a savings comparable to the reduction in demand will be achieved in those areas.

The following analysis assumes a 4.45% reduction in demand on the girders in negative moment areas only, and an equivalent reduction in structural steel cost in those areas.

Table 2. Comparison of initial cost and life cycle costs of new bridges with various types of deck reinforcing.

Reinforcing type	ECR, galvanized	MMFX-II™	FRP	Solid stainless	EnduraMet 32®
Deck cost (compared to total initial cost of "base" structure)	38.00%	39.14%	39.90%	42.18%	39.90%
Steel cost (compared to total initial cost of "base" structure)	31.00%	31.00%	30.50%	30.50%	30.50%
Foundation cost (compared to total initial cost of "base" structure)	25.00%	25.00%	25.00%	25.00%	25.00%
Earthwork, etc. cost (compared to total initial cost of "base" structure)	6.00%	6.00%	6.00%	6.00%	6.00%
Total initial cost of structure	100.00%	101.14%	101.40%	103.68%	101.40%
Estimated Life (yrs.)	40	50	65	100	100
Present worth of deck replacement at end of life	9.89%	6.88%	3.93%	1.05%	1.00%
100-year life cycle cost as a percentage of initial cost of "base" structure	111.48%	108.02%	104.88%	104.74%	102.40%

Design assumptions:

1. DL of structural steel is 50% of DL of concrete (std. deck)
2. Deck cost is 38% of the cost of the "base" structure
3. Steel cost is 31% of the cost of the "base" structure
4. Foundation is 25% of the cost of the "base" structure
5. Earthwork & misc. is 6% of the cost of the "base" structure
6. DL of concrete reduced 10.5% by reduction of deck thickness
7. Cost of deck is reduced 1.0% by reduced thickness
8. Total DL is reduced by 7.0%
9. Total DL + LL + I is reduced by 4.45%
10. Demand on girders in negative moment areas is reduced by 4.45%
11. Flange thickness of girders in negative moment areas is reduced by 4.45%
12. Self weight of steel in negative moment areas is reduced by 4.0%
13. Negative moment areas represent 40% of entire structure
14. Total weight and cost of structural steel is reduced by 1.6%
15. No reduction in foundation costs from reduced DL
16. Other assumptions same as Table 1

Table 2 shows that a bridge using EnduraMet 32® in the deck will have an initial cost only 1.4% higher than the same bridge using ECR, when the savings in structural steel are computed. Higher savings in structural steel could actually reduce the higher initial cost for EnduraMet 32®, but it is unlikely that the net initial cost difference could be reduced to zero, unless other savings can be found.

4.2 *Reduced foundation costs*

Table 2 assumes that there are no improvements in foundation design available from the reduction in dead load. In many cases, that is a valid assumption. However, for structures in poor soils, especially where high foundations are used, the reduction total dead load plus live load will provide savings in foundation design, especially where the foundation is governed by seismic loads.

A reduction in dead load of a superstructure supported by a tall pier can substantially reduce the seismic demand on that pier. This reduction can reduce the size of the pier column and can also reduce the size and cost of the footing or pile cap. The number of piles can sometimes be reduced.

Table 3 assumes that the 4.0% savings in superstructure cost is achieved in foundation cost also. This is obviously an arbitrary assumption: foundation savings in many structures will be very

Table 3. Comparison of initial cost and life cycle costs of new bridges with various types of deck reinforcing.

Reinforcing type	ECR, galvanized	MMFX-II™	FRP	Solid stainless	EnduraMet 32®
Deck cost (compared to total initial cost of "base" structure)	38.00%	39.14%	39.90%	42.18%	39.90%
Steel cost (compared to total initial cost of "base" structure)	31.00%	31.00%	30.50%	30.50%	30.50%
Foundation cost (compared to total initial cost of "base" structure)	25.00%	25.00%	24.00%	24.00%	24.00%
Earthwork, etc. cost (compared to total initial cost of "base" structure)	6.00%	6.00%	6.00%	6.00%	6.00%
Total initial cost of structure	100.00%	101.14%	100.40%	102.68%	100.40%
Estimated Life (yrs.)	40	50	65	100	100
Present worth of deck replacement at end of life	9.89%	6.88%	3.93%	1.05%	1.00%
100-year life cycle cost as a percentage of initial cost of "base" structure	111.48%	108.02%	103.88%	103.74%	101.40%

Design assumptions:

1. Foundation cost reduced by 4.0% where DL is reduced by 7.0%
2. All other assumptions same as Tables 1 and 2

small, while a structure with tall column piers in very poor soil may achieve savings in the range of 5% to 8%. When designing structures in these conditions, designers should consider various methods of reducing weight, including non-corrosive reinforcing, lightweight concrete, etc.

Table 3 is identical to Table 2, except for the reduced foundation costs for the FRP, Solid Stainless, and EnduraMet 32® options. For solid stainless steel (AISI 316 or 2205) a 15% reduction in foundation costs would actually reduce the total initial cost of a structure using solid stainless tell rebar below the "base" structure. While this is unlikely, except possibly in extremely poor soil conditions, the reduction in superstructure dead load can provide substantial reduction in cost for the entire structure. For EnduraMet 32®, a 7% reduction in foundation costs will reduce the total initial cost of the structure below the initial cost of the "base" structure using ECR in the deck. While this reduction in foundation cost will not be available on the average highway bridge, it could be achieved in some cases.

5 CONCLUSIONS

The use of carbon steel reinforcing bar has been common for over 100 years. Recent advances in materials will provide superior durability and reduced life cycle costs, compared to carbon steel, even when epoxy coated or galvanized. Some more modern materials, such as solid stainless steel reinforcing bar, will actually provide a reduced total cost of a new bridge structure in specific cases, while providing longer life, at no additional cost.

The various relative costs and percentages given above are based on specific assumptions, which the authors believe are representative of typical bridge projects. These assumptions will obviously not be valid for all cases. This paper is intended to illustrate that the more expensive material does not always make a more expensive project. The economic savings available from the use of better materials can frequently offset the higher initial cost of those materials, when one employs the use of full life cycle cost analysis.

Bridge designers should evaluate different reinforcing materials, during the design of major rehabilitation projects, as well as any new bridge project. A project involving deck replacement and steel repair on a deteriorated bridge could use the design advantages of corrosion resistant reinforcing bar to reduce the cost of steel repairs. The weight savings can substantially reduce the cost of a seismic upgrade for an older bridge which is being rehabilitated. The methodology used here can be used by designers to determine the economic value of various design options on many bridge projects.

REFERENCES

Burke, D.F., 1994. Performance of Epoxy-Coated Rebar, Galvanized Rebar, and Plain Rebar with Calcium Nitrite in a Marine Environment, pub. Naval Facilities Engineering Service Center, reprinted by CRSI

Clemena, G.G. & Yirmani, Y.P.; 2004. Comparing the Chloride Resistances of Reinforcing Bars, Concrete *International,* Nov. 2004, pp. 39–49

Cui, Fushuang & Krauss, P.D.; 2006. Corrosion Resistance of Alternative Reinforcing Bars: An Accelerated Test, Pub. By CRSI

Darwin, D.; Browning, J.; Nguyen, T.V.; & Locke, C.; 2002. Mechanical and Corrosion Properties of a High-Strength, High Chromium Reinforcing Steel for Concrete, FHWA report SD2001-05-F

Fanous, F.; Wu, H.; & Pape, J.; 2000. Impact of Deck Cracking on Durability, Iowa DOT Project TR-405

GangaRao, H. 2007. Verbal communication at Polymer Composites Conference IV

Hartt, W.; Lysogorski, D.; & Leroux, V.; 2004. Characterization of Corrosion Resistant Reinforcement by Accelerated Testing

Humphreys, S.R.; 2004. Improving the Quality of Epoxy-Coated Steel Reinforcing Bars through CRSI's Epoxy Coating Applicator Plant Certification Program, pub. CRSI

Lee, S.-K. & Krauss, P.D.; 2004. Long-Term Performance of Epoxy-Coated Reinforcing Steel in Heavy Salt-Contaminated Concrete, Report No. FHWA-HRT-04-090

McDonald, D. B.; Pfeifer, D. W.; & Sherman, M. R.; 1998. Corrosion Evaluation of Epoxy-Coated, Metallic-Clad and solid Metallic Reinforcing Bars in Concrete, Publication FHWA-RD-98-153

Samples, L.M. & Ramirez, J.A.; 1999. Methods of Corrosion Protection and Durability of Concrete Bridge Decks Reinforced with Epoxy-coated Bars – Phase I, Report FHWA/IN/JTRP-98/15

Smith, J.L. & Yirmani, Y.P.; 1996. Performance of Epoxy Coated Rebars in Bridge Decks, Publication FHWA-RD-96-092

6 Bridge inspection, monitoring & condition assessment

Chapter 19

Use of structural health monitoring techniques for a forensic study of bridge accidents

Hae-Bum Yun, Reza D. Nayeri, Raymond W. Wolfe, Sami F. Masri, Mazen Wahbeh, Farzad Tasbihgoo & John P. Caffrey
University of Southern California, Los Angeles, CA, USA

Li-Hong Sheng
California Department of Transportation (Caltrans), Sacramento, CA, USA

ABSTRACT: This paper presents an overview of a real-time web-based continuous monitoring system for the Vincent Thomas Bridge. An effective multi-thread bridge monitoring system architecture is shown. Using the bridge monitoring system, the bridge response to earthquakes, bridge-ship collision and ambient vibration was measured, and the bridge modal frequencies were successfully determined with vibration-based identification methods.

1 INTRODUCTION

Interest in the field of Structural Health Monitoring (SHM) has been growing at a fast pace in the recent past due to the great developments in the efficient fabrication of innovative sensors, the ease of deploying sensor networks, and the associated high growth in the computational power that is becoming readily available with PC's. Furthermore, the development of sophisticated digital signal processing tools for the analysis of vibration signatures of dispersed civil infrastructure systems has generated a lot of interest in the application of such analysis tools, in conjunction with real-time monitoring approaches, in order to perform virtually continuous condition assessment (of limited scope) of any instrumented structure. Some representative publications that include information about applications of SHM to full-scale bridges include the work of Housner et al. (1997), Nigbor and Diehl (1997), Aktan et al. (2003), Wu (2003), IABMAS (2004), Chang (2004), Masri et al. (2004), Wahbeh et al. (2005) and Yun et al (2007).

A real-time web-based continuous structural health monitoring system for the Vincent Thomas Bridge (VTB) has been developed by researchers at the University of Southern California (USC). The VTB is located in the larger metropolitan Los Angeles region connecting the Los Angeles Harbor operations on Terminal Island to the mainland. For last several years, the monitoring system on the bridge has successfully captured the dynamic response of the bridge under the ambient and special vibration conditions, such as earthquakes and a ship-bridge collision. In this paper, selected results of a forensic study of the bridge characteristics for different vibration scenarios are presented. Further information on the forensic study discussed in this paper can be found in Smyth et al. (2003), Wahbeh et al. (2005), and Yun et al. (2007).

2 DESCRIPTION OF THE BRIDGE

The VTB is located in the metropolitan Los Angeles region. This bridge was a toll bridge before 2000, and is considered a major bridge in California. It connects two main harbors in this region, the

Figure 1. The Vincent Thomas Bridge located in San Pedro, CA.

Port of Los Angeles and the Port of Long Beach. These two ports are among the busiest in the U.S. The bridge handles approximately 39000 cars and trucks daily. The VTB is a cable-suspension bridge, approximately 1850-m long, consisting of a main span of 457 m, two suspended side spans of 154 m each, and two ten-span cast-in-place concrete approaches of 545 m length on both ends. The roadway is 16 m wide and accommodates four lanes of traffic. The bridge was completed in 1964 with 92000 tons of Portland cement, 13000 tons of light-weight concrete, 14100 tons of steel and 1270 tons of suspension cables. It was designed to withstand winds of up to 145 kilometers per hour. A major seismic retrofit was performed between 1996 and 2000, including a variety of strengthening measures, and the incorporation of about forty-eight large-scale nonlinear passive viscous dampers. The photo of the VTB is shown in Figure 1.

3 REAL-TIME WEB-BASED BRIDGE MONITORING SYSTEM

The developed monitoring system is based on a multi-threaded software design. This highly efficient software architecture allows the system to acquire data with multiple channels, monitor and condition this data, and distribute it, in real-time, over the Internet to various remote locations. The software has three main threads: (1) data acquisition thread (publisher), (2) data transceiver thread (server) and (3) local monitoring thread (clients). The sensor locations on the bridge and a schematic of the system architecture of the developed bridge monitoring system are shown in Figure 2.

4 BRIDGE IDENTIFICATION WITH SEISMIC VIBRATIONS

4.1 *Description of the earthquake*

In the early morning hours of 22 February 2003, a relatively small earthquake (magnitude M = 5.4) occurred in the vicinity of the city of Big Bear, California (Figure 3). The epicenter was located

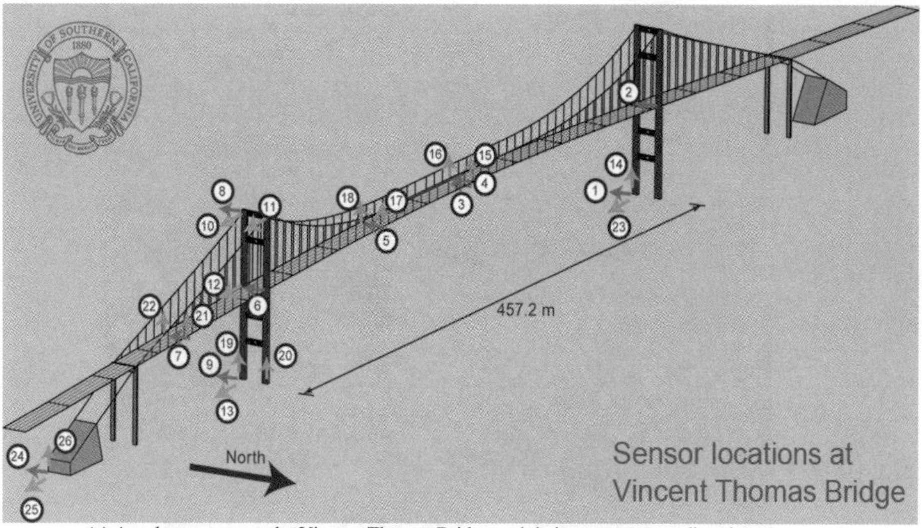

(a) Accelerometers on the Vincent Thomas Bridge and their measurement directions.

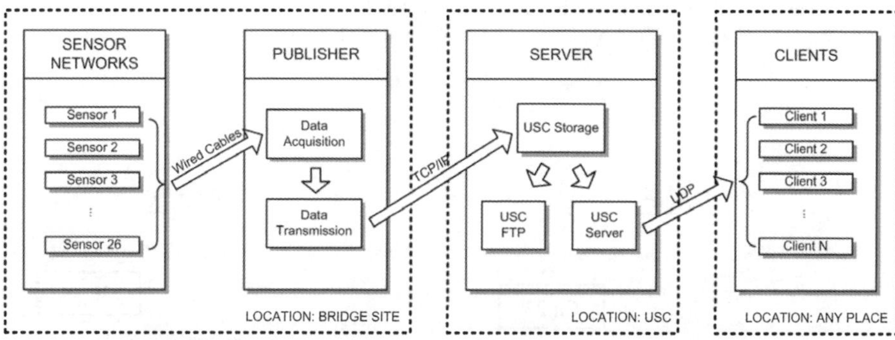

(b) A schematic of high-level system architecture of the real-time bridge monitoring system.

Figure 2. The real-time web-based continuous monitoring system for the Vincent Thomas Bridge (Yun, et al. 2007).

about 180 km from the Vincent Thomas Bridge. All acceleration channels were triggered on the VTB, and a complete data set with twenty-six channels was obtained by the real-time monitoring system.

A sample of two acceleration and displacement components at the base of VTB, as well as three response components on the bridge deck are shown in Figure 4.

4.2 *Bridge identification results*

The VTB was identified with a reduced-order multi-input/multi-output (MIMO) nonlinear model expressed as

$$M_{11}^{-1}C_{11}\dot{x}_1(t) + M_{11}^{-1}K_{11}x_1(t) + M_{11}^{-1}M_{10}\ddot{x}_0(t) + M_{11}^{-1}C_{10}\dot{x}_0(t) + M_{11}^{-1}K_{10}x_0(t) = -I\ddot{x}_1(t_1) \tag{1}$$

where M_{11}, C_{11} and K_{11} are the mass, damping and stiffness matrices, which characterize the forces associated with the unconstrained degrees of freedom of the system. The matrix dimension

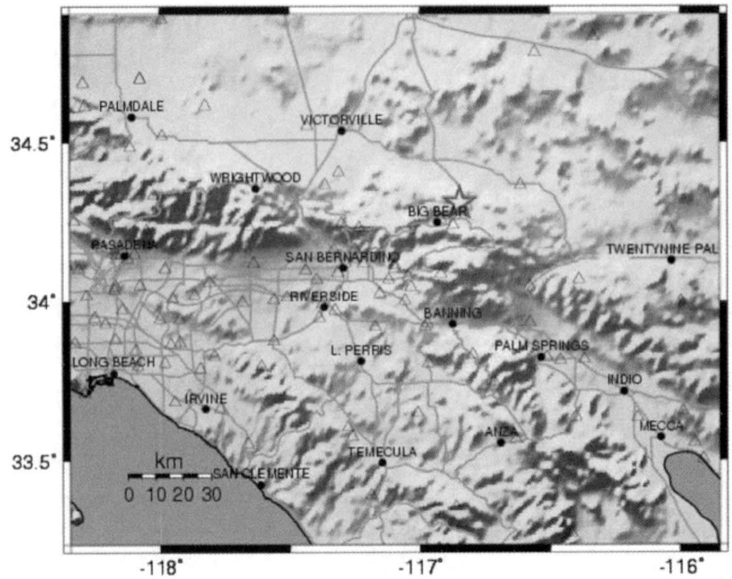

Figure 3. Geographical location of the Big Bear earthquake (M5.4). The epicenter was 3.1 miles north of Big Bear City, CA. (N34.31°, W116.5°, Depth 1.2 km) (Smyth et al., 2003).

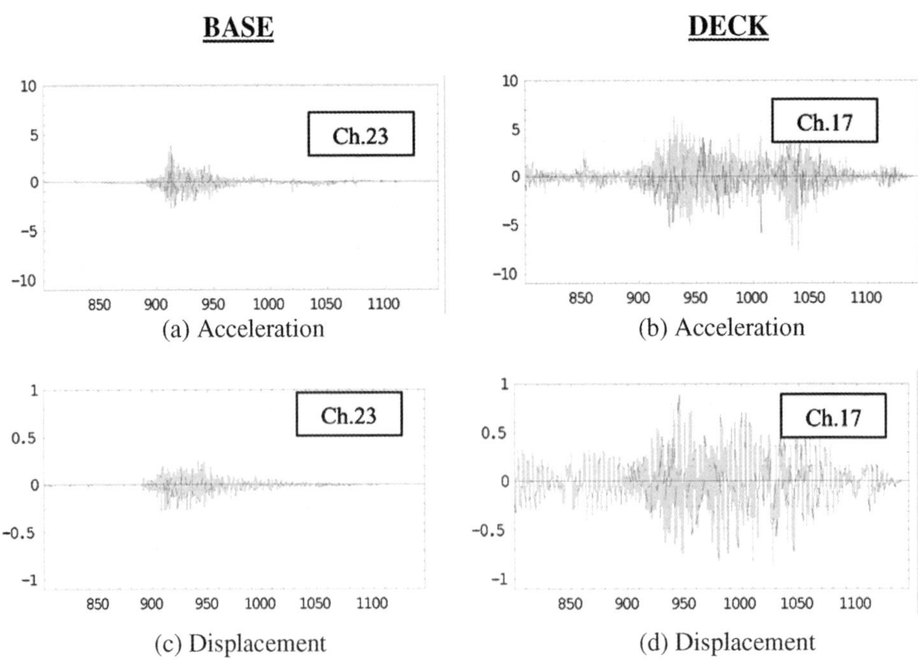

Figure 4. Sample displacement time histories measured at the VTB base and its deck during the Big Bear earthquake on 22 February 2003. The displacements were obtained through numerical integration of the measured accelerations (Wahbeh et al., 2005).

Table 1. A comparison of the identified natural frequencies of the bridge for 1987 Whittier, 1994 Northridge and 2003 Big Bear earthquakes (Wahbeh et al., 2005).

Smyth et al.(2003) Dominant modes (all directions)		Lus et al. (1999) Vertical component modes		Wahbah et al. (2005) Dominant modes (all directions) (f_c = 0.15 Hz)
Whittier	Northridge	Whittier	Northridge	Big Bear 2003
0.212	0.225	0.234	0.225	0.194
0.242	0.240	0.388	0.304	0.234
0.317	0.358	0.464	0.459	0.283
0.531	0.390	0.576	0.533	0.361
0.570	0.448	0.617	0.600	0.515

is ($n_I \times n_I$). M1, C10 and K10 are constant matrices, which are associated with support or input motions. $x_1(t)$, $\dot{x}_1(t)$ and $\ddot{x}_1(t)$ are vectors of order n_I of the active degrees of freedom (measured response) displacement, velocity and acceleration, respectively. $x_0(t)$, $\dot{x}_0(t)$ and $\ddot{x}_0(t)$ are vectors of n_0 order of the support displacement, velocity and acceleration, respectively. Detailed discussion of this method can be found in Masri et al. (1987).

Using this method, the bridge response was successfully identified. A comparison of the identified natural frequencies for three different earthquakes (1987 Whittier, 1994 Northridge and 2003 Big Bear earthquakes) is shown in Table 1.

5 FORENSIC STUDY OF BRIDGE-CARGO SHIP COLLISION

5.1 *Factual information of the incident*

The *Beautiful Queen* is a 189 m (620 ft) 32000 ton cargo ship, owned by the Pasha Hawaii Transportation Line. The cargo ship is a bulk carrier, not a container ship, commonly hauling rolled steel, coal or grain. The ship is equipped with onboard cranes for freight loading. On Sunday, 27 August 2006, the ship departed from the Los Angeles harbor via one of the channels in the harbor district. At 16:40 (Pacific Daylight Time), the ship was sailing under the Vincent Thomas Bridge, linking San Pedro and Terminal Island. When the ship passed under the bridge, the ship operators miscalculated the tide, and one of the onboard cranes scraped a guide rail of a maintenance scaffold secured at the bridge center span, which was about 56 m (185 ft) above the water. No injuries were reported during the incident, but the guide rail of the maintenance scaffold was damaged during the collision.

About thirty minutes after the collision, the vehicular traffic across the bridge was stopped by Caltrans to investigate potential damage. Vessel traffic was also stopped under the bridge by the Los Angeles Port Police and the U.S. Coast Guard. Two incoming cargo ships were delayed due to the vessel traffic shut-down. After investigating the incident for a period of about two hours, Caltrans engineers declared that the bridge was sound and that the damage was limited to the maintenance scaffolding. Both vehicle and vessel traffic were re-opened at 18:55 the same day. An independent investigation was also conducted by the U.S. Coast Guard on the colliding cargo ship.

5.2 *Measurements of the real-time monitoring system*

The VTB vibration during the cargo-ship incident, and the two-hour traffic shut-down afterward, were successfully captured by the real-time monitoring system. Sample acceleration time history data are illustrated in Figure 5. The figure shows twenty-four hour displacement (lateral) time history of the bridge deck when the cargo-ship accident occurred. The displacements in Figure 5 were obtained through numerical integration of the measured accelerations.

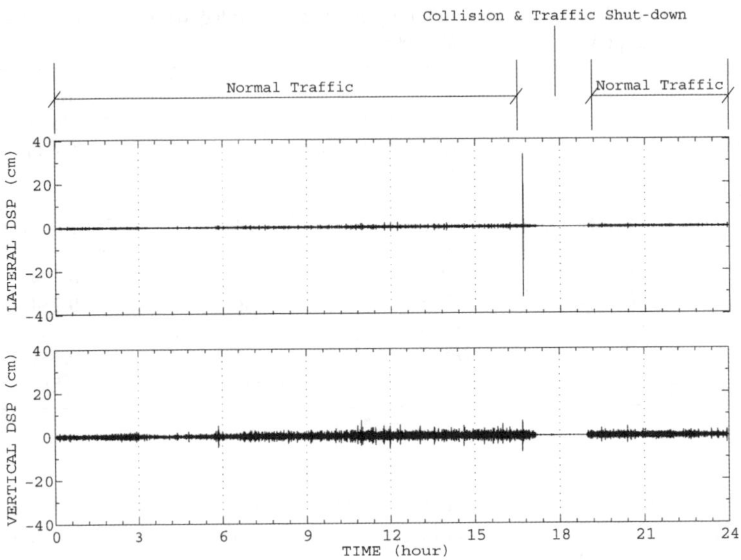

Figure 5. Twenty-four hour displacement (lateral) time history of the bridge deck on 27 August 2006 when the cargo-ship accident occurred (Yun et al., 2007).

Table 2. A comparison of the identified natural frequencies of the bridge for 1987 Whittier, 1994 Northridge and 2007 bridge-ship collision (Yun, et al., 2007).

Smyth et al. (2003)				Lus et al. (1999)				Yun et al. (2007)					
Vertical direction				All directions				All directions					
Whittier		Northridge		Whitter		Northridge		Impact		w/o traffic		w/traffic	
f	ζ	*f*	ζ	*f*	ζ	*f*	ζ	*f*	ζ	*f*	ζ	*f*	ζ
0.212	1.2	0.225	0.1	0.234	1.5	0.225	1.7	0.150	4.0	–	–	–	–
0.242	1.7	0.240	8.2	0.388	38.2	0.304	28.6	0.233	2.7	0.244	2.5	0.2.5	1.9
0.317	−4.3	0.358	−4.7	0.464	9.7	0.459	1.8	0.536	0.8	0.543	0.6	0.534	0.6
0.531	10.2	0.390	4.2	0.576	9.9	0.533	4.0	1.394	1.5	1.392	1.3	1.400	1.7
0.570	0.6	0.448	−0.7	0.617	14.5	0.600	26.2	1.869	1.3	1.893	1.2	1.867	1.9
0.636	4.2	0.478	1.3	0.617	76.8	0.632	13.7						
0.672	0.1	0.522	1.4	0.769	29.7	0.791	15.6						
0.734	2.4	0.587	−0.1	0.804	1.4	0.811	1.0						
0.818	1.9	0.625	7.4	0.857	11.6	0.974	2.7						
0.958	2.9	0.733	1.2	0.947	4.3	1.110	0.6						
1.027	−1.9	0.837	5.0										
1.111	1.3	0.935	−1.8										
1.159	1.7	1.036	1.6										
1.391	2.3	1.110	1.7										
1.554	−1.3	1.136	1.4										

5.3 *Identification results*

Using the measured bridge accelerations, the bridge was identified using the NExT-ERA method for the conditions during-collision, post-collision with no-traffic, and with regular-traffic flow. The detailed discussion of the Next-ERA method can be found in Nayeri et al. (2006a, b). A comparison of identified natural frequencies for the 1987 Whittier, 1994 Northridge, and 2007 bridge-ship collision is summarized in Table 2.

6 SUMMARY AND CONCLUSION

This paper gives an overview of a real-time web-based continuous monitoring system for the Vincent Thomas Bridge. The monitoring system is based on a highly efficient multithreaded software design that allows the system to acquire data from a large number of channels, monitor and condition this data, and distribute it, in real-time, over the Internet to multiple remote locations.

Through examples of different excitation conditions, it was shown that the developed bridge monitoring system can provide a robust long-term bridge condition assessment tool. Using the obtained bridge response data, the bridge was successfully identified for earthquakes, bridge-ship collision, as well as ambient vibration. Recovery of such data allowing detailed computational investigation of the structure is otherwise infeasible given the current practice of post-event visual inspection that is prevalent not only in California, but elsewhere as well.

REFERENCES

Aktan, E, Ciloglu, K, Grimmelsman, K, Pervizpour, M and Qin, X., (2003), "Monitoring Operations, Security and Structural Health of Major Long-Span Bridge Hyper-Systems," Proc. of *Structural Health Monitoring 2003*, (Edited by F-K Chang), DEStech Publications, Lancaster, Penn, pp. 188–196.

Chang, F.K., (2003), (editor), *Structural Health Monitoring: Current Status and Perspectives*, Proceeding Fourth International Workshop on Structural Health Monitoring, Stanford University, September.

Housner, G.W., Bergman, L.A., Caughey, T.K., Chassiakos, A.G., Claus, R.O., Masri, S.F., Skelton, R.E., Soong, T.T., Spencer, B.F., and Yao, J.T.P., (1997), "Structural Control: Past, Present and Future," ASCE *Journal of Engineering Mechanics*, (Special Issue), 123(9), Sept 1997, 897–971.

IABMAS, (2004), Proceedings of *the Second International Conference on Bridge Maintenance, Safety and Management,* Kyoto, Japan, 18–22 October 2004.

Lus H, Betti R, and Longman RW., (1999), "Identification of Linear Structural Systems using Earthquake-Induced Vibration Data," *Earthquake Engineering and Structural Dynamics*, vol 28: pp. 1449–1467.

Masri, S.F., Miller, R.K., Saud, A.F., and Caughey, T.K., (1987), "Identification of Nonlinear Vibrating Structures: Part I – Formulation," *ASME Journal of Applied Mechanics*, Vol. 54, December, pp. 918–922.

Masri, S.F., Sheng, L-H, Caffrey, J.P., Nigbor, R.L., Wahbeh, M., and Abdel-Ghaffar, A.M., (2004), "Architecture and Utilization of a Web-Based Real-Time Structural Health Monitoring System", *Smart Materials & Structures* September (12), pp. 1269–1283.

Nigbor, R L., and Diehl, J G, (1997), "Two Year's Experience Using OASIS Real-Time Remote Condition Monitoring System on Two Large Bridges," Proc of *Structural Health Monitoring 1997*, (Edited by F-K Chang), Technomic Publishing, Lancaster, Penn, pp. 410–417.

Smyth, A.W; Pei, J-S., and Masri, S.F., (2003), "System Identification of the Vincent Thomas Suspension Bridge using Earthquake Records," *Intl Jnl Earthquake Engineering and Structural Dynamics*, Vol 32, pp. 339–367.

Wahbeh, M., Tasbihgoo, F., Yun, H., Masri, S.F., Caffrey, J.P., and Chassiakos, A.G. (2005). "Real-time earthquakemonitoring of large scale bridge structures," Proceedings of *the international workshop on structural health monitoring.*

Wu, Z.S., (2003), Proceedings of the *First International Conference on Structural Health Monitoring and Intelligent Infrastructure* , Tokyo, Japan, 13–15 November 2003.

Yun, H., Nayeri, R., Tasbihgoo, F., Wahbeh, M., Caffrey, J., Wolfe, R., Nigbor, R., Masri, S. Abdel-Ghaffar, A., Sheng, L-H. (2007), "Monitoring the collision of a cargo ship with the Vincent Thomas Bridge," *Structural Control and Health Monitoring* (accepted).

Chapter 20

Bridge management and inspection system for Montgomery County, Maryland

Jeremy K. Shaffer & Michael C. Schellhase
InspectTech, Pittsburgh, PA, USA

ABSTRACT: Montgomery County is Maryland's largest county, containing nearly 20% of the state's population, and owns over 300 bridges covering a wide range of types and lengths. Its prior methods of utilizing bridge data for its management and inspection processes were typical of that of many other cities and counties: a variety of spreadsheets and databases with a large paper based component. For its latest inspection cycle the county adopted an advanced bridge inspection and management system to organize all of its bridge information.

This paper covers an overview of the county's needs and the solutions that have been developed to significantly improve both the inspection and management processes. Electronic forms were created to meet the county's requirements, the most rigorous in Maryland, and allow for entry of all information from the inspection. County personnel are now able to review all inspection data via a secure web-based browser and utilize powerful tools to instantly track changes occurring, highlight problem areas, and generate a wide range of needed reports. This new approach has allowed the County to significantly streamline the entire process from field to office and integrate all data into a single application from which all personnel can access needed information.

1 INTRODUCTION

As Maryland's most populace county with nearly a million residents [Census 2006], Montgomery County has been a state and national leader in the technology and techniques used to manage and inspect its over 300 bridge structures. The county's bridges represent a wide diversity in structure types and functions with structures located along the border of Washington, D.C. within major suburban communities, to rural areas with minimal usage. To cope with the challenge of organizing all bridge information from field inspections to office information in 2006 the County's Department of Public Works and Transportation sought the implementation of a state of the art computerized inspection and management system. The system replaces a difficult to manage process that had become very paper intensive in which it had become difficult to quickly obtain needed and useful information.

The County selected InspectTech's BridgeInspect™ software suite as the solution to meet its needs for an integrated end-to-end system. The core system was customized to correspond to the County's exact specifications with interfaces, work/process flow, maintenance items, additional county specific data/inspection forms, and security settings. The software contains two primary parts an inspection component and a management system. The inspection software has both a field/standalone version that runs on tablet/laptop computers for use by inspectors while at the bridge site and a web-based office version for integrating the field data and finalizing the reports. This software contains all digital versions of the necessary forms needed to generate a complete county inspection report (typically 20–30 pages in length). The Management system is a completely web-based program that can be accessed securely from any county computer or from home with a correct username and password. County personnel are able to access all information on bridge structures from current and historical inspection reports, pictures, sketches, memos, and maintenance needs along with having numerous tools such as GIS/mapping, full searching, and cost estimating components.

This paper is organized in the following sections. Section 2 provides Background information on the County and its former system of handling inspections and management. Section 3 provides an overview of the overall project goals and a high-level view of the main system components. Section 4 discusses the inspection process and software used to create a final report while Section 5 presents the details of the software used to enable the bridge management process. Section 6 ends this paper with conclusions and remarks on the overall project implementation.

2 BACKGROUND

2.1 *Bridge system information*

Montgomery County manages a diverse set of bridges with widely varying needs. The total inventory of structures is over 300. This is composed of nearly 200 large structures (greater than 20′ in length), over 100 small structures (less than 20′ in length) as well as an additional 20 pedestrian bridges [MONTCO 2006]. These bridges span a wide variety of features from rivers to major interstate highways (some up to 12 lanes) and from railroads to the DC Metro transit system. In the more urban areas of the county the high volume of usage can often require significant maintenance of traffic to perform even minor tasks or inspections. Inspections of structures spanning railroad tracks or the Metro system require significant coordination with outside agencies that makes even the simplest tasks far more complex to achieve.

The County implements a regular maintenance and capital planning program to achieve its minor and major repair needs. Information obtained from the bi-annual inspection program forms the basis for identifying bridge deficiencies and obtaining the necessary data for prioritization and funding requests.

2.2 *Federal and state requirements*

The county's bridge inspection and management program is designed to meet and exceed the requirements specified by the United States Government's FHWA and Maryland State Highway Agency (SHA). FHWA's National Bridge Inspection Standards (NBIS) require the collection of over 200 pieces of inventory and appraisal data on all 20′ or larger structures [FHWA 1995]. The regulations additionally specify requirements (training and experience) needed for personnel performing the inspections. Maryland's SHA administers these NBIS details as well as adds an additional layer of requirements [MD 2004]. As with most other states, Maryland, collects element level inspection data [MD 2002]. Bridges have been divided into their primary elements and the evaluations of each of these elements are defined by requiring inspectors to quantify the amount of each element based on a varying number of condition states. In addition to element level data Maryland SHA has added on several additional inventory related fields that are not present in the NBI data and are required to be collected.

2.3 *Inspection process*

The County utilizes an engineering consulting firm to perform all of its bridge inspections. In Maryland the State Highway Administration provides a set of pre-selected consultant firms from which the Counties can chose/be assigned a firm to perform the inspections. For the latest inspection cycle the firms selected were a joint venture of Greenhorne & O'Mara and Kennedy Porter Associates with sub-consultants of Alvi Associates and Tuhin Basu Associates. The consultants perform an in-depth hands on inspection of the structures and when necessary will perform new load rating calculations. The traditional deliverables prior to this inspection cycle were paper inspection reports for each structure along with updating a Maryland SHA Access File containing NBI and Element Level data.

2.4 *Original system*

Prior to implementation of the integrated inspection and management system the County relied on a variety of differing formats in order to meet its needs. The County utilized the state's Microsoft Access file to store basic inventory data, condition rating information, and element level coding. In addition several Excel spreadsheets were kept for each bridge to keep track of coating conditions and ratings, guardrail and approach data, as well as maintenance information. The bulk of the narrative data of each inspection report is stored in printed reports kept in file cabinets and shelves for review and reference. In addition many of the files composing these reports are saved as PDF or Word documents under individual folders saved on a network drive. When questions arise on a bridge or bridges information must be either manually retrieved from a hard copy of a report or compiled from the correct Excel, Access, or Word files. The user must be located physically within the primary administration building of the County in order to electronically or physically access any of the information.

3 SYSTEM OVERVIEW

3.1 *System goals*

The County chose to implement an integrated inspection and management system for a number of key reasons. First, the County desired a system that would enable considerable efficiencies and time-savings on both the inspection and management components. Starting with the inspection process the County sought to establish a high-level of quality and consistent format for all consultants to use. The County sought to eliminate all errors from inconsistent data and dramatically increase the reliability of the information obtained via inspections and utilized to make critical maintenance and capital planning decisions. Finally, the County desired to unify all of the bridge information (inspections, as-built drawings, load ratings, work orders, etc.) into one place available from all computers securely via the internet from any county office or individual's home. This integrated bridge information and management system is able to serve a variety of user needs from maintenance to bridge managers to higher-level executives. Each group requires different permissions and functions to meet their unique needs and prevent one from inadvertently corrupting data they should not have access to. Overall, the goal is to dramatically streamline the inspection and management process for all of the County's bridges and as a result save time, money, and increase the operating efficiency and safety of the bridge network.

3.2 *System structure*

In order to meet the County's goals InspectTech's core BridgeInspect™ Collector and BridgeInspect™ Manager software was chosen and quickly customized per the County's specifications. Figure 1 shows the overall system architecture and how the various components fit together from field module to office based management system. The system is composed of three primary components. The first component shown in the upper-left corner of Figure 1 is the field inspection software. This software runs on individual laptop or tablet computers and is taken to the bridge site in order to start the inspection report. Once the field work is done and the laptop is back in the office or at another location with an internet connection the data collected is uploaded to the online office inspection module (Figure 1 – bottom-left). Once the data is submitted to the online inspection module it is available for continued edit and final review by any office based personnel. Detailed sketches or load rating analysis sections can be added to the report with multiple users working on different sections of the same bridge report. Once all sections have been added the report can go through internal quality assurance and quality checks by the consultant project manager. When the consultant project manager is satisfied with the report he submits it to the owner entity (County) for review and final approval. The county accesses the report using their bridge

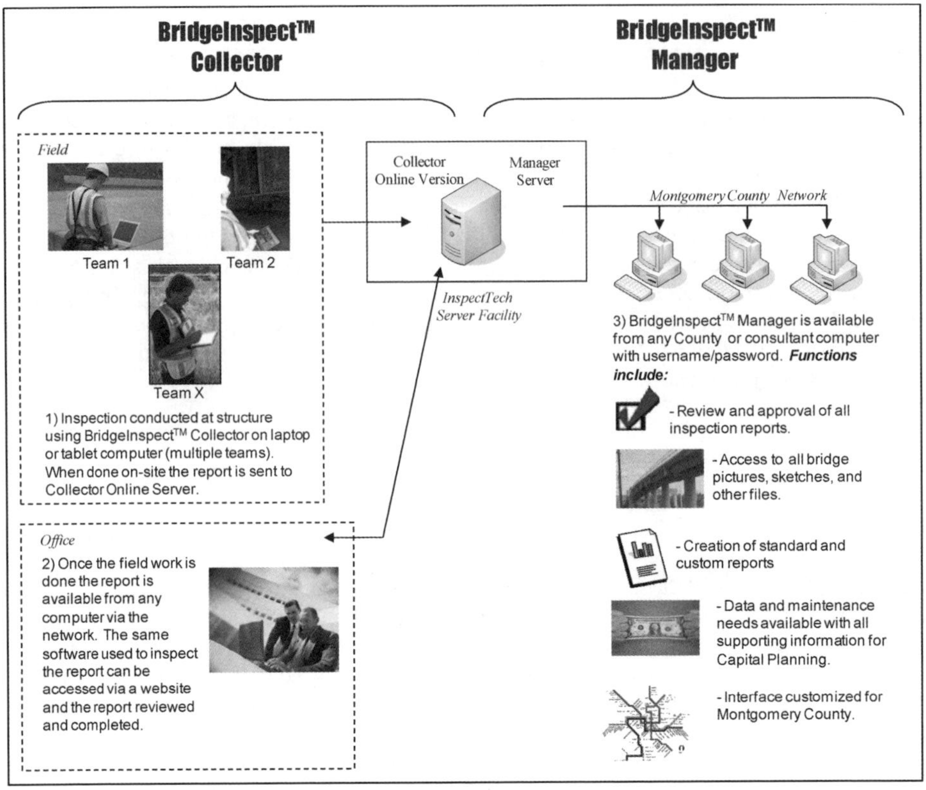

Figure 1. Software components by functions.

Table 1. Field computer hardware and software requirements.

Component	Requirement
Processer Speed (CPU)	800 MHz*
Operating System	Windows XP, 2000, Vista (all versions)
RAM	256 MB*
Screen Resolution	1024 × 768 pixels (or higher)
Free Disk Space	1 GB (or more for storing pictures and sketches)
PDF Reader	Adobe Acrobat Viewer 5.0 or higher

*Minimum requirements additional resources allow for faster performance.

management software via the internet. (Figure 1 – right side). The county bridge manager can approve the report and then data is available for usage in all of the various bridge management modules. The bridge management software runs entirely off a server computer and accessed securely through encrypted connections from any internet computer.

3.3 *Technical requirements*

The system software is designed to run on the typical computers that consultants and office based managers already have. To access the online inspection or management software the consultants needs only to be using Internet Explorer 6.0 or higher browser along with an Adobe Acrobat Acrobat Reader program in order to view the completed reports which are generated as PDF files. The standalone inspection software used in the field has more additional requirements as shown in Table 1. It is also recommended that for optimal performance the software be run on a computer designed for outdoor usage. Standard laptop or tablet computers are not designed for usage in outdoor weather or lighting conditions. On sunny days glare or washout can render the screen unreadable on a regular computer. Additionally, rain or accidental drops could also severely damage a standard laptop. It is recommended that a semi-ruggedized or full ruggedized computer with outdoor optimized screen be used for field inspection work. Many of the input screens utilize drop down lists for entry and value selection which make the added feature of a pen-based touchscreen optimal but not necessary for the field computer.

4 BRIDGE INSPECTION SOFTWARE

4.1 *User interface overview*

The bridge inspection software has been constructed to be very user-friendly software (Figure 2). Significant care in the development of the interface was done for allowing for inspectors to literally "pick-up and go" with the software with minimal training. The main interface used for an individual inspection report uses a tab-based approach that functions well on traditional laptops and pen-enabled tablet/laptop computers. The top of the screen shows a two-layered tab menu. The main tabs are on the very top and provide the main sections by which the individual inspection forms are grouped. The main window in the screen below shows the digitally enabled inspection form.

4.2 *Tab menus*

There are two rows of tab menus across the top of the screen. For the County the main tabs on the top row are: "SI&A", "Pontis", "Condition Forms", "Mont. Co. Forms", and "Report Sections". The SI&A main tab has nine sub-tabs which each correspond to a traditional Maryland/FHWA inspection form. The Pontis main tab when clicked enables sub menu tabs that show the elements associated with the bridge and additional sub-tabs that allow new elements to be assigned or incorrect ones to be removed. The Condition Forms main tab when clicked shows sub-tabs that link to greatly expanded forms that each correspond to the main condition state ratings (58 – Deck, 59 – Superstructure, 60 – Substructure, etc.). These forms allow for considerable detail on these main bridge components and correspond to traditional forms used in the state of Maryland. The Mont. Co. Forms main tab enables sub-tabs that correspond to several unique forms that the County has developed for their own use independent from FHWA or SHA requirements. Some examples of these forms include a Coating Evaluation & Rating form, a Guardrail form, and a detailed bridge description and summary. Finally, the Report Sections main tab provides links to sub-tabs that allows for tasks for complete report compilation. Other of the sub-tabs allow for picture insertion and viewing the current sections that make up a printed report. The software is extremely flexible in that it allows the input of any PDF file as a new section. Both pictures and sections can be re-ordered easily by the inspectors.

In addition to the main and sub-tabs other information in the top rows includes the bridge number or name in the top left corner. In the top right corner of the main menu bar are two links to take the inspector back to the county bridge list or to the main menu.

4.3 *Main screen*

The main window of the inspection software displays the interactive inspection form that corresponds to the selected sub-tab. Most of the forms are constructed to look identical to the paper

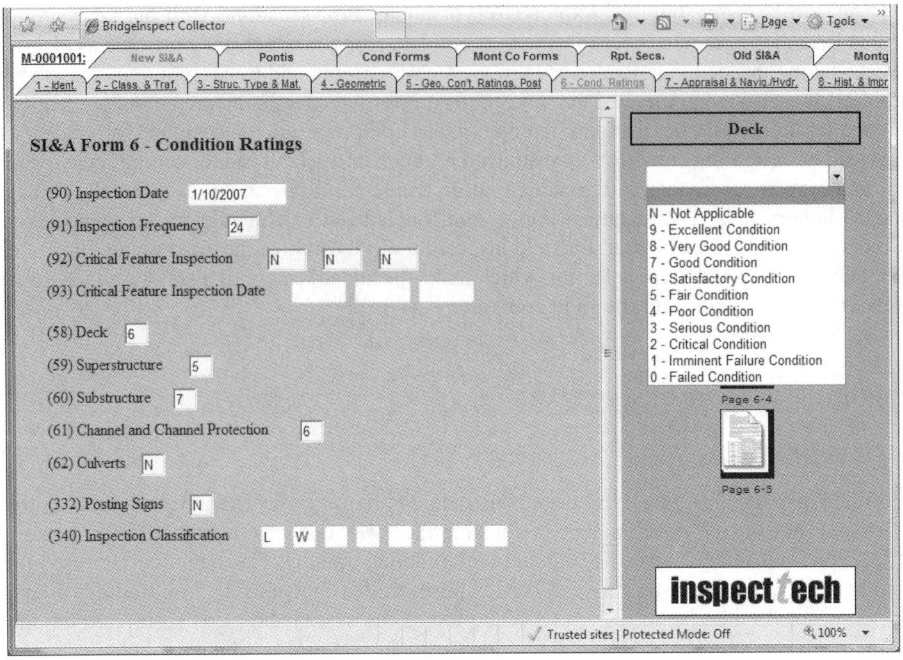

Figure 2. Inspection software screen showing: (1) Tabbed Menu across the top, (2) Main Entry area on the left and bottom (3) Information Sidebar on the right.

forms allowing inspectors to utilize a familiar format and thus not having to spend time learning a new layout. Fields within the form come in a variety of sizes and functionalities from large text areas to check boxes. Text areas can grow and most support up to eight pages of text within a single box effectively allowing for unlimited data entry. The color scheme of the input screen can be toggled between an outdoor optimized input screen (black background/white text) or an indoor screen of (light background/black text). Past values are automatically loaded into the report saving the inspector valuable time in not having to retype entire descriptions when not necessary but only altering those parts that have changed. Colors are used extensively within the backgrounds of the individual fields to show what has been loaded in from a past report and not changed (yellow) to what is new or has changed (white). This feature allows for quickly identifying fields that have changed and those that have not. Fields on the form can be rapidly moved between by using one of several options the tab key, the mouse, or a pen (on touch screen computers). The main screen can scroll both vertically and horizontally whenever it the forms size exceeds the normal viewing area. Scroll bars will only appear when necessary for each axis.

4.4 *Information sidebar*

The information side bar located on the right portion of the program's screen is an extremely useful tool for the inspectors. Whenever a field is selected or clicked on the main screen the content of the information sidebar automatically updates. The sidebar displays several critical pieces of information. On the top of this area is first the full field name and description of the selected main form field. Immediately below this are described any restrictions on that data entered in the field, i.e. "Maximum Character Length of 20". These restrictions often are based on limitations/requirements from external sources such as the Federal NBI [FHWA 2002]. The next feature that appears is a drop-down list for fields that have choices available. For example, the 58 Deck field has a 0–9 list of options available with corresponding descriptions for each field (Figure 2 – right side).

Next, the inspectors have direct links or text messages from the relevant coding guides (FHWA or Element Level). Only the relevant description and links to specific pages are shown on the sidebar for the specific field selected. By clicking on a page link the inspector will immediately be shown the full page in the coding guide corresponding to the field they are attempting to enter information on. This feature effectively has integrated in over 600 pages of manual data into an extremely user friendly and accessible format for field inspectors to instantly access when needed onsite. Additional, advanced information can also be shown on the sidebar such as example pictures or sketches demonstrating how to take measurements or what a specific rating for the element looks like. These features serve both to avoid common mistakes and help to take subjectivity out of the rating process.

5 BRIDGE MANAGEMENT SOFTWARE

5.1 *User interface overview*

The bridge management software is accessible via a secure website. County personnel can reach the website from any internet enabled computer with their secure username and passwords. Based on their login information they are given varying permissions to access different modules within the management system. The manager software uses a drop down menu based system that provides much the same interaction and experience of a standalone program. At any time users can select individual bridges by a drop down list in the upper right corner or enter a specific bridge number. Like the inspection software the management program is extremely intuitive and easy to use.

5.2 *Bridge detail page*

When an individual bridge is selected or clicked on within a search result or from the interactive map users are taken to the bridge detail page. This page serves as the integration of all information on the bridge structure. If a user within the organization has any question on the structure they can come here to find out current conditions, historical conditions, trending, past and future work that has been scheduled, drawings, pictures (current and historical) and any digital file that has been attached to the bridge (memos, CAD, etc.). The page is composed of many sections each of which provides a snapshot of information. Also on the page are links to full documents available. A user is able to retrieve complete inspection reports (latest or any historical) as a PDF file by clicking on a link. Summaries of all of the current maintenance needs for the structure are also present. Since pictures can be linked directly to the maintenance needs by the inspectors, managers are also able to drill down into the maintenance needs for specific pictures and detailed descriptions. A popular item present on the detail page is the load rating summary and link to the full detailed calculations. The detail page serves as a powerful index that allows the County management personnel to quickly reference information that could fill a file cabinet on a specific bridge.

5.3 Full searching

The management software provides the capability to search across any field or combination of fields from within the inspection reports. This includes all inventory data, condition ratings, and maintenance information. If a Manager needs to quickly identify all prestressed adjacent concrete box beam bridges with superstructure ratings of 5 or less and that were built before 1970 they can do that with a couple clicks. Queries from very simple to extremely complex nested logic strings can be created and saved for future usage. Results are displayed in tabular form and indexed by bridge. Users can click on the bridge name/code to go directly to the bridge detail page for more information or click on a mapping link to plot all the bridges on the mapping interface to visually determine where the structures are located.

5.4 *Document storage*

In order to serve as the single source for all bridge information. The management system allows for the uploading and direct storage on the server of any digital file. Files can be associated with dates and given descriptions during the submission process. As some examples, the Manager software can store Word documents describing agreements with utilities running on the structure, emails regarding actions to be done, CAD drawings, or a variety of other information. For another user to access this information stored on the server they have a direct link to download the file onto their computer. In order for the file to be opened the user accessing the data will need that specific software application to read a file of that type (i.e. Word file requires Microsoft Word to be installed).

5.5 *Mapping interface*

The County's version of the management software integrates directly with Google maps. Using latitude and longitude coordinates already present in the inventory data all bridges can be plotted directly onto an interactive map of the county. No additional software needs to be installed on the user's computer. Users can select between the roadway layer, a satellite layer, or a combination of both. The Google maps allows for the county users to zoom in and out and see a very high level of detail in aerial photos of the structure. In addition to displaying the entire set of the bridges the software can also display only an individual bridge or other subset of bridges that are outputted from reports or that meet a specific search criterion.

5.6 *Administrative features*

The software allows the administrative features necessary to handle the addition, editing, and deletion of bridges as well as new users. A user who is an administrator in the software can create accounts for others with specified permissions and correct access by structure or module. Additionally, accounts can be setup for the consultant inspectors to allow them permissions to retrieve data for performing inspections automatically from their laptop computers. This is especially useful when starting a new inspection cycle. A variety of tasks can be done to allow for the full system maintenance and importing/outputting of data to external programs.

6 CONCLUSIONS

Bridge owners of all sizes can significantly benefit from adoption of a customized and integrated inspection and management system. The traditional processes of inspection with disjoint or incomplete databases and extensive usage of static paper reports leads to considerable inefficiencies, is prone to errors, and lacks the flexibility or function necessary for bridge managers or inspectors. The Montgomery County implementation demonstrates how both consultant inspectors and bridge owners can utilize the software to facilitate better communication, quicker results, and much more in-depth and usable information. Bridge inspection is far more than just collecting data for storage in a file cabinet with little practical usage. Effective bridge inspection software helps to highlight and provide quick and easy access to turn data into useful information. Problems can be quickly identified, documented, and action plans developed. The management software will creates a unified location for the full documentation and plans on all bridges to be accessible throughout an organization helping to prevent internal communication errors. The user-friendly nature of the software has required little training and fits well with the standard inspection flow. It has been demonstrated that the software does not create an extra burden for the inspectors but is instead a powerful tool that allows them to do their job without having to focus on tedious clerical work in organizing a report and entering duplicate information. Overall, the system has been extremely well received and has allowed the County to achieve its primary goal of integrating all bridge information in a single location for quick and easy access with tools to prioritize and highlight problem areas.

REFERENCES

Maryland Bridge Inspection Coding Manual and Inspection Forms: Maryland SHA Office of Bridge Development, 2004, Baltimore, MD.

Maryland Element Level Inspection Guide: Maryland SHA Office off Bridge Development, 2002, Baltimore, MD.

Montgomery County Department of Public Works and Transportation, Bridge Condition Summary Report, 2006, Rockville, MD.

United States Census Bureau – Maryland Quick Facts: July, 2006.

United States Federal Highway Administration: Recording and Coding Guide for the Structure Inventory and Appraisal of the Nation's Bridges, 1995, FHWA-PD-96-001.

United States Federal Highway Administration: Bridge Inspector's Reference Manual, 2002, FHWA-NHI 03-001, 002, 004.

United States Federal Highway Administration, 1999, *Asset Management Primer*, Office of Asset Management, Washington, D.C.

Chapter 21

Objective condition states for concrete bridge deck assessment

M.L. Knight
Middle Tennessee State University, Murfreesboro, Tennessee, USA

ABSTRACT: Inspection of bridge decks generally relies on visual inspection and use of basic non-destructive testing techniques. Assessment typically involves comparison of observed condition with pre-defined condition states. Current condition states require little quantitative data and must apply across many different material types and bridge elements. Use of these types of subjective techniques may lead to uncertain assessment of structure condition. This is particularly true when comparing different structures or structures assessed by different personnel. One improvement that may be considered to reduce the uncertainty or subjectivity of the current process is the introduction of quantitative measures within the condition states. The study presented herein discusses condition states developed for assessment of concrete cast-in-place bridge decks. The proposed condition states include basic quantitative information and address specific forms of deterioration consistently identified during inspection.

1 INTRODUCTION

1.1 *Motivation*

Current bridge inspection methods rely heavily on subjective assessment based on visual inspection and comparison with pre-defined condition states (Chajes et al. 2000). These condition states, or definitions of bridge condition, are generally quite broad and do not provide a definitive identification of the current condition of the bridge and the type of deterioration that is present. Also, these condition states do not allow for utilization of the different types of quantitative information that may be obtained during inspection of a bridge structure. Therefore, condition states are proposed to address a portion of the inadequacies of the current system. Basic quantitative information is integrated into the proposed condition states to provide more precise, objective, and comparable assessment of bridge deck condition.

1.2 *Need for improvement*

In regard to current condition states, at least one research study has identified the need for condition states to be quantifiable to provide a more accurate assessment of the structure at hand (Phares et al. 2001). The need exists for inspectors to have the capability of identifying actual amounts of damage or change in damage from prior inspections or initial construction conditions. Also, current condition states do not provide opportunity for integration of quantitative data such as that obtained from testing and or monitoring of a structure. Improper identification of the condition rating of the bridge, or components thereof, has been associated with compromising public safety, inefficient allocation of public funds, and major difficulties with heavy truck traffic (Chajes et al. 2000).

Several studies have noted that the current system of visual inspection relies on the inspector's subjective assessment of bridge condition at the time of inspection. Additionally, the reliability of inspectors choosing the correct condition state has been investigated through actual inspection and condition assessment of structures with known deterioration. This study revealed that routine inspections and condition assessments are completed with significant variability and that typically an average of four different condition ratings were given for the same component. This study also found that inspectors participating in the study successfully identified large widespread deficiencies such as corrosion or section loss on steel girders but rarely identified deficiencies that would typically call for more in-depth inspections such as fatigue cracks in steel girders. Inspectors also found difficulty locating and estimating areas of concrete bridge decks experiencing delamination (Graybeal et al. 2001, Phares et al. 2000 & 2001).

These studies have shown that the reliability of the condition ratings assigned during routine visual inspections, as well as the results from in-depth inspections, are not providing accurate assessments of current bridge condition and or deterioration and that the current condition state definitions do not provide adequate opportunity for inspectors to properly classify each of the bridge components. Proper identification and assessment of bridge deterioration is a major key to assuring public safety.

Non-destructive evaluation techniques (other than visual inspection) are being increasingly utilized in bridge inspection (Rolander et al. 2001). Integration of these techniques supports more accurate identification and assessment of bridge deterioration. However, to improve the inspection process, the understanding of actual bridge condition, and the link between inspection results and planning or modeling, the condition states must be organized in a manner that accepts quantitative data.

An earlier study has investigated the use of condition states that integrated different or additional inspection and testing procedures as the elements transition from one condition state to the next (Hearn & Shim 1998). This study was primarily focused on the integration of non-destructive testing methods into bridge inspection, condition states, and bridge management systems. This was accomplished through the development of augmented condition states.

An additional study was interested in the inspection of highway bridges using segmental inspection, a technique that breaks each element into several segments rather than evaluating the element as a whole (Hearn 1999). This study suggested that more information may be obtained about the deterioration patterns of a given bridge through this type of inspection, as well as relative and causative deterioration among groups of elements. This type of inspection may also provide more repeatable results, calculation of quantities, more accurate location of deterioration for future inspection and repair considerations, aid in selection of repair options, and the ability to track the effect of repairs through the remaining life of the bridge. In addition to these advantages, this type of inspection may allow better communication between administrative and field personnel responsible for inventory, assessment and repair of bridges.

1.3 *Current condition states*

The most recent edition of the "*Recording and Coding Guide*", which provides guidance on assessment of bridge structures, identifies two different sets of condition states for the individual characteristics of a typical bridge structure, a set of ratings for use in the appraisal of the entire bridge, and an additional set of ratings concerned with the vulnerability of the bridge due to scour (FHWA 1995). These condition states provide the framework for the information maintained in the National Bridge Inventory (NBI). Generally, as is the case in Tennessee, bridge inspectors utilize these condition state definitions during the inspection and appraisal process as do many research studies concerned with bridge inspection, deterioration or modeling (Mauch & Madanat 2001, Dunker & Rabbat 1995, Chase & Gaspar 2000).

The most frequently utilized set of condition states is shown in Table 1. These condition states are utilized when assessing the deck, superstructure, and substructure of a typical bridge. As shown, these condition states identify, in general terms, the amount of damage present thus providing an opportunity for the inspector to match the actual condition of the bridge to the condition state that is most similar.

These condition states provide the inspector the opportunity to classify each bridge component or characteristic based upon a short, non-quantitative definition. As previously discussed, these definitions are quite subjective and do not provide many distinct transition points between the different ratings.

Table 1. Current condition states for deck assessment.

NBI code	NBI condition	Description
N	Not applicable	–
9	Excellent	–
8	Very good	no problems noted.
7	Good	some minor problems.
6	Satisfactory	structural elements show some minor deterioration.
5	Fair	all primary structural elements are sound but may have minor section loss, cracking, spalling or scour.
4	Poor	advanced section loss, deterioration, spalling or scour.
3	Serious	loss of section, deterioration, spalling or scour have seriously affected primary structural components. Local failures are possible. Fatigue cracks in steel or shear cracks in concrete may be present.
2	Critical	advanced deterioration of primary structural elements. Fatigue cracks in steel or shear cracks in concrete may be present or scour may have removed substructure support. Unless closely monitored it may be necessary to close the bridge until corrective action is taken.
1	Imminent failure	major deterioration or section loss present in critical structural components or obvious vertical or horizontal movement affecting structure statbility. Bridge is closed to traffic but corrective action may put back in light service.
0	Failed	out of service – beyond corrective action.

Also, when considering a component or characteristic of interest, the inspector must generalize the rating for the entire component. In example, a few specific portions of a bridge deck may be in very poor condition, while the rest of the deck remains in satisfactory condition. In this instance, the inspector must combine, or average, these characteristics to obtain a single condition rating. Inspection results obtained utilizing only these condition states provide little information that can be utilized in calculation of load capacities or accurate identification of actual repair requirements.

2 PROPOSED CONDITION STATES

2.1 *Augmentation*

The proposed condition states were developed through augmentation of the existing condition states in Table 1 and are intended for use during visual bridge inspections. Augmentation accomplished two main goals including the integration of basic quantitative data into the rating procedure and the provision of definitive transition points between adjacent condition states. These condition states are typically compatible with segmental inspection and allow more objective comparison of inspection results completed by different personnel or from different structures. In conjunction with notes, sketches, and measurements taken during field inspection, the proposed condition states will help provide a more clear understanding of the actual condition of the structure and improved information to support repair planning and load capacity analysis.

The proposed condition states represent the typical types of deterioration found during routine visual inspection of highway bridges throughout the Tennessee bridge inventory. Similar condition states may be developed for bridge deterioration found in other inventories or other elements of interest.

Augmentation was accomplished through study of inspection reports generated utilizing visual inspection and consultation with Tennessee Department of Transportation (TDOT) personnel responsible for bridge inspection and repair planning. Transition between adjacent condition states was defined based upon historical application throughout the Tennessee inventory and the amount of deterioration thought to be representative of the current condition state definition. Therefore, these quantitative transition points do not represent an exact relationship between a type of deterioration and a quantified reduction in the structural capacity of the component in question.

Many of the proposed condition states rely upon measurements taken or estimated in the field during inspection. The transition points utilized by many of the condition states require quantitative measurement such as percent of deck area affected. Although these transitions are exact in nature in the proposed condition states, the actual measurement must typically be estimated by the inspection team. However, careful estimation and partitioning of the component of interest will provide results deemed accurate enough to be utilized with the proposed condition states.

In contrast to existing condition states, the augmented condition states include separate condition states for each different type of deterioration typically found during visual inspection of decks representing significant portions of the bridge inventory in Tennessee. During typical inspection, each deterioration type will be assessed using the applicable proposed condition state, with the final assessment equal to the minimum of all assessments completed.

2.2 *Tennessee bridge decks*

The distribution of deck types throughout the Tennessee bridge inventory was investigated. Approximately sixty percent of the inventory utilizes concrete cast–in–place decks. Due to the frequency of use of this particular bridge deck type, proposed condition states were developed only for concrete cast-in-place decks. Six major types of deterioration were identified as the typical reasons resulting in degradation of the deck and subsequent reduction in assessment. These included partial depth deterioration, full depth deterioration, scaling, structural cracks, non-structural cracks, and chloride contamination. New condition states were created for each identified type of deterioration. Typical assessment is based upon the percentage of deck area (entire deck or segment) deteriorated with assessment of chloride contamination based upon the maximum contamination identified throughout the entire deck or particular segment.

2.3 *Partial depth deterioration*

Partial depth deterioration reaches a maximum depth equal to either layer of reinforcing steel when compared to the respective nearest face of the deck. This type of deterioration represents a structural concern due to loss of section and opportunity for additional deterioration to occur if not repaired. Delamination, spalling, and exposed reinforcing steel are indicative of partial depth deterioration. The mechanisms at work causing partial depth deterioration may include corrosion of reinforcing steel, overstress from traffic loading, and environmental loading such as frost action. Poor quality control during initial construction may also play a role in this form of deterioration. The proposed condition states for partial depth deterioration are shown in Table 2. Partial depth deterioration is not allowed in condition states 9 thru 7 and alone cannot reduce the assessment to ratings below 3. Condition states 6 thru 3 represent different levels of deterioration ranging from less than five percent to greater than fifty percent of deck area, respectively.

2.4 *Full depth deterioration*

Full depth deterioration is defined as deterioration of the deck that penetrates to a level extending beyond either layer of reinforcing steel resulting in deterioration of a majority of or the entire depth of the deck. The deterioration mechanisms for full depth deterioration are similar to that of partial depth deterioration only differing in the extent to which the deck is damaged. The proposed condition states are shown in Table 3.

Due to the loss of capacity associated with full depth deterioration, condition states 9 thru 6 do not allow full depth deterioration. Condition states 5 thru 2 represent full depth deterioration from

Table 2. Proposed deck condition states for partial depth deterioration.

NBI code	NBI description	%Deck/segment
N	Not applicable	Not applicable
9	Excellent	None allowed
8	Very good	None allowed
7	Good	None allowed
6	Satisfactory	<5%
5	Fair	5% to 20%
4	Poor	20% to 50%
3	Serious	>50%
2	Critical	Not applicable
1	Imminent failure	Not applicable
0	Failed	Failed

Table 3. Proposed deck condition states for full depth deterioration.

NBI code	NBI description	%Deck/segment
N	Not applicable	Not applicable
9	Excellent	None allowed
8	Very good	None allowed
7	Good	None allowed
6	Satisfactory	None allowed
5	Fair	<15%
4	Poor	15% to 50%
3	Serious	50% to 75%
2	Critical	>75%
1	Imminent failure	>75% in critical area
0	Failed	Failed

less than fifteen percent to greater than seventy five percent of deck area. Decks with greater than seventy five percent deterioration in critical areas such as maximum positive moment regions are assessed with condition state 1.

2.5 *Scaling*

Scaling generally deteriorates the top of a concrete bridge deck. This particular deterioration mechanism by itself is typically not a structural concern however it may provide an additional opportunity for other forms of deterioration to initiate or accelerate. Scaling may also result in decreased functionality of the deck through reduced ride quality. The proposed condition states developed for scaling of concrete bridge decks are provided in Table 4. Condition states 9 and 8 do not allow scaling, and due to the non-structural nature of this type of deterioration, scaling alone cannot reduce the assessment of the deck below a rating of 4. Condition states 7 thru 4 represent different levels of deterioration ranging from less than two percent to more than fifty percent of deck area affected, respectively, with states 6 and 5 representing intermediate levels of deterioration.

2.6 *Structural cracks*

Tension and shear cracks in concrete bridge decks are typically considered structural cracks. Shear cracks are typically found near points of support and typically run diagonally across the section affected, whereas tensile cracks are typically found in areas of maximum flexure. These cracks are

Table 4. Proposed deck condition states for scaling.

NBI code	NBI description	%Deck/segment
N	Not applicable	Not applicable
9	Excellent	None allowed
8	Very good	None allowed
7	Good	<2%
6	Satisfactory	2% to 25%
5	Fair	25% to 50%
4	Poor	>50%
3	Serious	Not applicable
2	Critical	Not applicable
1	Imminent failure	Not applicable
0	Failed	Failed

Table 5. Proposed deck condition states for structural cracks.

NBI code	NBI description	%Deck/segment
N	Not applicable	Not applicable
9	Excellent	None allowed
8	Very good	None allowed
7	Good	None allowed
6	Satisfactory	None allowed
5	Fair	<5%
4	Poor	5% to 50%
3	Serious	>50%
2	Critical	Not applicable
1	Imminent failure	Not applicable
0	Failed	Failed

typically caused by dead and live loads, and in extreme cases, are a result of restricted thermal movement.

For this study, cracks with a width of one sixteenth of an inch or greater are considered structural in nature. Cracks may be evident due to corrosion stains and efflorescence. When inspecting for cracks, either structural or non-structural, use of segmental inspection may be advantageous due to the possible difficulty in identifying percentages of deck deteriorated due to cracking. The deck may be broken into segments and each segment rated based purely on the existence of cracks. Results from all of the segments can be combined to gain an overall picture of the entire deck. Selection of the appropriate segment size and layout are important as is the use of the same combination during future inspections to facilitate an improved understanding of the change in deterioration from one inspection or repair to the next. The proposed condition states for structural cracks are shown in Table 5. Due to the serious nature of this type of deterioration, structural cracks cannot be present in condition states above 5. Ratings of 5, 4, and 3 indicate less than five percent, five to fifty percent, and more than fifty percent of the deck deteriorated by structural cracks. Condition states below 3 are not utilized.

2.7 *Non-structural cracks*

Non-structural cracks are typically initiated by stresses due to temperature and shrinkage. These cracks are fairly common in reinforced concrete bridge decks, and alone do not represent a great risk to the structure. However, these cracks may allow the intrusion of deleterious elements, such

Table 6. Proposed deck condition states for non-structural cracks.

NBI code	NBI description	%Deck/segment
N	Not applicable	Not applicable
9	Excellent	None allowed
8	Very good	None allowed
7	Good	<10%
6	Satisfactory	10% to 50%
5	Fair	>50%
4	Poor	Not applicable
3	Serious	Not applicable
2	Critical	Not applicable
1	Imminent failure	Not applicable
0	Failed	Failed

Table 7. Proposed deck condition states for chloride contamination.

NBI code	NBI description	Measured level
N	Not applicable	Not applicable
9	Excellent	<1.2 kg/cubic meter
8	Very good	<1.2 kg/cubic meter
7	Good	<1.2 kg/cubic meter
6	Satisfactory	<1.2 kg/cubic meter
5	Fair	>1.2 kg/cubic meter
4	Poor	>1.2 kg/cubic meter
3	Serious	>1.2 kg/cubic meter
2	Critical	>1.2 kg/cubic meter
1	Imminent failure	>1.2 kg/cubic meter
0	Failed	Failed

as water and chlorides, which may initiate or accelerate deterioration. Non-structural cracks are typically identified through visual inspection and may be present on the top or bottom of the deck. Map cracking is one example of non-structural cracking. The proposed condition states for non-structural cracks are provided in Table 6. Condition states 8 and above do not allow non-structural cracks and due to the non-structural nature of these cracks, states 4 and below are not utilized. Decks are assessed condition states 7, 6, and 5 when non-structural cracks affect less than ten percent, ten to fifty percent, and greater than fifty percent of deck area, respectively.

2.8 *Chloride contamination*

Chloride contamination has been identified as a major factor in concrete bridge deck deterioration. The presence of chlorides in the bridge deck does not necessarily represent deterioration or damage, but does indicate that favorable conditions exist for deterioration to begin. Regardless of the source, a threshold contamination level of between one and two kilograms of chloride per cubic meter of concrete has been linked to the possible initiation of corrosion. Typical bridge inspection in Tennessee requires a bridge deck survey that includes the identification of the chloride concentration present at a depth equal to the top layer of reinforcing steel. The condition states proposed for this predictor of deterioration are shown in Table 7, which identifies a transition from a rating of 6 to 5 as the chloride level increases above the assumed threshold. This rating will generally support (will not increase or decrease) other ratings identified during an inspection unless the chloride level is found to be above the threshold level with all other ratings remaining at 6 or above. In this instance the chloride level would reduce the overall rating to a 5. Otherwise, the final assessment of the structure will be the minimum of the other five condition states.

3 CONCLUSIONS

Condition states were proposed in this study for each major type of deterioration found to reduce condition ratings for concrete cast-in-place bridge decks in the Tennessee inventory. Each condition state is to be utilized during a deck inspection, with the overall assessment equal to the lowest single assessment. The proposed condition states allow the integration of quantitative data into the inspection process and the use of segmental inspection. The proposed condition states were developed to improve the accuracy of bridge deck assessment, repair planning, monitoring of specific conditions or repairs, and communication of actual field conditions between field and administrative personnel. Although improvement in assessment and communication of actual condition may result through the use of condition states similar to that proposed, the transition from assessment based on visible damage to actual reduction capacity of a structure may ultimately provide the most useful information and allow the most efficient allocation of repair and replacement funds.

REFERENCES

Chajes, M. J., Shenton, H. W., and O'Shea, D. (2000). Bridge-condition assessment and load rating using nondestructive evaluation methods. *Transportation Research Record*, 1696, 83–91.

Chase, S. B., and Gaspar, L. (2000). Modeling the reduction in load capacity of highway bridges with age. *Journal of Bridge Engineering*, 5(4), 331–336.

Dunker, K. F., and Rabbat, B. G. (1995). Assessing Infrastructure deficiencies: the case of highway bridges. *Journal of Infrastructure Systems*, 1(2), 100–119.

Federal Highway Administration (FHWA) (1995). Recording and coding guide for the structural inventory and appraisal of the Nation's bridges. FHWA-PD-96–001.

Graybeal, B. A., Rolander, D. D., Phares, B. M., Moore, M. E., and Washer, G. A. (2001). Reliability and accuracy of in-depth inspection of highway bridges. *Transportation Research Record*, 1749, 93–99.

Hearn, G. and Shim, H.-S. (1998). Integration of bridge management systems and nondestructive evaluations. *Journal of Infrastructure Systems*, 4(2), 49–55.

Hearn, G. (1999). Segmental inspection for improved condition reporting in BMS. Proceedings of the 8th International Bridge Management Conference, Denver, Colorado, B-3/1–B-3/8.

Mauch, M., and Madanat, S. (2001). Semiparametric hazard rate models of reinforced concrete bridge deck deterioration. *Journal of Infrastructure Systems*, 7(2), 49–57.

Phares, B. M., Rolander, D. D., Graybeal, B. A., and Washer, G. A. (2000). Studying the reliability of bridge inspection. *Public Roads*, November/December 2000, 15–19.

Phares, B. M., Graybeal, B. A., Rolander, D. D., Moore, M. E., and Washer, G. A. (2001). Reliability and accuracy of routine inspection of highway bridges. *Transportation Research Record*, 1749, 82–92.

Rolander, D. D., Phares, B. M., Graybeal, B. A., Moore, M. E., and Washer, G. A. (2001). Highway bridge inspection/state-of-the-practice survey. *Transportation Research Record*, 1749, 73–81.

Chapter 22

The 2006 rope access inspection of the Brooklyn Bridge towers: A new view of an old bridge

J.C. Schmidt
DMJM Harris, Philadelphia, Pennsylvania, USA

ABSTRACT: The Brooklyn Bridge stands as a monument to bridge engineering, and while easily accessible to the public, access for structural inspection is difficult. As part of the 2006 biennial inspection of the Bridge, a detailed masonry inspection of the Manhattan and Brooklyn towers was conducted using rope access techniques to examine areas previously investigated only through remote visual methods. This paper discusses the access methods employed for a detailed inspection of the bridge tower masonry. Challenges included performing this work without adding anchors to the towers, registration of inspection findings on a massive masonry structure in a repeatable format, and providing tactile inspection access in stone overhangs, beneath steel walkways and within recesses. An overview is included of the relevant safety and fall protection specifications and how they were employed in conjunction with DMJM Harris' own safety requirements. Inspection methods are presented and findings are summarized.

1 INTRODUCTION

Mention the name of the Brooklyn Bridge to people around the country, or even the globe, and you begin to understand the impact this engineering monument has had on the world. It has faithfully served New York City since construction was completed in 1883, and is crossed daily by thousands of motorists and pedestrians, and yet the structural inspection of such iconic towers has proved problematic for the engineering community. The "Great Bridge" towers stretch 276′ above the mean high water of the East River and are constructed of solid un-reinforced masonry. The historic nature of the bridge, the limited access below the deck, the turbulent East River, and the use of back-beveled stone overhangs at the top of the piers have prevented direct access to large sections of the towers. Previous inspections relied on visual inspection techniques and/or rope-access descents off the edge of the bridge cables. These techniques proved adequate for visual inspection, but have not provided the level of detail necessary to fully evaluate each bridge tower. The 2006 inspection by B&H Engineering, P.C. in conjunction with DMJM Harris chose a new approach to look at this structure through the use of industrial rope access techniques. The inspection was directed by the New York State Department of Transportation in conjunction with the New York City Department of Transportation and the East River Bridge Maintenance Group who provided invaluable support throughout the project and daily access to the bridge itself. B&H Engineering provided support services and worked directly with the client to provide shielding of traffic, access to the structure, and oversight regarding the inspection methods and recording of deficiencies. As a sub-consultant to B&H Engineering, DMJM Harris has been involved in the practice of structural climbing inspections since 1999 and provided the experience, climbers and equipment necessary for successful completion. The inspection team consisted of 4 highly trained engineers, each with experience in working at heights, rope-access, bridge climbing, bridge inspection, structural bridge design, and stone masonry construction. Three members of the DMJM Harris team are registered professional engineers, and all members participated in rope-access training as dictated by company safety policies. Due to the size of the project, safety risks, and complexity of access, Ropeworks, an industrial

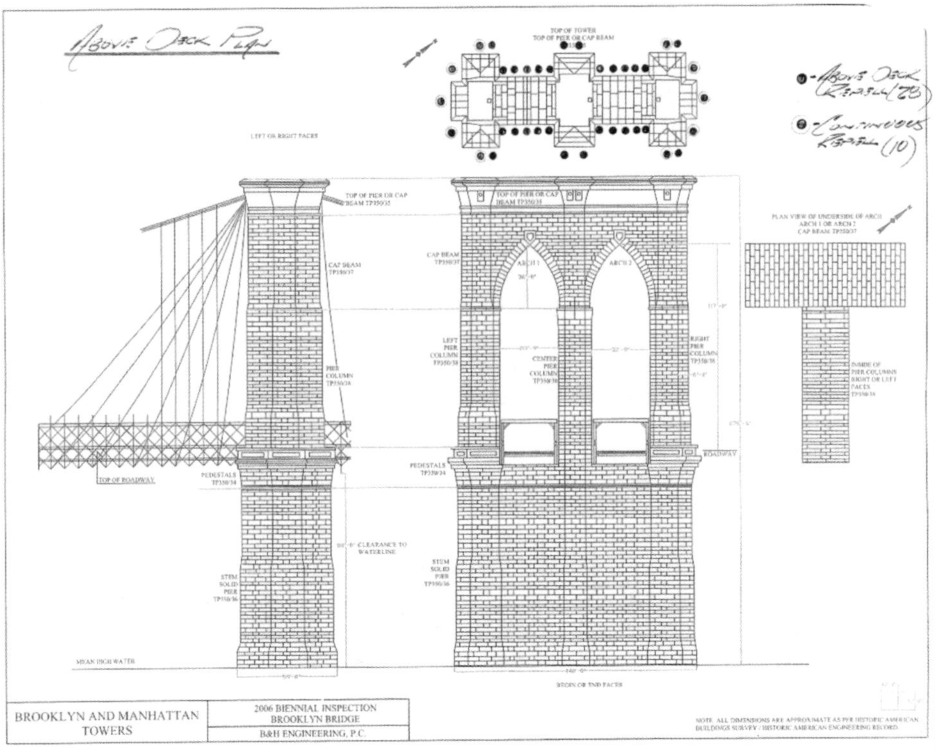

Figure 1. General Brooklyn Bridge Tower configuration and inspection plan.

access specialty company who had provided training and equipment on several other bridge inspections for DMJM Harris, was engaged to provide rigging services, specialized equipment, safety and rescue services, and access expertise. During the inspection, Ropeworks served to set additional lines and monitor the climbers, allowing the engineers to focus on the inspections at hand. This arrangement was essential to completing the job safely within the scheduled 15 day window. The resulting inspection provided access to areas of the bridge never before inspected by hands-on-methods and shed new light on the condition of the towers.

The tower inspection occurred from October 16 to November 2, 2006 and included the detailed inspection and "sounding" of the stone and mortar of both towers. Loose stone fragments and loose or deteriorated mortar encountered during the inspection were removed by the inspectors. Photographs were taken documenting the typical condition of the structures and localized conditions and defects. The intrados of the arches, cables and eyebars were inspected during the course of the 2006 biennial bridge inspection, but were not included in this project's DMJM Harris scope of services.

All access to the towers was through vertical rappels with descent spacing selected to provide access for a detailed visual inspection. The average visual inspection distance was 6′ or less while maintaining a maximum site distance of 10′. Localized suspect areas were sounded and the rappel spacing adjusted as needed to focus on areas of concern. Figure 1 shows a general view of the towers and approximate locations for rappels from the top of the tower.

A site visit prior to the beginning of the inspection revealed the top of the cap beam of both towers to be the area of greatest deterioration, and therefore was the focus of the inspection. Since the arches span the roadway and pedestrian Promenade, an increased number of rappels were used over the face of each arch to provide a more detailed inspection and remove any loose stone fragments and mortar. For rappels not interrupted by roadway or Promenade, a long rappel was implemented that covered the entire height of the pier. As the inspector progressed down the tower the lateral movement allowed by the rope increased, and was utilized to reduce the number of rappels needed below deck

2 OSHA AND CONSULTANT SAFETY PROCEDURES

Safety is a prime and critical element of any bridge inspection, whether the inspector remains on the ground or elevated involving the use of inspection vehicles or equipment. While the goal of the inspection is to investigate and document the physical condition of the structure, the number one priority is always the safety of the inspection team and public. Rope access inspection is no different with regards to the importance of safety practices, but does pose a variety of unique challenges. Standard safety issues, such as the use of personal protective equipment, were integral to the inspection, but will not be elaborated upon in this paper which focuses on rope access and associated procedures. The U.S. Department of Labor Occupational Safety & Health Administration (OSHA) has developed Regulation 29CFR1926 Subpart M-Fall Protection governing tasks while working at heights. The regulation makes no distinction between inspection and other types of work at elevations. The sections most relevant to this project are summarized as follows:

- OSHA 1926.500 describes the scope of the standard and in what conditions it is to be applied. A glossary of terms is included that define words and concepts specific to fall protection and as used in the standard's sections that follow.
- OSHA 1926.501 provides the requirements of an employer to provide a safe working environment for employees via fall protection systems. The most basic requirement as found in 1926.501(b)(1) describes the duty to have fall protection. It states "Each employee on a walking/working surface (horizontal and vertical surface) with an unprotected side or edge which is 6 feet or more above a lower level shall be protected from falling by the use of guardrail systems, safety net systems, or personal fall arrest systems."
- OSHA 1926.502 discusses requirements for fall protection systems and how they are utilized. The systems discussed include guardrails, net systems, and personal fall arrest systems. These elements were pertinent to the tower inspection of the Brooklyn Bridge and are discussed later herein as well as anchorages, safety monitoring and protection from falling objects.
- OSHA 1926.503 focuses on the needs for an employer to provide adequate training via a training program for all employees exposed to falling hazards. This includes certification and retraining of employees to maintain a safe work environment and monitor operations.

The DMJM Harris Safety and Health Program and Procedures Manual is a framework for employees to meet or exceed the Company's safety standards in planning and executing all work tasks in a safe manner. Included are sections on fall protection in general and specific sections on structural inspection by climbing and rope access techniques. The document addresses the responsibilities, development of site-specific safety plans, training requirements and climbing procedures. Specifics will be discussed further relating to actual conditions experienced during the inspection.

Prior to beginning the inspection, a site-specific Climbing Inspection Safety Plan was developed by DMJM Harris staff and coordinated with the Ropework's Job Hazard Analysis to ensure all risks were assessed and safety requirements met. This document identifies the key personnel on the job and their roles in maintaining the safe working environment. By assessing safety risks in advance, surprises in the field were limited. The safety plan also dictates the requirement for climbing practice, equipment and techniques, and provides the following daily safety routine during the inspection.

Each day began with the team of DMJM Harris and Ropeworks meeting in the B&H Engineering field office near the bridge for a review of the client and project procedures and plans. These meetings focused on inspection access, anticipated work for the day, weather conditions, and emergency contact numbers and planned rescue response. In several instances, high winds were forecasted and directly influenced the type of work for that day. Alternate locations and tasks were then opted for. In these cases, long rappels were avoided, or conducted to keep the wind at the climbers back. An assessment was made prior to each rappel to verify that conditions were still safe. At the daily meetings, issues, concerns, safety, access, equipment, or inspection findings from the previous day could also be discussed and addressed as necessary.

Following the daily office meeting, the team would report to the bridge and proceed to equip themselves with a harness and related gear. At this time, all personal protective gear was inspected to ensure that it was functioning properly and in good condition before donning it and beginning the ascent up the tower cables. Two-way radios were used throughout the inspection to communicate findings of the inspection to a team member below for documentation and to assure ongoing contact between the rigging crew and climbers. Another equipment check by another team member was conducted prior to each rappel. Ropes, anchorages and other rigging equipment were also checked on a daily and ongoing basis during the inspection.

Upon reaching the top of the tower, several safety considerations had to be resolved to work within safety parameters. Since no catwalk or railings are present at the edge of the towers, a chalk line was sketched 6′ beyond to point where the masonry starts to slope towards the edges. This established line was a warning line demarcation. Climbers were instructed that at no time and for no reason was an inspector permitted to step beyond the line without being secured. Following a thorough inspection to confirm condition, the anchored railing in the center of the bridge tower was used as the attachment point for a restraint system while ropes were being anchored. Once secured, the ropes were dropped in preparation for the actual inspection descents. While the railing in general appeared as an adequate anchor location, it was deemed generally unacceptable by the inspection team for use as a rope anchor point. OSHA 1926.502(d)(23) states "Personal fall arrest systems shall not be attached to guardrail systems. . ." and OSHA 1926.502(e)(2) "Positioning devices shall be secured to an anchorage capable of supporting at least twice the potential impact load of an employee's fall or 3,000 pounds, whichever is greater.

Several embedded anchors in the masonry were present on each tower, but no record was available concerning the anchor material, age or embedment depth. Since the Brooklyn Bridge is a historic monument, it was agreed that no new anchors would be added to the bridge. This condition then necessitated the inspection team to work with the existing available anchor locations. Each anchor was therefore inspected by two professionally licensed engineers on the team. The inspection involved hammer sounding of the anchor for a clear ringing sound, and of the surrounding stone and mortar to ensure a solid anchorage. The eye of the anchor was also twisted and pried and visually investigated for any signs of cracking, or distress. If an anchor caused any concern, it was marked as unusable and eliminated from service. Calculations based on conservative estimates of material strength and anchor depth verified the adequacy per the above requirements. The final element of anchorage safety was "redundancy", which will be discussed in the section on access methods.

One safety consideration particularly challenging for the Brooklyn Bridge tower inspection, was the proximity and constant presence of the public for virtually all phases of the inspection. There were no vehicular traffic restrictions and only limited disruptions to pedestrian traffic during the inspection. Great care was taken to provide a safe environment for those using the bridge. Rovi Construction Corp. was brought onto the team by B&H Engineering to provide both pedestrian traffic control on the Promenade and shielding of the roadways.

The Promenade is a timber deck consisting of one bike lane and one pedestrian lane that runs the length of the bridge and splits around the central column of the tower and under each arch providing pedestrians, runners and bikers a scenic view as they cross between Brooklyn and Manhattan. To protect pedestrians, one side of the split was closed to visitors while inspectors sounded stone and removed loose debris over the walkway. To accomplish this, traffic cones and safety tape were placed well in advance of the tower, directing the two lanes down into one. A flag-person was stationed at either end of the closure with a stop sign, to control the flow of bikers and walkers and to ensure the area beneath the inspector remained clear.

Vehicular traffic lanes lie on either side and slightly beneath the Promenade and extend under both arches. To shield this traffic, netting was used in conjunction with corrugated steel decking supported by the stiffening truss of the bridge. The netting stretched from the outside edge of the

Promenade to the inside edge of the arch, covering the entire walkway. The decking also allowed for the climbers to land safely in this zone for rappels along the face of the arch. The original configuration employed netting beneath the arch and for one bay of the truss on either side. During the inspection, it was decided that this might not be sufficient, and an additional bay was added to either side. Supplemental netting was also provided by Rovi Construction when it was identified that some of the loose mortar and stone could be larger than anticipated. While climbers made every attempt to prevent debris from falling, by collecting loose mortar and stone chips in pockets or small bags clipped to the climber, these shielding measures ensured complete safety to the public and kept open access to this historic bridge.

A further issue of public safety was the need to notify police and emergency response personnel of the inspection. In today's environment of heightened security, a climber on the Brooklyn Bridge may be viewed by the public as either a threat or a fallen worker in need of rescue. To allay these concerns, a press release was issued describing the type of inspection, and local authorities notified each day the team was on the bridge.

At the end of each day, all ropes and rigging equipment would be collected and secured within one of the saddle chambers on top the tower and protected from weather. This prevented the team from the arduous task and additional materials movement risks of getting hundreds of pounds of ropes and gear up and down each day, and ensured the tools would stay safe and dry. A debriefing at the conclusion of each day's work provided the opportunity to discuss any issues that arose, options and resolutions, and verify that all safety requirements were being met.

3 ACCESS CHALLENGES AND SOLUTIONS

The Brooklyn Bridge Towers are imposing structures of solid masonry supporting the superstructure of cable and structural steel. The techniques required to access them can be simplified into four major elements. These elements are anchorage, placement of lines, "getting on rope" and the full decent/return cycle. The following paragraphs will discuss how each was addressed for each type of rappel.

Access to the towers was provided via the 15-$^{3}/_{4}$ inch diameter main suspension cables. Two of these cables can be accessed from the Promenade and thus each cable has a locked gate about 20′ up that prevents access by the public. Two smaller cables serve as hand-rails and support points for attachment by a full body harness and double-locking device worn by each climber. At the top of the cable, a caged ladder, approximately 10′ high, provides access to the top of the towers.

In all rope-access inspections, two separate lines are used. One operates as the working line, while the other exists solely as a redundant "backup" and rescue line. Due to the unknowns associated with the anchors structural integrity, it was decided that a fully redundant anchorage would be used for each of the individual lines at all times. This required each line to have two separate anchors, resulting in a total of at least 4 anchors used for each climbing operation (as shown in Figure 2). Due to the small number of anchors, the arrangements of the ropes became quite complex, but ensured a safe support system for all climbers.

Below deck, transverse walkways adjacent to the masonry and hatches in the Promenade provided access to the tower faces below the arches. Unlike the top of the towers, the large amount of steel truss work below the deck provided many possible anchorage points. In these locations, only one anchorage was required for each rope of the two rope system.

With the question of anchorage addressed, the subsequent issue becomes where to place lines to provide adequate coverage while minimizing the number of required rappels. The limited number of anchors and their placement was not always convenient to the location of the descent. Thus, the tubular railing components were used as a diversion point to redirect the rope from an anchor to the location of descent. Often the ropes would have to reach to anchors on the opposite side of the tower to ensure that each anchor was only used once. In addition, protective nylon sleeves were used at all abrasion points to prevent damage to the lines, along with "good housekeeping" to avoid tripping and fouling while working atop the tower.

Figure 2. Anchor redundancy for each climbing line.

The uniformity and shear size of the tower masonry made it difficult to accurately locate where on the structure a defect was found, especially for an inspector suspended on the face of the tower. Thus, B&H Engineering developed a grid system and provided marked ropes to be dropped alongside the climber. These ropes were tagged at 10 foot intervals and used in conjunction with inspection forms to identify locations both vertically and horizontally (see Figure 3). The resulting inspection could pin-point findings and comments within a 10′ × 10′ square grid on either tower.

A total of 38 rappels were used from the top of each tower, using a combination of short (28) and long (10) rappels to provide the most efficient inspection (see Figure 1 for rappel layout). The short rappels ended at the Promenade, while the long rappels over the side of the tower extended the full height of the tower ending at the water line. These long rappels utilized the increased lateral movement of the inspector as he descended to reduce the number of descents necessary below deck level. This required the inspector to traverse more of the masonry and led to lengthier rappels, but proved to be more efficient than additional rappels due to the time it took to get back to the top and set for another descent.

Once the ropes were in place, the climber had to "get on rope" through a number of methods depending on the location. Perhaps the simplest involved accessing the rope from the cables themselves. This method has been applied by inspectors on the Brooklyn Bridge in the past, but was limited to areas adjacent to the main cables. The ropes are anchored as previously described, but the climber will attach himself to the ropes below the overhang at the level of the cable. A re-direct of the rope was used via a carabineer attached to cable-stays to pull the ropes over and hold the climber up against the stone (see Figure 4).

Descents from the top of the tower proved more difficult as an edge negotiation was required and compounded by the downward slope at the edge of the tower. A total of 26 rappels on each tower required this edge negotiation that provided the necessary access to the face of the arches and the

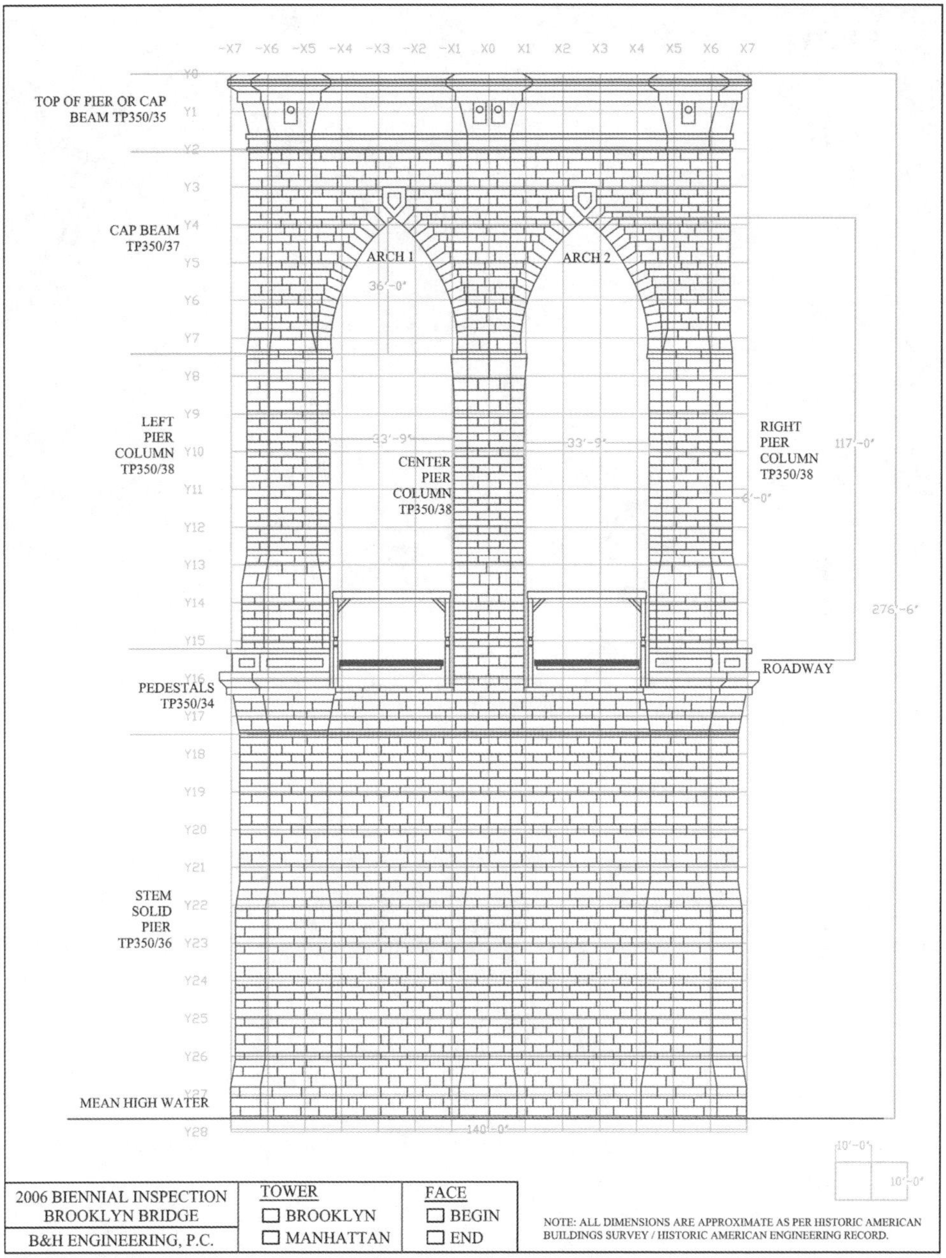

Figure 3. General Brooklyn Bridge Tower configuration and inspection plan.

outside faces of each tower. To complete the maneuver, a rope ladder was anchored to the top of the tower, allowing the climber a means to lift his weight from the ropes as necessary for equipment to clear the stone edge (see Figure 5). Once over the edge, the climber had direct access to the masonry. The remaining obstacle facing climbers descending from the top to the tower was the back-beveled masonry in the top of the pier. This area in conjunction with step-backs in the stone resulted in the climber hanging several feet off the face of the stone. Were the inspection to be a

Figure 4. Access from cable with re-direct of rope.

Figure 5. Access via edge-negotiation with rope ladder.

Figure 6. Access beneath deck through rope transfer.

visual only, this would have been acceptable, but a hands-on inspection required the climber to be in reach of the stone. The solution developed used a "trolley line" to pull the inspector closer to the stone. The trolley line is an additional rope pulled taut through the tower arch and then anchored at deck level. The inspector can either clip into the line with a carabineer to be pulled closer to the masonry, or use the trolley line as an anchor. When used as an anchor point, the climber could use a swinging motion to increase mobility and range during a traverse of the tower face.

To improve the climbers reach, a 2′ section of rebar was used for sounding the stone. A hole was drilled into one end of the rebar and fitted with a rope to allow the tool to be hung from the climber. This unique solution significantly increased the reach of each climber, allowing for greater access with less rappels.

The under deck rappels, while directly accessible by platforms, provided their own unique problems. The transverse walkways are set only a few inches off the face of the stone. This allows the ropes to be set behind the railing, but provides an opening that is far too narrow for a climber. To overcome this and still maintain access to the stone, two sets of ropes were used. One set was placed along the stone between the back of the railing and the face of the stone. The second set was hung on the open side of the walkway and used by the inspector to get below the walkway. The secured inspector would climb over the railing and lock his legs against the bottom kickplate (see Figure 6.) By beginning the descent in this position, the body would rotate about the feet until the climber was hanging upside down. By releasing one's legs, the inspector would rotate into an upright position with adequate clearance of his head and body. With movement, the inspector could maneuver beneath the walkway, grab the second set of ropes and transfer over. After completing the transfer, the inspector would be adjacent to the stone for full access.

The final challenge to be mitigated by the inspector having completed a rappel was how to finish the cycle and get back in position for the next rappel. For short rappels, the inspector would simply land on the Promenade or roadway shielding, disengage from the ropes and return to the main cable access location. The process was uncomplicated, but added a significant amount of time to the cycle. On the Manhattan tower, the long rappels and those below deck terminated at the waterline and required the use of a North East Marine boat. This vehicle served as a landing platform and shuttle to get the inspectors to a "sky-climber" (suspended scaffold) which served as an elevator back to the walkways beneath the deck. For the long rappels, the inspector would continue up through the access hatch in the Promenade and then back to the main cable entrance point. For rappels below deck, the climber could continue from the walkway with the next rappel. This system provided a safe and efficient method to quickly get climbers back into place for additional rappels. On the Brooklyn Tower a pier was constructed around the tower base, eliminating the need for a boat. Originally, the cycle to the top required a half mile walk back to the approach spans of the bridge at the origination point of the Promenade. This access was utilized several times, but was less preferable due to the time required to complete the cycle. The skyclimber was then implemented on the Brooklyn tower as well.

4 REPORTING FINDINGS & CONCLUSIONS

Having solved the issues regarding bridge access, the last problem to be conquered was how to record the results of the inspection. As previously mentioned, a grid system was used to located areas of interest or damage. Notes were then relayed by radio to an engineer, typically at deck level, who would record the note and the location on Bridge Inspection forms. In some cases as many as three independent rappels could occur simultaneously, requiring intense concentration and coordination between climbers and recorder. The inspector on the line could then focus on sounding the stone, taking photographs and navigating the ropes. Note pads were carried by each climber to allow for sketches when needed. Figure 7 shows an inspector sounding mortar beneath the cornice stone using the rebar tool. Note the adjacent flagged rope to define the climber's location vertically on the tower.

The resulting inspection revealed the masonry to be typically in good condition. As anticipated, the top 20 feet of the pier exhibited the greatest deterioration in the mortar and efflorescence in the stone. Particularly the first row of mortar between the top stones was found to be loose and deteriorated. In many locations vegetation was found in these joints, and the mortar could be readily removed as much as 3″ into the joint (see Figure 8). These results are consistent with leaking in the joints on the top of the tower due to a failure of the joint sealer. The cap beam of each arch was of particular interest since is covers the roadway and Promenade. The inspection found isolated locations of laminated and loose stone chips in this area that were removed during the inspection. This is typical of the structure, and not uncommon for granite masonry.

Once below the top 30 feet of masonry, the tower's general condition was good. The inspection team was able to isolate areas where the laminated stone fragments and/or mortar were loose or deteriorated and removed this debris. Efflorescence was seen in a number of joint locations, and several cracks were identified in the stone. Areas around the arches and cap stone tended to show increased efflorescence, open joints and cracking.

It was also noted that the mortar had been repaired in numerous locations. In some areas the repair mortar was tooled such that it stuck out from the stone. This convex exposed mortar was weathering, and tended to be loose and deteriorating much faster than original mortar, or mortar tooled in a concave fashion.

Several recommendations were made concerning needed repairs and maintenance as follows. Among our recommendations were that the joint sealer at the top of the tower had begun to degrade, and should be removed and replaced with new pourable sealer. The spalled base of the handrail on the Brooklyn Tower was to be cleared and the handrail reset. Joints with missing mortar should be re-pointed and areas of vegetation removed and cleaned. Several cracks were also identified that require sealing, especially in the cap under the roadway. None of these repairs were deemed critical and could be later addressed during routine maintenance.

Figure 7. Inspector sounding mortar on Brooklyn Tower.

Figure 8. Mortar joint deterioration on Manhattan Tower.

The successful completion of the rope access tower inspection of the Brooklyn Bridge has shown the feasibility of this inspection method to achieve a detailed level of inspection, without significantly impacting public access to the bridge. While climbing inspections may not always be warranted for every structure, they do provide a unique tool for bridge inspectors. The future of climbing inspection will depend on the needs of the client but should not be overlooked. With careful coordination, a diverse team of experts, innovative techniques and equipment and the appropriate focus on and detail to safety, no bridge is too difficult, too large, or too famous to inspect.

REFERENCE

The U.S. Department of Labor, Occupational Safety & Health Administration. Standard 1926: Safety and Health Regulations for Construction. Subpart M: Fall Protection.

7 *Bridge history & aesthetics*

Chapter 23

Walkway over The Hudson (historic bridge to Northeast recreational destination)

Peter Melewski
Bergmann Associates, Albany, NY, USA

Malcolm McLaren & James Green
McLaren Engineering Group, West Nyack, NY, USA

David Thurnherr
Bergmann Associates, Rochester, NY, USA

ABSTRACT: The Poughkeepsie Railroad Bridge, opened in 1888, holds tremendous historic significance. It was the longest bridge in the world when the first train crossed it. As the first bridge constructed across the Hudson River between New York City and Albany, the bridge had an enormous impact on transportation throughout the Northeast United States. After a long history of ownership and uses, the bridge suffered damage from a fire in 1974 that rendered it unusable for railroad traffic. A comprehensive study has begun to certify structural integrity and to produce a plan to establish it as a public park and walkway, as well as a bridge engineering educational resource. The paper will provide a brief historic overview, discuss the objectives of the comprehensive study and the findings of the late 2006 underwater inspections.

1 INTRODUCTION

The Poughkeepsie-Highland Railroad Bridge is a 19th Century engineering marvel on the National Historic Register. The cornerstone was laid in 1873, and when it was completed in 1888, five years after the Brooklyn Bridge, it was the longest bridge in the world (6,767 feet). It had the largest cantilever spans ever built, and the four river piers were supported on massive concrete filled timber crib foundations over ten stories tall. As the first bridge spanning the Hudson River between Albany and New York City, it had an enormous impact on the transportation of freight in the Northeast, and today's transportation network. In 1974, a fire brought an end to its useful life as a railroad bridge.

After years of discussion and studies by various agencies and private groups on how to make the best use of the structure, Walkway over the Hudson, a 501(c) (3) not-for-profit membership organization, has obtained full ownership of the bridge. In the fall of 2006, they hired the Bergmann Team (Bergmann Associates; McLaren Engineering Group; Ulazewicz, Melewski, and Greenwood; and Howard/Stein-Hudson) to certify the structural integrity of the bridge and to produce a comprehensive plan for its use as a public park and walkway.

In addition to providing breathtaking views of the Hudson River Valley from 212 feet in the air, the historic bridge will serve as an educational resource on bridge engineering and construction. The goal of the organization is to have the public enjoy the view in 2009, the 400th Anniversary of Henry Hudson's ship Halfmoon sailing up the Hudson, an American Heritage River. The authors wish to acknowledge Pulitzer Prize winning author Carleton Mabee and his book *Bridging the Hudson* from which some of the historical information in this paper was derived. It is a fascinating book, and a must read for anyone who has a love for bridges.

Figure 1. Historic etching of the bridge and river.

Figure 2. West shore view today of bridge.

2 BUILDING MOMENTUM FOR A CROSSING

Starting in the mid 19th Century, many plans and ideas were developed and discussed in the Hudson Valley on the merits of constructing a bridge over the Hudson River between Albany and New York City to facilitate the movement of raw materials from the west to the urban centers in the east. In order to cross the river, trains had to be disassembled and the locomotives and cars carried by ferry boats to the opposite shore where they were to be reassembled. The demand for items such as coal was driving the need for a more efficient method of crossing the Hudson River. A cornerstone was laid in 1873; however, as with many large bridges of that era, and even today construction stopped after several false starts. In 1875, a group of Boston investors visited several potential sites in the Hudson Valley and reaffirmed that Poughkeepsie-Highland location was preferred due to the presence of existing railroads on both sides of the river, the steep slopes which facilitated clearance for shipping. Upon their return, they created a map that demonstrated the viability of the location (Figure 4).

Figure 3. A current view of bridge surface looking east.

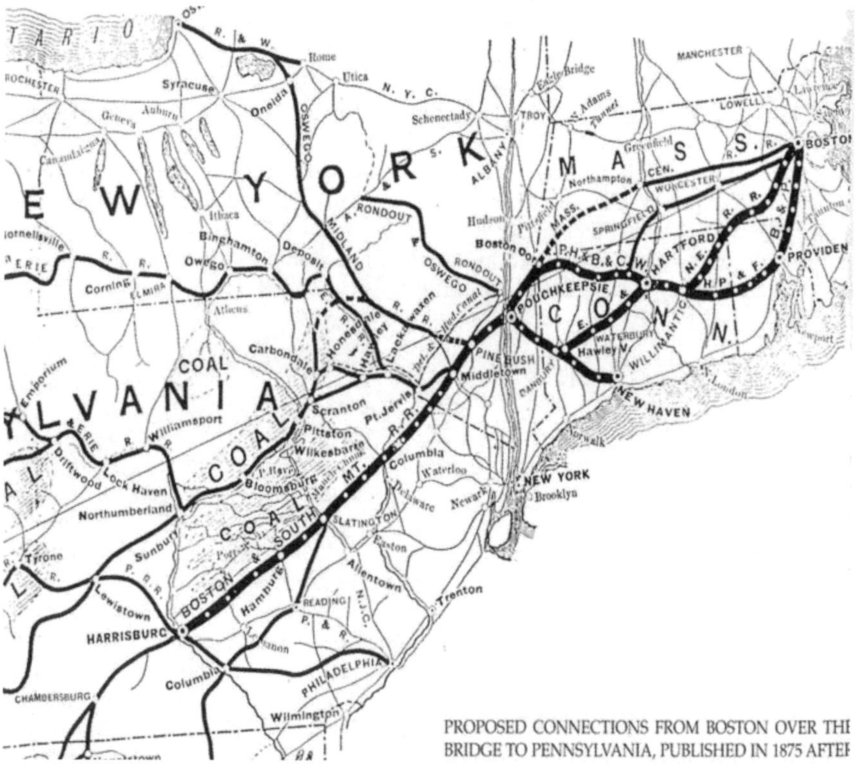

Figure 4. Proposed connections from Boston over the bridge to Pennsylvania.

When work began for a second time in 1876, plans called for 4 piers in the water, and a superstructure comprised of rectangular trusses which were common at that time. Construction of the 4 piers would be a considerable challenge given the depth of water and silt, as well as the concern over the use of timber cribs with air pressurized chambers that had been used in the construction of the St. Louis and Brooklyn Bridge. To avoid the concerns associated with the dangerous air pressurized chambers, the Chief Engineer for the American Bridge Company, W.G. Coolidge, chose to create giant timber cribs without the use of air pressure.

Each pier crib would be about 50 feet wide by 100 feet long. The first two were constructed along the shore to a height of approximately 30 feet and then launched into the river, where once in place, their height was increased. Each crib would require over one million feet of Pennsylvania hemlock. Two large barges were constructed from Florida yellow pine, and were equipped with two steam-operated derricks. The cribs were designed with pockets that derricks would fill with rocks in order to weigh the crib down and allow it to settle down through the river's silty bottom. Other open crib pockets were used to dredge out sediment to allow the cribs to continue settling until they reach bedrock. Once it was determined that the cribs were near bedrock, divers were sent down through the dredging pockets with crow bars to verify that the cribs had reached bedrock. The derricks were then used to dump concrete mix into the water filled crib pockets. The result was a wooden crib filled with concrete and/or stone depending on the pocket, which provided a solid foundation for the stone pier work to follow.

Unfortunately, due to various financial issues in the industry, work was stopped in 1878 with two piers partially constructed. The top of the piers were above the high water mark (one foot and twenty feet, respectively). Lights had to be place on the two piers to protect navigation. Nothing occurred on the bridge for the next 8 years, with company directors failing to meet for years at a time. However, interest was rekindled due to growing amount of rail traffic crossing the Hudson via rail-car ferries in Newburgh. The primary cargo was Pennsylvania coal. The amount of traffic demonstrated that a rail bridge in Poughkeepsie would be financially feasible.

In addition, there is nothing like competition to spur people to act. Promoters were discussing the possibility of building a bridge near Storm King Mountain. The New York Times claimed that it was the Storm King Bridge activity which "led to a movement among the dry bones at Poughkeepsie." Brooklyn railroad contractor John Clarke Stanton in discussions with Philadelphia utility executive William W. Gibbs, was able to convince Mr. Gibbs that a bridge at Poughkeepsie would facilitate the marketing and delivery of Pennsylvania coal to its Northeast customers. The pace accelerated once again, with the Manhattan Bridge Company contracting with the Union Bridge Company to design and construct the bridge.

3 DESIGN & CONSTRUCTION

Two of the five partners in the Union Bridge Company played a major role in the bridge's redesign. Thomas Clarke was the president, a Harvard graduate, and was "one of the best – known civil engineers in America" per the New York Times. Charles Macdonald was a graduate of Rensselaer Polytechnic Institute. Both were well known for their accomplishments with iron and steel bridges.

In the two previous attempts at building the bridge in 1873 and 1876, the intent was to provide a rectangular truss bridge utilizing 4 river piers. However, due to recent construction success of a cantilever bridge over the Niagara River Gorge by other partners in the firm, it was decided to use cantilever spans. The New York Herald stated that it would be the longest cantilever bridge in the world. Clarke and Macdonald brought their knowledge of the construction of the Brooklyn Bridge to the project. Macdonald was a classmate of Washington A. Roebling, who was the chief engineer for the Brooklyn Bridge after his father died. Clarke and Macdonald served as trustees of the Brooklyn Bridge from at least 1881 until it opened in 1883, and they were on a three-man committee reviewing how to increase live load traffic on the Brooklyn Bridge while the Poughkeepsie bridge construction was underway. Union Bridge appointed John F. O'Rourke to be its chief engineer for the bridge. O'Rourke directly supervised construction under the guidance of Clark and Macdonald.

The decision to use cantilever spans was driven in large part by the fierce resistance by the shipping industry regarding any type of obstruction in the river. Cantilevered spans improved vertical clearance for sails, and they could be built without scaffolding in the water, significantly reducing interference with navigation, as well as reducing costs. The shipping industry also wanted the piers

Vital Statistics	
Length over water, anchorage pier to anchorage pier	3094 ft.
Total length of bridge, including approach viaducts	6768 ft. (1 ¼ mi.)
Height from water to base of rail track	212 ft.
Vertical clearance for shipping: under connecting truss spans under center cantilever span	 130 ft. 160 ft.
Depth of water and sediment to the hard gravel on the rock bottom	130 – 140 ft.
Width of top of bridge	35 ft.
Usual speed limit on bridge	12 mph
Steel: "mild" with a strength in thousands of lbs per sq. inch	60-65 psi
Weight of one yard of track rail: originally by 1907 by 1954	 70 lbs. 100 lbs. 131 lbs.

Figure 5. Vital statistics.

as narrow as possible. Narrower piers, however, would not be able to provide adequate anchorage and support for cantilevered spans across the entire river. Therefore, it was decided to provide alternating deck truss spans to help the smaller piers anchor the cantilevers. The Poughkeepsie Bridge would demonstrate that cantilevered bridges were a viable design for more than just deep river gorges.

The Poughkeepsie Bridge set many bridge records. While the short approach spans over land were made of conventional iron, the very large river piers spans and piers were designed of mild steel at time when the use of steel in bridge construction was in its infancy. According to the New York Times, just one of the 525 foot long steel spans would be "the largest and heaviest steel truss in the world." Although under considerable pressure to keep costs down, the designers wanted the bridge to be able to handle two of the heaviest coal trains at the same time, as well as high winds strong enough "to blow trains off the bridge." Movements due to temperature expansion and contraction, and live load deflections were addressed by substantial use of pin and eye bars, as well as sliding bearings.

The pace of pier construction accelerated in 1887, with the work force going from approximately 200 in January to 1500 men in July of that year. Similar to the 1876 approach, cribs were floated out to the site, and carpenters then continually increased the height of the cribs while they sank. During crib construction, American Society of Civil Engineers was holding a major conference at the Hotel Kaaterskill in the nearby Catskill Mountains. A field visit for 150 attendees was arranged using trains (lobster lunch included), steamers, and tugs to get them to the site. The visitors were very impressed with the site. More information on the crib and caisson construction is provided later in this paper.

Steel for the river piers and trusses was produced in Pittsburgh. However, so much steel was required that some of the steel for the bridge was produced in England to keep the project on schedule. The two truss spans necessitated the construction of massive, temporary wood trestles, which were impressive in their own right. In a concession to navigation, only one trestle between piers would be in the river at any one time. The trestles were supported by hundreds of wood pile clusters, in row after row, bolted together. The piles were 130 feet long or more, and were made of yellow pine and spruce. Once the trestle was completed, a work deck was built on top, along with tracks to accommodate traveling derricks and riveters.

Figure 6. Temporary scaffolding for permanent October 6, 1887.

Figure 7. Poughkeepsie shore construction truss span.

In an interesting side note, as the truss spans were being erected, locals inquired about the accommodation of carriages underneath the permanent rail tracks, as was called for in the original charter. Ownership was not inclined to add the carriageway despite the language in the charter due to cost and operational concerns. They were supported by local ferry operators who did not want to see carriages using the bridge. The matter made it to the state legislature for a vote to require the carriageway, it failed. Upon completion of the truss spans, and removal of the temporary trestles, including the removal of the pile clusters, work was able to begin on the cantilevered spans. Derricks were used to lift pieces of steel from barges on the river and extend the cantilevers until they were completed. Eight men lost their lives in the construction of the bridge.

When the bridge opened in January of 1889, it was the longest bridge in the world. However, that distinction only lasted a year until 1890, when the Firth of Forth Bridge opened in Scotland. Engineering News called the Poughkeepsie Bridge "colossal," and claimed that, among all the structures in the world, it was "one of the grandest." Scientific American called it "a monument of engineering." Its promoters celebrated it as "one of the wonders of the world," providing "the grandest" views "to be seen from any railroad line in the world," and built to "last forever."

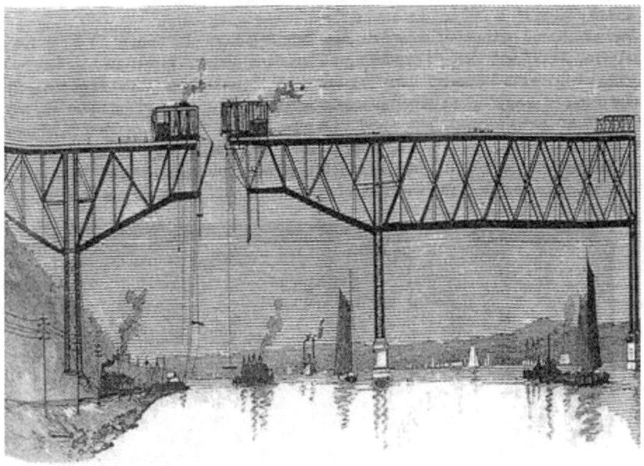

Figure 8. Cantilever construction.

Figure 9. 1974 fire on east approach.

4 OPERATIONS, STRENGTHENING AND FIRE

The bridge saw dramatic increases in the volume and weight of train traffic in the years following its opening. In 1905, 400 freight cars a day were crossing on the bridge. During World War II, almost 3,500 cars a day were traveling over the bridge. The usage continued to rise after World War II despite the increasing competition from trucks. Volume in the early 1950s increased to 17 million tons per year, with many trains stretching all the way across the river. As was typical with many east-west trail routes, significantly more weight and tonnage headed eastbound than westbound, primarily due to coal and oil. Many westbound cars were empty.

To address increasing loads, the New Haven Railroad strengthened the bridge in 1906–07 by inserting a new center steel truss between the two existing trusses. The strengthening changed the aesthetics of the bridge from being light and airy where many people wondered how it could handle the trains, to one resembling a spider's web of steel members. Trains continued to get heavier in the years that immediately followed the strengthening, with locomotives such as the Santa Fe weighing as much as 295 tons. The New Haven Railroad was concerned about the stresses an uneven weight distribution on the bridge with very heavy trains heading eastbound, and lighter or empty trains using the westbound tracks, could have on the structure.

Figure 10. View of Mid-Hudson Bridge and Sloop Clearwater from the walkway.

Consequently, the bridge was strengthened again in 1917–18. Workmen raised the tracks onto oak blocks to allow trains to run while a new steel floor system was installed. To distribute weight more evenly, the tracks were overlapped (gauntleted) so in effect, the bridge was a single track bridge, directing trains to run down the center of the bridge, thus providing better load distribution. The approaches and steel piers were again strengthened, with the viaducts transitioning from iron to primarily steel. By allowing the new Santa Fe engines to run on the bridge, the new strengthening of the 1.25 mile long structure paid for itself in less than two years. The late 1950s and 1960s saw the nation relying more on the movement of goods and services by truck instead of trains due to the development of the Interstate highway system. Less traffic was also going over the bridge due to the opening of the Castleton-On-Hudson railroad bridge near the Selkirk rail yards just south of Albany, New York. Dedicated funds and labor to maintain the bridge waned as the years went on. The bridge's one time permanent maintenance staff had been eliminated. In 1974, sparks from either the brakes or engine exhaust of a diesel engine ignited the creosoted railroad ties and the neglected wooden maintenance walkway on the east approach of the bridge.

There were no Penn Central guards or maintenance men on duty at the time the fire broke out. Once firemen arrived, they tried to open the fire fighting water line that ran the length of the bridge. Unfortunately, due to the lack of maintenance, the pipe had not been drained the previous winter and it had burst in numerous locations. Penn Central was aware of the problem, but never repaired it. Firemen were forced to try and get water to the fire from far below on ground level. The fire destroyed about 700 feet of deck and track, and had warped the steel girders. While the damage was minor considering the overall size of the bridge, Penn Central showed little interest in repairing the bridge, and was satisfied with sending traffic over the Castleton-On-Hudson Bridge.

5 STUDY OBJECTIVES

5.1 *Vision*

The project objectives can be summarized as: 1) Save and restore an historic landmark; and 2) Building a park and walkway for people to enjoy the view for generations to come. (Figure 11) Short term, the goal is to make the view available to the public. The longer range goal is to create a National Icon and restore the bridge's place in history. Most people do not realize what an important structure this was from an engineering and transportation perspective. When completed, the Walkway will be the only dedicated park and walkway across the Hudson River. It will serve once

Figure 11. Future walkway visualization.

Figure 12. Overall view of the river spans of the bridge, looking north.

again as a "great connector" – only this time it will be for 30 miles of rail trails, rather than railroad tracks. With proper planning, funding and enthusiasm, the Walkway over the Hudson will be a "must see" destination.

The bridge is already located near other major tourist destinations such as the home of Franklin D. Roosevelt, the historic Hudson Valley mansions of the Vanderbilts and others, as well as institutions such as the Culinary Institute of America. It is within walking distance of the Poughkeepsie Amtrak and Metro North rail station, providing easy access to weekend recreationists who will come and walk or bicycle the extensive trail network.

Access to the top of the bridge on the Poughkeepsie side of the river, will be provided by elevators. An elevator near the east end of the bridge will be convenient for Poughkeepsie residents. A more ambitious elevator, reminiscent of the elevators that used to be on the Palisades escarpments, is envisioned for the pier along the east shoreline. The ride 212 feet in the glass elevator to the top will be an attraction on its own. Many will come just to ride the elevator and enjoy the view from the top. Amenities such as a café and gift shop will be considered.

Figure 13. Pier 2 west and north elevation.

5.2 *Making it happen*

Based on very preliminary estimating, construction values range from \$10 M to \$35 M, depending on the width of the Walkway, amenities provided, types of materials used, scope of the rehabilitation, the level of donations received, and other variables.

Phase 1 of the comprehensive plan was completed in the fall of 2006. Phase 1 was comprised of a Strategic Planning Session and Underwater Inspection of the river piers. At the Strategic Planning Session, the Walkway Board, the consultant team, and a representative from the National Park Service discussed strategies on how to progress the project. On a unique, ambitious project such as this, extensive community outreach, fundraising, and informational meetings with key elected officials, and community/business leaders is imperative. The underwater inspections discussed in section 6, were important to assure that the "project is on solid footing".

Phase 2 is expected to start in the summer of 2007. Under Phase 2, an environmental assessment will be conducted following NEPA and SEQRA standards. A cost/benefit analysis determining local and regional economic benefits derived by the project will be conducted. Extensive community outreach and fundraising already underway will be enhanced. It is the firm belief from all those involved that as the comprehensive plan progresses, and details are released, the momentum for the project will increase exponentially.

Our goal is to develop practical, cost effective and perhaps innovative solutions to meet the project objectives. An example of this would be the design loading for the Walkway, which is less than 1/3 the original design loading of the bridge. While certain levels of steel section loss are acceptable, and do not require repairs, one of our objectives will be to predict the rate of steel corrosion if the steel is not painted, to estimate when repairs will be necessary. The extent of painting required, and timeframe, will significantly impact the budget. Since current recommendations include providing a pedestrian and bicycle path on the bridge rather than a freight railroad, it is anticipated that the live loading will be significantly less than the coal locomotives that the bridge was initially designed for (approximately 1/3 of the design loads). The designers will be able to accept some section loss, thereby, reducing the amount of superstructure repairs required.

Various deck systems will be evaluated for the structure based on cost, ease of construction, structure weight and future maintenance. It is anticipated that a prefabricated and panelized system will be utilized. The team will evaluate typical concrete, steel and composite systems as well as innovative materials such as fiber reinforced polymer systems. Other elements of the comprehensive study include permitting evaluations (including railroad, utility, environmental, and historic permits), and design recommendations for incorporating an historic railway while meeting code

Figure 14. Pier 2, south and east elevation.

requirements for safety. Also, ADA accessibility for an elevated structure needs to be addressed. We hope to provide cantilevered overlooks in certain locations on the bridge to capitalize on the spectacular panoramic views of the Hudson River Valley. Speaking of the view, non-operational catenary lines owned by Central Hudson are carried on the south fascia of the bridge and impact the view to the south. These lines will be removed as part of the project. Finally, typical trail amenities such as kiosks for way-finding and interpretive signage, restrooms and parking will be addressed.

6 PRELIMINARY UNDERWATER INSPECTION FINDINGS

6.1 *Conventional inspection methods and results*

All in-water pier substructures received a visual inspection from Mean High Water line (MHW) to the mud line. Water depth were measured and recorded at eight points around the perimeter of the piers along the mud line. Probes were taken into the timber cribbing and grillage. Mud line elevation and composition were recorded. Additional investigation was performed using a "DIDSON" Sonar Camera on November 21, 2006 at Pier 2 to determine the extent of the void area observed during the initial investigation.

6.2 *Methodology*

The dive team conducting the investigation was composed of a Professional engineer diver, a diver and a diver tender. Dive operations were conducted from either a thirty-foot aluminum boat or a twenty-two foot fiberglass dive boat. Diving was performed using surface supplied equipment with constant two-way radio communication and real-time video recording. The divers' visibility during the investigation was limited to three inches or less, making visual observation very limited. Most of the inspection was performed using tactile investigation skills. Due to high current velocity, inspections time was limited to hours of slack current.

6.3 *Skills necessary for demanding tidal environment*

The inspection diver must have a unique skill set to meet the challenges of a demanding environment. A few of the qualifications include a high level of physical fitness, mental alertness and safety awareness. These qualities must be combined with proper training and years of experience to accurately and effectively conduct an inspection. Additionally the diver must have training and experience in structural systems and behavior in order to distinguish between structural and non-structural deterioration and the effects this has on the structure.

Ordinary underwater inspections pose many hazards and difficulties. The especially challenging facets of this particular inspection are as follows:

High water current exacerbated by the tidal fluctuation and river flow present a problem for maneuvering and mooring the boat, and handling the divers' umbilical. Additionally high current can quickly fatigue a diver. It is difficult to move underwater, and the diver can be swept off of the structure by the current.

Low water temperature, approximately 42F, can quickly lower a divers body temperature and hasten fatigue. Specialized undergarments and dry suits are required to combat the cold temperatures. The fully dressed diver can be carrying up to 80 lbs in addition to his own body weight when fully equipped with dry suit, weight, bail out, harness, dive helmet, lights, video/sonar unit and inspection tools.

Due to lack of visibility, caused by high turbidity, the inspection must be accomplished by hand. This requires a high degree of skill to accurately locate and recognize deficiencies by feel rather than sight and determine the degree of deterioration. There is however a limited degree of visibility, approximately 3 in. to 5 in. Visibility will vary the tide (flood or ebb), amount and time of a recent rainfall, water temperature, time of year (melt runoff).

It is extremely important that the diver knows his exact location at all times. This is so the top-side crew can quickly respond during an emergency and so the inspection notes and corresponding deterioration is noted in the proper location. This requires the diver to have the ability to keep a visual image of the plan of the structure in his mind and coordinate it with his physical location while relaying the inspection information and location to topside personnel.

Inspection divers must contend with debris on a regular basis on almost every underwater inspection of all structures inspected. Some types of debris typically encountered are debris from initial or rehabilitation construction on bridges and include steel cables used for temporary mooring of work barges, temporary suspended catwalks and general rigging, chain link fencing for temporary safety nets, catwalks and netting. Additionally steel and timber beams, scaffolding, ladders and other items used in falsework construction are usually strewn about the pier area. Other types of debris include concrete debris from over pours and demolition, and pile stubs from false work or previous structures. All of these types of debris not only hamper the inspection, but also pose a hazard for snagging and entrapment of the divers' equipment or umbilical.

Some types of debris and hazards are due to the deterioration of the structure itself such as knife edged steel piles and bracings that can severe the umbilical or the diver himself. Timber bracing and stay lathing are commonly found wedged between the typically closely spaced timber pile bents and make for quite a circuitous route when traveling in and out of the pile bents. Other debris found around piers and pile bents include fishing nets, abandon cables, boats, crab and lobster pots all of which can be a potential entrapment hazard for the diver or his equipment and umbilical.

Debris typically encountered in this underwater inspection included fishing nets, timber stub piles and most frequently spaces in exposed timber cribbing and caissons with exposed steel spikes. Although the timbers were originally installed pinned tight against each other, deterioration of the outer surface of the timber has left 1 in to 2 in gaps between them after 100 years. These gaps catch and hold divers umbilical making it necessary for him to return to the point on entanglement, free the umbilical, inspect it for damage such as cuts, pull additional slack into the water and then return to his last inspected point with limited visibility before continuing with the inspection.

Safety is paramount in any diving operation. Special emphasis is always placed on accident avoidance through training and hazard analysis planning, but if a mishap should occur, it is of vital importance that an emergency response plan be in place and acted upon without hesitation. The planning locations and contact numbers for local emergency responders (ambulance, fire, hospital and police) and specifically the location of the nearest decompression chamber and method of transportation to that location should a diver be accidentally exposed to conditions requiring hyperbaric treatment.

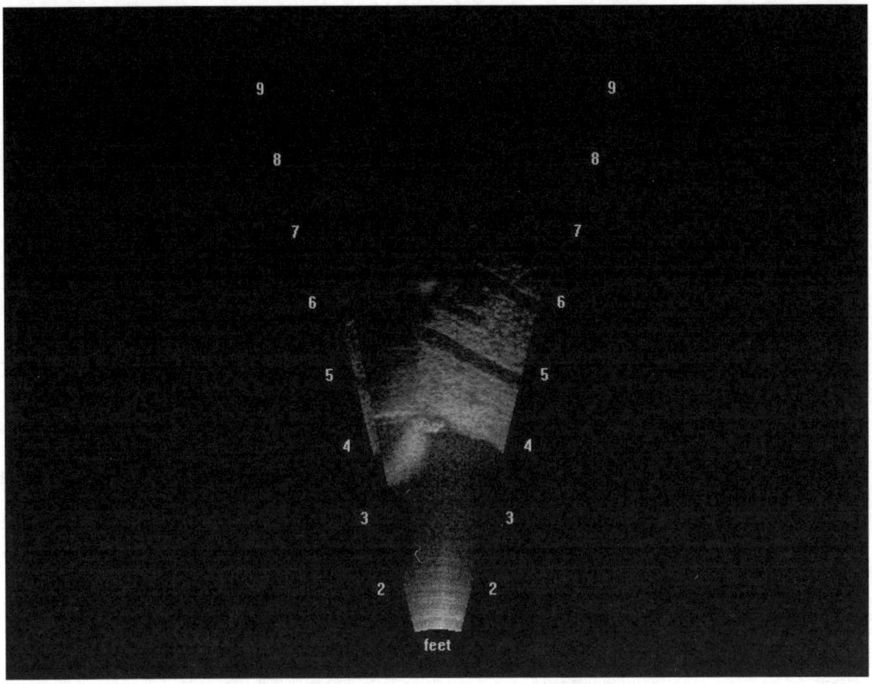

Figure 15. Sonar inspection reveals stone fill.

Several specialized techniques are used to work in high current situations. Mooring locations around pier are carefully selected when conducting the dive inspection. The direction of the current and eddys formed around the pier, as well as the time and predicted current change, must be considered. The dive is arranged to avoid maximum current by working during the slack current (time when tide is changing direction. Diving inspection is generally conducted from upstream/current to downstream/current to eliminate the need for the diver to swim against the current.

6.4 *Information on crib and caisson construction*

Typical substructure components of the piers consist of a heavy timber crib structure that was used as a deep dredging system. The cribbing is essentially a bottomless box and was constructed of several layers of 12 in. by 12 in. timbers positioned horizontally and fastened with steel pins, typically the exterior was then sheathed with vertical timber planking to complete the structure. The cribbing structures measure approximately 60 ft wide by 100 ft long at the bottom (see Figure 19), and tapers along the east and west faces to 50 ft wide by 100 ft long at the top. The timber cribbing was partially built on land then floated out in the river where it was positioned, sunk by filling the weighting pocket with stone and built up to the required elevation as dredges worked to remove material from the interior dredging wells until the bottom of the crib structure was founded on firm soil at approximately 130 ft below the surface of the river. The dredging wells were then filled and leveled with concrete to provide a stable foundation for the timber grillage layer.

When the concrete filled timber cribbing installation was complete the next step was constructing and placing a floating caisson over the cribbing. The caisson was designed with the bottom serving as a mat of timber grillage and was constructed of six layers of 12 in. by 12 in. timbers (see Figure 20). Once the caisson was floated into position over the previously placed cribbing, construction of the masonry faced concrete pier began within the floating caisson. As the pier construction progressed the weight of the pier gradually sunk the caisson until it rested on the cribbing.

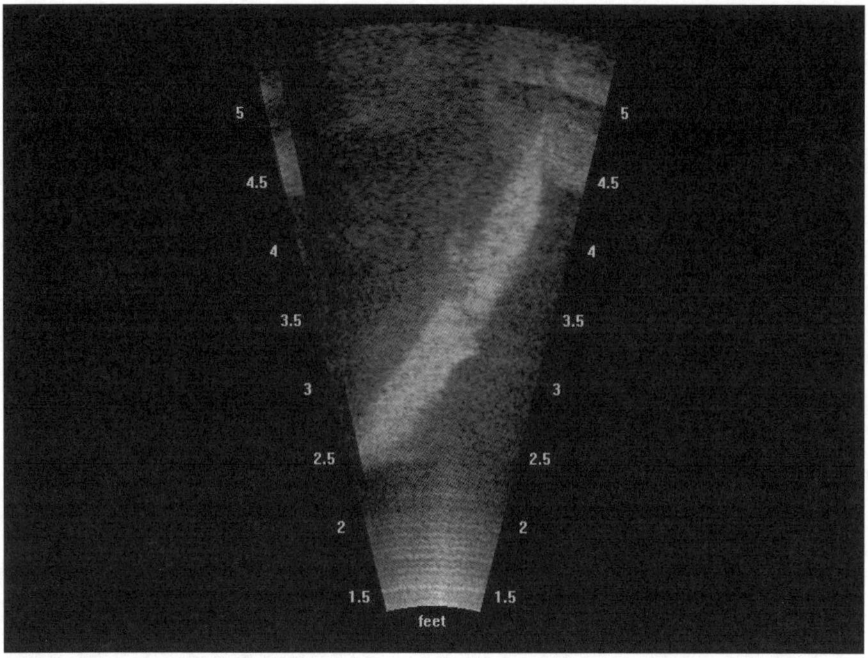

Figure 16. Sonar view of inner timber wall and tie.

Once the masonry and concrete pier was completed the caisson was flooded and the sides were removed leaving the completed masonry and concrete portion of the pier resting on the timber grillage. The finished overall dimensions of the concrete filled masonry piers are approximately 25 ft wide by 90 ft long.

The Pennsylvania Railroad officially started the initial construction of the Poughkeepsie Railroad Bridge in 1873. However, actual bridge construction started in 1876 by the American Bridge Company who built Pier 2, installed the cribbing for Pier 3, and partially completed construction on the timber cribbing for Pier 4. Construction was suspended from 1876 to 1886 when construction was resumed by the Union Bridge Company, who modified the design and reworked Pier 2 and completed the remainder of the bridge with the first traffic across the bridge on January 1, 1889.

6.5 *Previous inspections differing opinions*

Previous diving inspections were performed in 1950, 1969 and 2002 with differing observations and recommendations. Only excerpts from the 1950 inspection report that were included in a 1978 structural assessment were available for review. The 1969 inspection was conducted for Penn-Central Railroad. The 2002 report was compiled for the New York Bridge Authority.

The 1969 and the 2006 inspection generally agree on the type and location of deterioration. The most serious deterioration being at the west face of the cribbing of Pier 2 and the interface between the cribbing and grillage at the east face of Pier 3. The 2002 inspection indicates serious deterioration of the cribbing at the southeast corner and face of Pier 2.

Extensive plan and document research was conducted prior to the fieldwork. This provided a valuable insight into the substructure construction and provides the engineer with information to determine between structural and non-structural deficiencies.

6.6 *Use of DIDSON Sonar to see inside voids*

"DIDSON" is an acronym for Dual Frequency Identification Sonar. It provides a black and white image reproduced from the reflection of sound waves off of objects underwater. It can generally provide images from a distance of 1 meter to 30 over meters. The sonar camera used for this inspection consisted of a hand-held camera unit comprised of an acoustic beam projector/receiver, electronics package and battery power supply. This was connected through 150 ft umbilical to a topside computer unit where the engineer could control the camera settings and record sonar images.

Images obtained are black and white video images or stills. Since the sonar "illuminates" the scene by emitting high frequency sound wave, which produces an image, the images obtained have a high degree of shadowing making some images difficult to interpret, especially when imaged from an acute angle, unless the engineer is familiar with the structure.

6.7 *Underwater findings/recommendations*

6.7.1 *Pier 2*

Pier 2 is located nearest to the western shoreline of the river (Figure 13 and Figure 14). Construction of Pier 2 varies slightly from the other piers (see Figure 21). Construction of the substructure was partially completed when the project was postponed, redesigned and work eventually started again by a different Contractor. A bottomless timber caisson was constructed around the existing pier and pumped dry, then partial demolition of the existing pier was performed, the new masonry and concrete pier constructed over the remains of the previously pier foundation and then the interior of the caisson was filled with concrete.

The stone masonry portion of the pier is generally in fair condition. Typical deficiencies include intermittent areas of missing mortar from between the joints, with penetration from 4 in. to 12 in. deep, and moderately spalled and loose coping stones on the south face of the pier. Moderate efflorescence and rust staining were also evident on the face of the masonry.

The timber cribbing and grillage typically exhibit moderate rot and loss of cross sectional area with gaps between the timbers averaging 1 in. to 2 in. wide. Intermittent penetrations into the gaps of the cribbing were taken and typically varied from 12 in. to over 3 ft deep. The outer layer of vertical timber sheeting is missing from around the entire pier. Intermittent missing pieces of 12 in. by 12 in. timber cribbing were also observed. A significant horizontal void area was observed behind the outermost layer of the timber crib wall, extending along the east, south and west faces of the pier. The voids are located approximately 29 ft below the water surface and extend approximately 16 ft along the south face and 56 ft along the west face. The void in the timber cribbing is 2 ft high at the south west corner and tapers down to 2 in. as it progresses along the south and west elevations. Penetrations in the void varied from 3 ft to more than 6 ft deep. The maximum height of the void on the inside of the timber cribbing at the southwest corner is unknown.

Additional investigation was performed to determine the depth of the void. A dive crew equipped with a "DIDSON" Sonar Camera performed a real time sonar survey of the void area. Interpretation of the sonar images revealed that only the outer layer of weighting pockets has been compromised revealing the stone fill (Figure 15). The inner timber wall and transverse ties appeared to be in place (Figure 16). It was not possible to visually inspect or probe the timber on the interior pockets to determine their condition.

The mud line generally consists of silt and sand over scattered rip rap stone with concrete and steel debris. No signs of scour were observed in the vicinity of the pier. Water depth varied from 39 ft to 66 ft around the pier.

6.7.2 *Pier 3*

Pier 3 also has a slightly different construction than the other river piers. The timber grillage mat was built up of 14 layers of 12 in. by 12 in. timbers in order to bring the foundation up to the required elevation (see Figure 23).

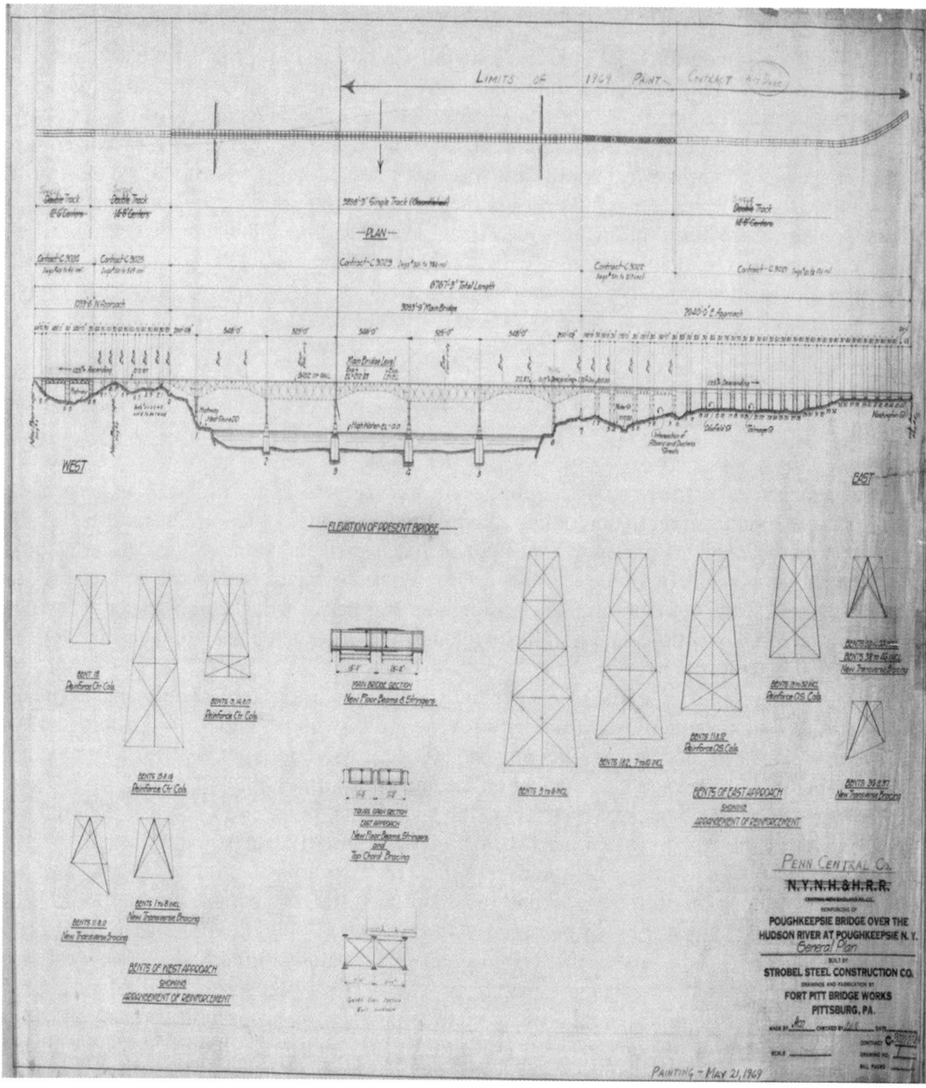

Figure 17. Bridge elevation.

The stone masonry portion of the pier is generally in fair condition. Typical deficiencies include intermittent areas of missing mortar from between the joints, with penetration from 2 in. to 14 in. deep. Moderate efflorescence and rust staining were also evident on the face of the masonry. The timber cribbing and grillage typically exhibit moderate rot and loss of cross sectional area with gaps between the timbers averaging 1 in. to 2 in. wide. Intermittent penetrations into the gaps of the cribbing were taken and typically varied from 12 in. to over 4 ft deep. The outer layer of vertical timber sheeting is missing from around the entire pier. Minor areas of intermittent missing pieces of 12 in. by 12 in. timber cribbing from the outer layer, up to 4 ft long were also observed.

A significant void was discovered along the east elevation at approximately 22 feet below the water surface. The void was located at the interface between the timber grillage and the cribbing structure. The void measured approximately 4 ft high at the southeast corner and tapered down to 2 ft high at the north end of the pier. Penetrations into the void varied and were approximately 2 ft deep

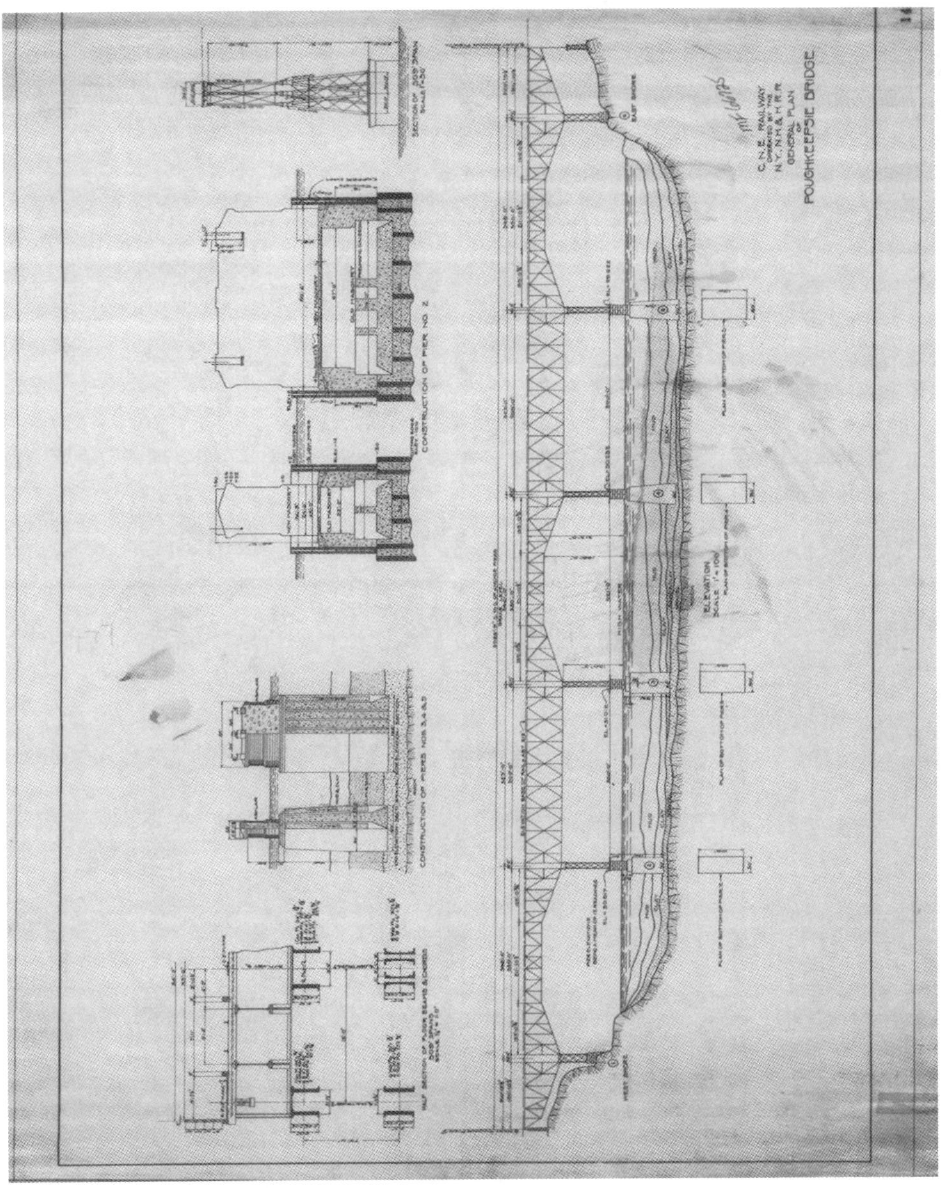

Figure 18. Construction of piers 2 through 5.

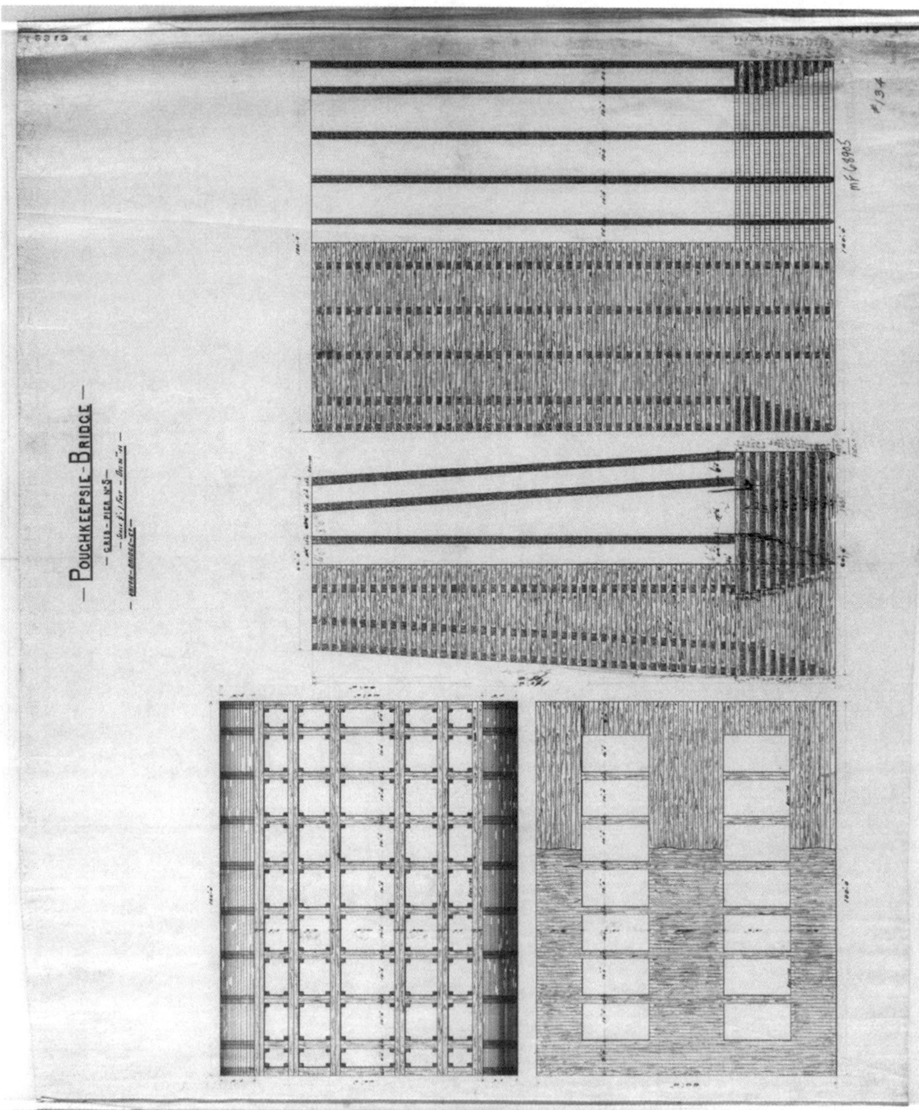

Figure 19. Pier 5 crib details.

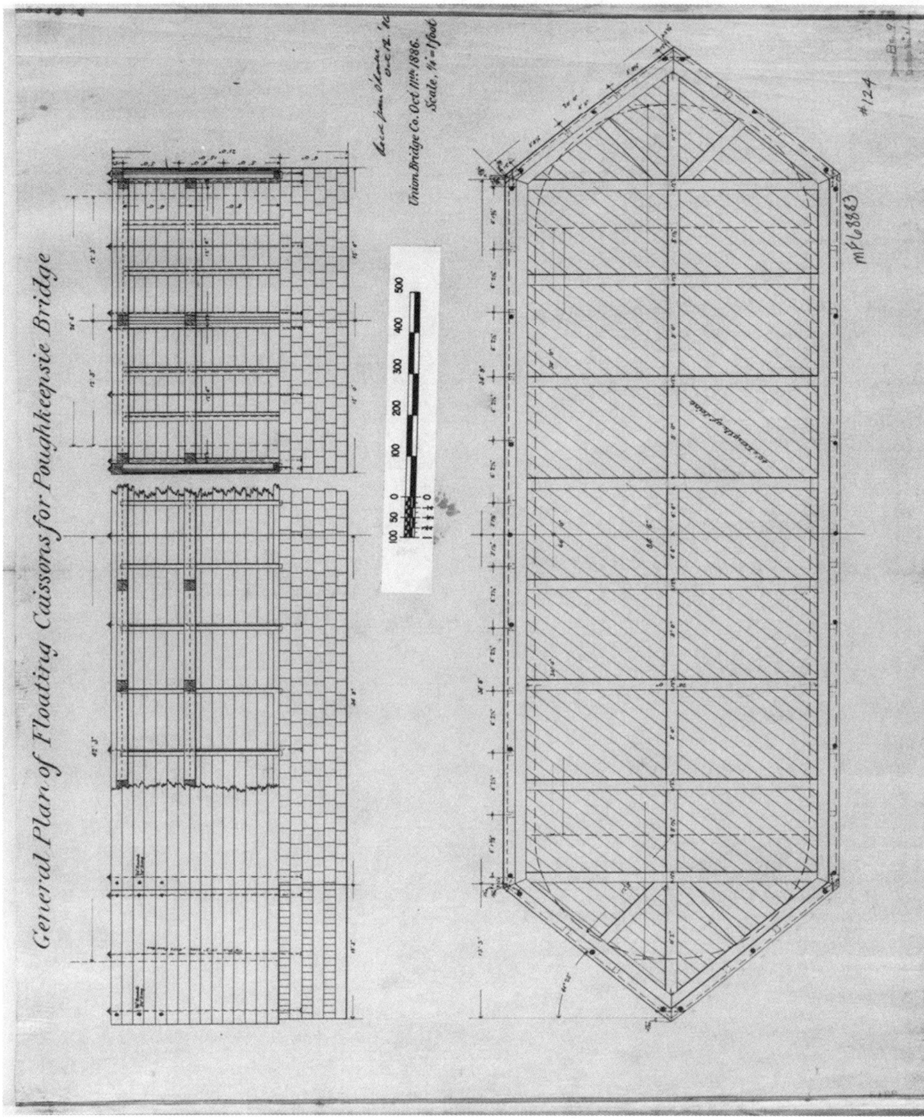

Figure 20. Floating caisson general plan.

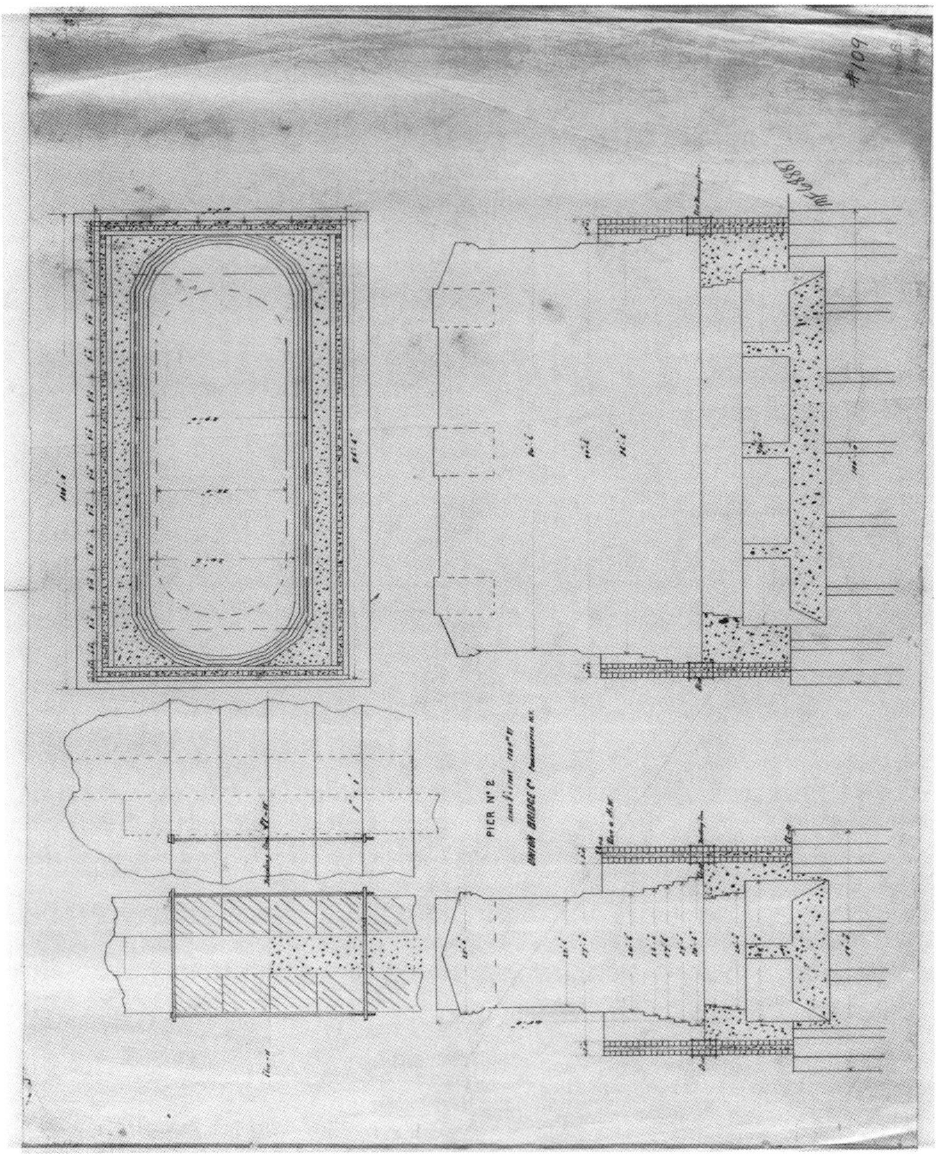

Figure 21. Pier 2 construction details.

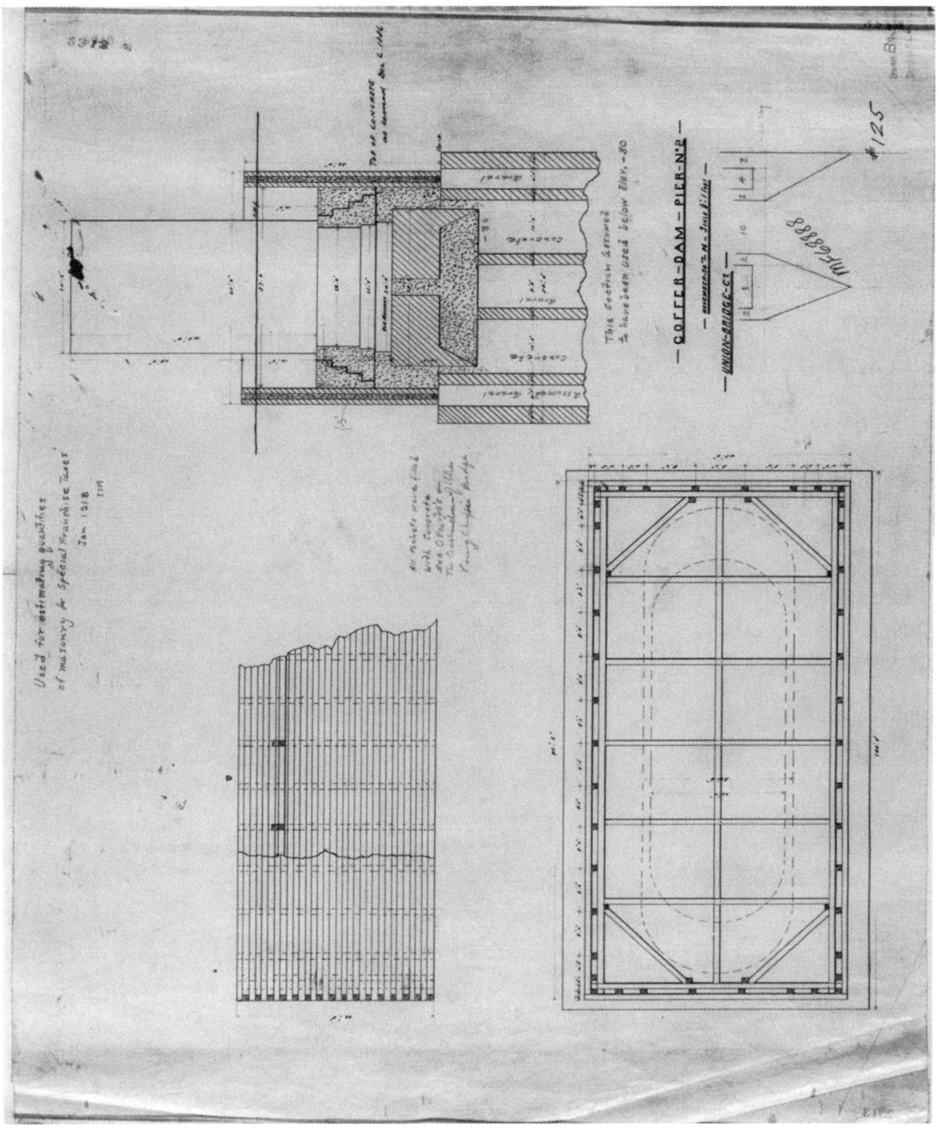

Figure 22. Pier 2 cofferdam.

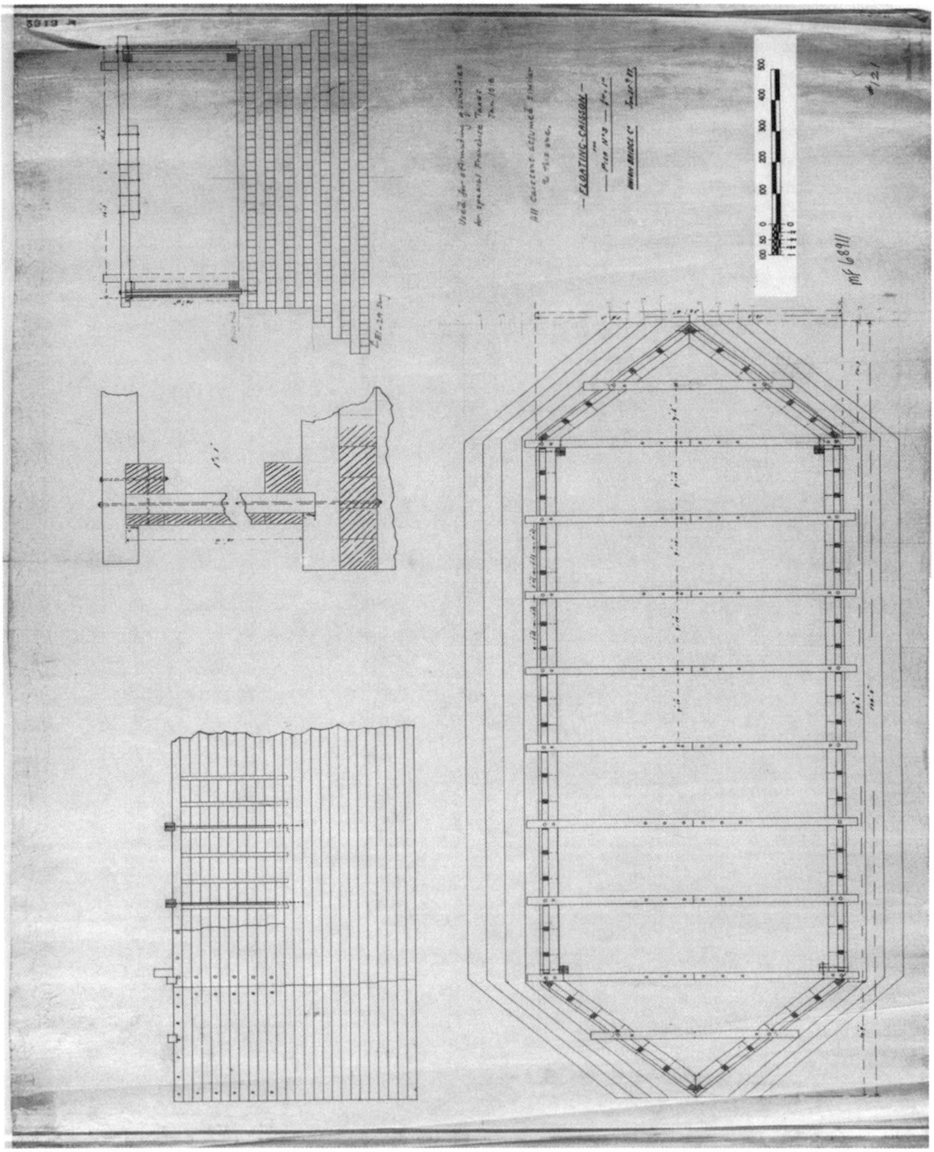

Figure 23. Pier 23 caisson.

at the northwest corner and up to 6 ft deep at the southwest corner. The mud line generally consists of silt and sand over scattered rip rap stone with concrete and steel debris. No signs of scour were observed in the vicinity of the pier. Water depth varied from 38 ft to 51 ft around the pier.

6.7.3 *Pier 4*

Constructions of the pier sub structural elements are as described above under "Typical Substructure Construction". The stone masonry portion of the pier is generally in fair condition. Typical deficiencies include intermittent areas of missing mortar from between the joints, with penetration from 4 in. to 16 in. deep.

The timber cribbing and grillage typically exhibit moderate rot and loss of cross sectional area with gaps between the timbers averaging 1 in. to 2 in. wide. Intermittent penetrations into the gaps of the cribbing were taken and typically varied from 12 in. to 2 ft deep. The outer layer of vertical timber sheeting is missing from around the entire pier. Minor areas of intermittent missing pieces of 12 in. by 12 in. timber cribbing from the outer layer were also observed. The mud line generally consists of silt and sand over scattered rip rap stone with concrete and steel debris. No signs of scour were observed in the vicinity of the pier. Water depth varied from 39 ft to 58 ft around the pier.

6.7.4 *Pier 5*

Pier 5 is located nearest to the eastern shoreline of the river. Constructions of the pier sub structural elements are as described above under "Typical Substructure Construction".

The stone masonry portion of the pier is generally in fair condition. Typical deficiencies include intermittent areas of missing mortar from between the joints, with penetration from 2 in. to 12 in. deep. Moderate efflorescence and rust staining were also evident on the face of the masonry.

The timber cribbing and grillage typically exhibit moderate rot and loss of cross sectional area with gaps between the timbers averaging 1 in. to 2 in. wide. Intermittent penetrations into the gaps of the cribbing were taken and typically varied from 12 in. to over 3 ft deep. The outer layer of vertical timber sheeting is missing from around the entire pier. Minor areas of intermittent missing pieces of 12 in. by 12 in. timber cribbing from the outer layer, were also observed.

The mud line generally consists of silt and sand over scattered rip rap stone with concrete and steel debris. No signs of scour were observed in the vicinity of the pier. Water depth varied from 42 ft to 54 ft around the pier.

7 RECOMMENDATIONS

It is recommended to repair the void areas at Pier 2 and Pier 3 to stop the loss of fill from within the cribbing and restore structural integrity. These deficiencies are not an emergency or a structural stability issue at this time; however, the repairs are needed to provide long-term protection and insure stability. It is recommended that these repairs be completed within the next five years to prevent accelerated deterioration of the substructures. We strongly recommend that the deteriorated portions of Piers 2 and 3 be inspected on an annual basis until repairs are made, to arrest or respond to any sudden change of these conditions. It is our intent to avoid any dramatic increase in rehabilitation costs due to lack of attention. It is also recommended to perform an underwater inspection just prior to repair construction to confirm that the extent of deterioration has not changed.

Typically, repairs to the voids involve sealing the outer surface of the void by installing formwork or grout bags. The void area is then pumped full with concrete. After repairs have been completed, the piers should be regularly inspected at five-year intervals to monitor the deterioration of the substructure elements and recommend any additional repairs.

Permits for the repair construction will be required from various state and federal agencies. Since the permitting process approval may take an unusually long time (over 1 year), it is recommended that work on submitting the permits begin immediately so as to not delay the repairs.

8 SUMMARY

After three decades, the historic Poughkeepsie Railroad Bridge is awakening from its long slumber to resume its role as a "great connector". The bridge community is encouraged to support the restoration of this engineering marvel. To learn more about Walkway over the Hudson's efforts, or to purchase *Bridging the Hudson*, Pulitzer Prize winning author Carleton Mabee's fascinating book on the history of the bridge, please log onto www.walkway.org.

ACKNOWLEDGEMENT

We wish to acknowledge Assemblyman Maurice Hinchey and all of the volunteers from the Walkway over The Hudson project for their dedication, persistence and vision.

REFERENCE

Mabee, Carleton. *Bridging the Hudson*. Purple Mountain Press, Fleischmanns, New York, 2001.

Chapter 24

Aesthetics and durability aspects in the realization of small and medium span arch bridges

E. Siviero & A. Zanchettin
IUAV University of Architecture, Venice, Italy

ABSTRACT: A bridge does not fulfil its function in the simple conjunction of two river banks or in the overcoming of an obstacle; it represents many times one of the main marks left on the territory by man and it determines the visual aspect of the surrounding areas. It is then important to pay great attention to the formal aspects of the bridge itself and to its good insertion in the environment. The use of the arch as a structural type is functional to dialogue with the surrounding environment thanks to its wide use in the past and its consistent presence in the territory. In more recent epochs, arches were penalized by the considerable cost of the provisional formworks. Some examples of bridge realizations are presented, which show the possibility of erecting a pleasant structure respecting times and costs expected by clients, guaranteeing also a long lasting durability.

1 INTRODUCTION

During the last fifty years, Italy has experienced a period of continuous economic development, peace and increasing international communications. These factors contributed to an important growth in number of vehicles circulating and to the consequent need for a suitable road network.

Especially after the Second World War and until the Sixties the need for a rapid reconstruction together with booming economy forced to face the problem of realizing many infrastructures in the quickest and most standardized way possible.

In the case of road intersections, flyovers have often been preferred to tunnels because of their limited cost and ease of construction. The most used structural system was the simply supported beam with two or three spans depending on the width of the underlying road. This solution guaranteed an extremely rapid time of erection, a minimization of costs, relatively unskilled workmanship needed, easy design procedures and limited possibilities of errors during realization. On the other hand, simply supported beams do not guarantee high standards in terms of durability and aesthetic aspects. Regarding durability, bearings and expansion joints are critical points, while regarding aesthetics, a simply supported scheme implies high structural sections and consequently more "impacting" structures on the surrounding environment.

Differently from other European countries, in Italy simply supported prestressed beams are still the structural typology largely prevalent in flyovers of small and medium span, due to the reasons explained above, but in particular for the very limited initial cost of construction.

This paper aims at presenting some feasible alternatives, paying particular attention to the aspects of aesthetics and durability but also considering the issues of competitive cost and limited time of realization.

2 AESTHETICS

Too often a flyover is considered by clients as a simple mean to connect two river banks or to overcome an obstacle. However, designers should be aware that a bridge represents an important

Figure 1. The erection of an arch bridge in the city of Padua; on the background an existing masonry bridge.

landmark on the territory where it is built. A careful study of its formal aspect and of its good insertion in the environment permits to obtain a functional and at the same time pleasant result. This is very important in particular for small span structures, usually scarcely considered, which represent the vast majority in number and characterise quite often the visual aspect of many urban and suburban areas.

The use of continuous beams instead of simply supported ones, for example, can be a relatively simple way to have thinner sections and to increase the visual permeability of the bridge. However, the authors believe that, considering the peculiarity of Italy's territory, designers should dare more: our land is characterised by one of the widest artistic heritages in the world and virtually any corner of the country can boost the presence of an historical finding. The use of the arch as a structural type is functional to dialogue with the surrounding environment thanks to its strong use in the past and its consistent presence in the territory. Arches resist by shape in a very natural manner and knew an extraordinary success when stone, masonry and to a smaller extent, wood, were the only known construction materials. Any new realization of flyovers in an Italian city will probably have to cope with the pre-existence of an arch in the neighbourhood.

In more recent epochs, the success of concrete and steel determined a progressive use of structures with a flexural behaviour instead of arch ones, which were penalized by the considerable cost of the provisional formworks necessary to sustain the arch during erection.

As it is widely known, an arch works properly only after the completion of the whole structure, while during realization it needs some form of provisional support.

A continuous centering is one of the possible solutions to sustain the arch during erection.

These very complex provisional structures, usually made with timber elements, became too onerous as the cost of workmanship started to rise, especially after the Seventies. This was particularly evident for small flyovers, which needed an easy and serialized erection process.

Designers who nowadays want to conjugate the needs for economy, velocity and aesthetics, have to rely on alternatives to traditional erection processes for arch constructions.

A careful study of the static behaviour for any stage of construction, joined with an efficient recourse to precast elements, easy to transport and to assemble, allows to reduce the economic gap with other structural systems and to make feasible an arch realization even for medium and small spans.

Figure 2. Centering and final aspect of the bridge over the river Aglio, G.Oberti, 1960.

In this paper we will present three examples of bridges realized in northern Italy during the last fifteen years. All the bridges, relatively small (spanning from 40 to 63 m), have a very simple shape, resembling the lowered arch bridges that Maillart, incomparable master, designed in Switzerland and throughout Europe.

The aim at the base of design is a simple construction process, allowing the contractor to save time and money and to employ a limited number of high skilled workers.

In all three cases soils has poor mechanic characteristic, situation very common in the alluvional plains of northern Italy. The horizontal thrust deriving from the arch is then partially borne by deep foundations and partially by the bridge deck, which is monolithic with the arch, forming a bow-string structure, as explained later.

3 DURABILITY

It is well known that the initial cost of an infrastructure, although very important, is only one of the several expenses that the client will have to withstand during its service life. Maintenance has an important weight in the total budget, especially because it brings a series of indirect costs, such as traffic interruption, which create discomfort among infrastructure's users. For this reason, in recent years road authorities and municipalities tend to accept slightly higher initial costs of construction in order to obtain a structure with an improved durability.

In bridges and flyovers, bearings and expansion joints represent the most critical points for durability; in fact, water flowing through the discontinuities of the structure bring many degenerative phenomena in concrete and in particular corrosion of reinforcement.

The easiest way to overcome the problem of joints is to reduce their number by means of a continuous scheme (see fig 5). A more effective solution, which implies a little more complicated design and construction process, is the integral bridge (see fig. 6). In an integral bridge the intermediate pier bearings are removed and the pier and deck act as a single unit. The maintenance of integral structures is easier and cheaper comparing to traditional ones; anyway great care has to be paid to soil-structure interaction, particularly as bridge expands due to thermal variation.

In the following pages, three examples of arch bridges are presented. In all of them the presence of joints is limited at minimum in order to create a monolithic structure scarcely affected by water flow.

The bridge is located in an historical and architectonic setting of great interest, close to the city of Padua. The municipality planned its construction as an alternative to an historic masonry bridge, which was inadequate for the growing traffic load of the area. The new structure was placed 50 m far from the old one, which has been maintained with the function of a footbridge.

The new structure spans 42.7 m and is 13 m wide; the deck has a trapezoidal shape ranging from 50 cm to 20 cm in height.

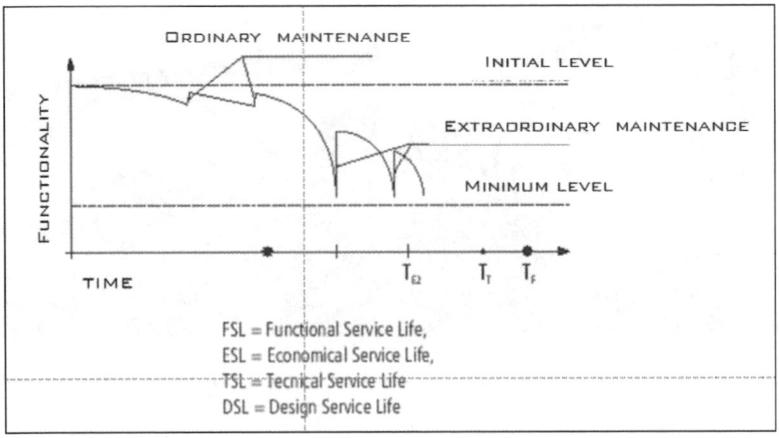

Figure 3. Functionality against time plot for an infrastructure. The importance of maintenance is evident.

Figure 4. Damaged concrete and concrete pull out in correspondence of the bearings of a flyover.

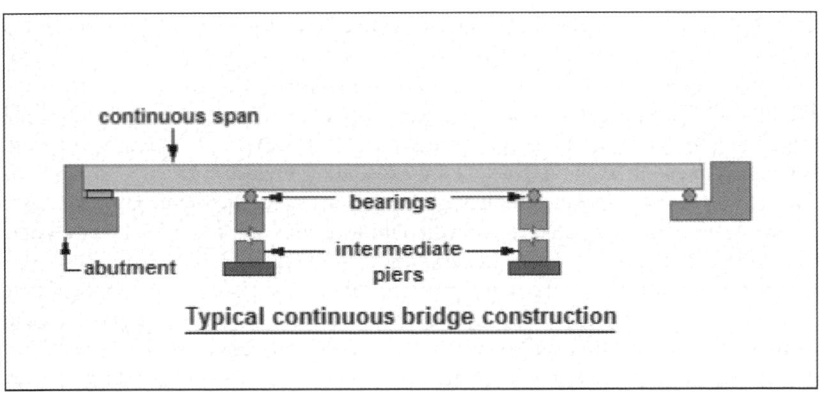

Figure 5. Continuous beam flyover scheme.

The arch, entirely made of precast members, spans 29.5 m and is composed of two series of six semi arches 1.2 m wide and 0.55 m high, disposed as in fig. 8, joined at the keystone and at foundations.

In order to reduce structural dimensions, high strength concrete has been used, employing additives enhancing fluidity to reduce water-cement ratio; the following addition of acrylic fibres

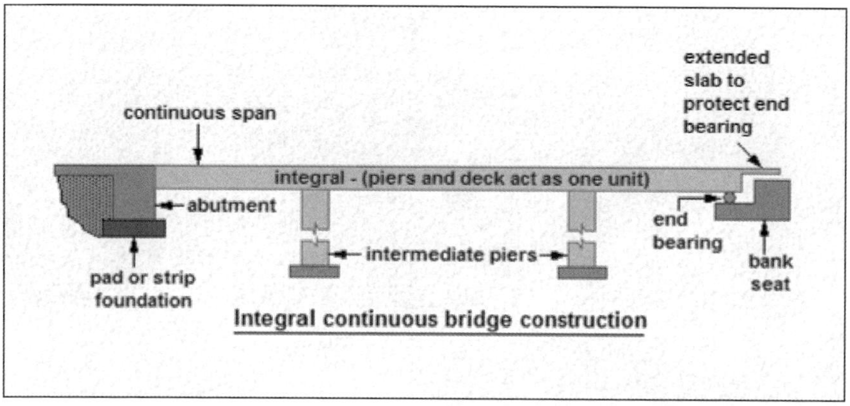

Figure 6. Integral flyover scheme.

Figure 7. Bridge over river Battaglia in Battaglia Terme, Padua.

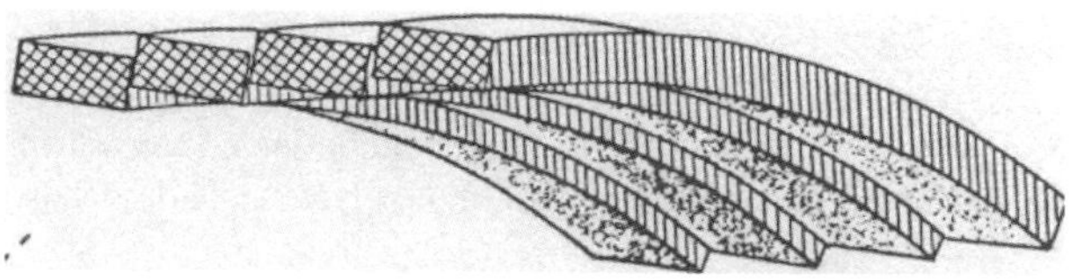

Figure 8. Scheme of the disposition of the semi arches.

permitted to contrast the long term plastic phenomena of concrete, which are one of the main causes of structural damaging. The absence of joints moreover reduces drastically maintenance needs and enhances the durability of the structure.

The realization process can be schematized in five simple stages, each one characterized by a different static scheme:

- realization of deep foundations.
- positioning of precast semi arches on a provisional pier; during this phase the semi arches behave as simply supported beams.

Figure 9. Building phases of the bridge.

- cast of concrete to form an arch monolithic with foundations; removal of provisional piers. During this stage in fact, the structure starts to behave as an arch working mainly in compression.
- cast of the upper deck; in the hypothesis of realizing a single cast, the structural scheme is analogous to phase 2, apart for the loads acting on the structure. During this phase the new loads are vertical reaction induced by the deck's supports and an uniformly distributed load at midspan where the deck bears directly on the arch.
- Completion of the bridge; removal of all provisional supports and prestression. All the remaining design loads are then applied.

In the following images it is possible to point out the main phases of construction, the overall lightness of the structure and the good insertion of the structure in the surrounding environment.

An important aspect is the rapidity of execution: the whole structure was built in only five months.

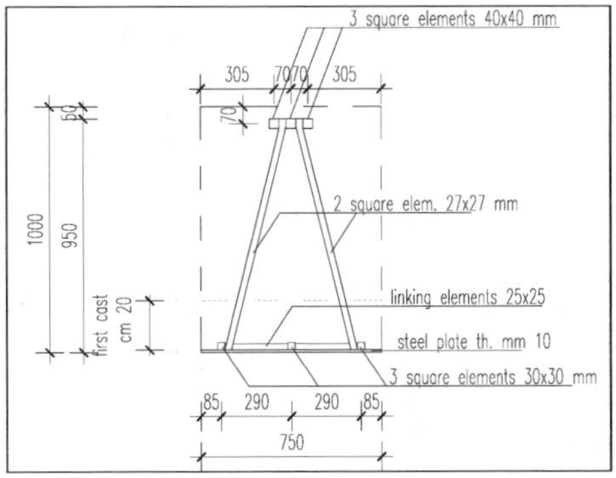

Figure 10. Transversal section of a REP beam forming the main arch.

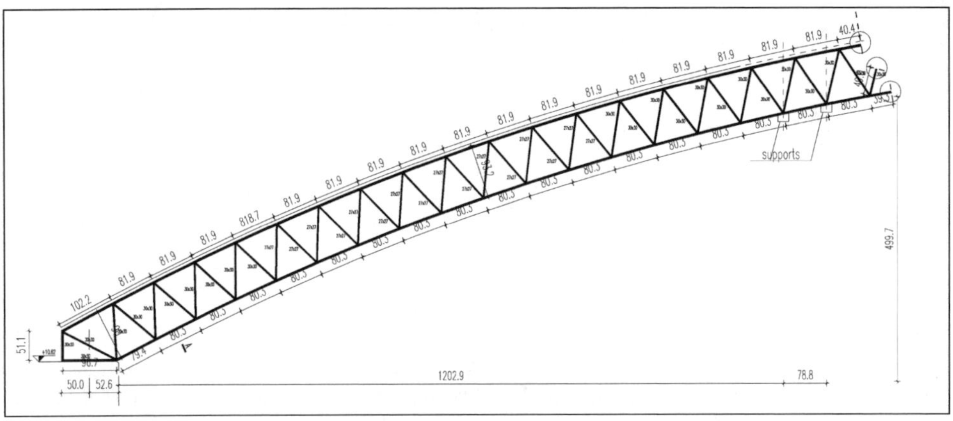

Figure 11. Lateral part of the REP beam forming the arch.

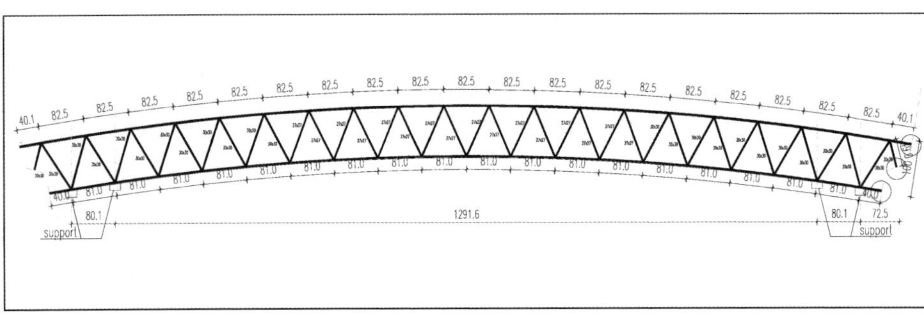

Figure 12. Central part of the REP beam forming the arch.

4 BRIDGE OVER THE RIVER PIAVE VECCHIA, S. DONÀ DI PIAVE, VENICE, 2002

S. Donà di Piave is a small city of about 100.000 inhabitants, characterised by a strong traffic due to its important industrial district and to its position, along the road connecting Venice to its beach resorts.

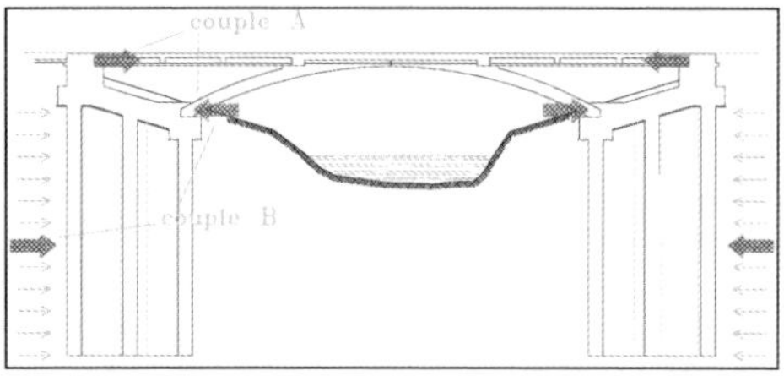

(a)

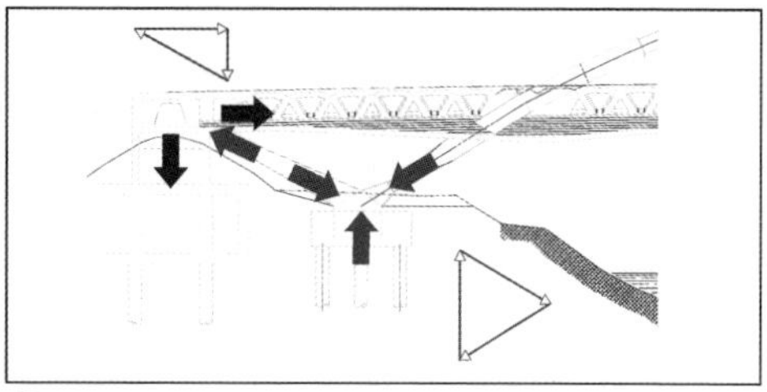

(b)

Figure 13. Bow – string configurations.

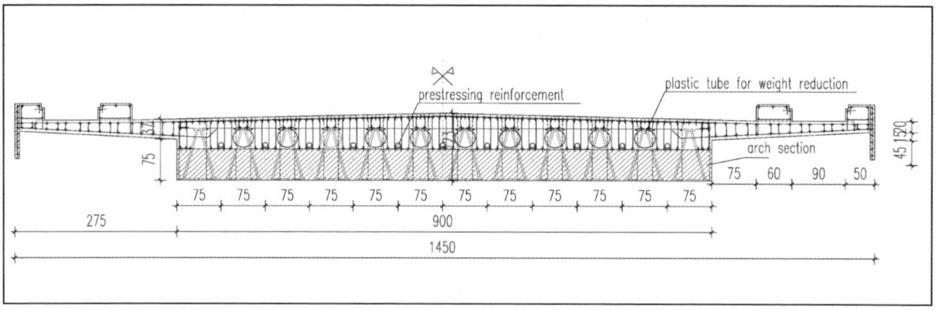

Figure 14. Section of the bridge at midspan.

It is crossed by an important river, the Piave, and its tributaries; until year 2002 there was only one bridge connecting the two river banks, in correspondence of the city centre. In order to reduce the amount of vehicles circulating, the municipality planned the realization of an alternative road system, permitting to reach the two sides of the river without passing through urban areas. Two bridges have been built as part of the new project: one on the river Piave, spanning about 500 m, and a smaller one, on the river Piave Vecchia.

The latter is an arch bridge, spanning 63 m and sustained by an arch of 45 m. The deck, 14.5 m wide, is made of a rectangular prestressed concrete section.

The interesting aspect of the structure is that the arch is realized with composite beams, patented as REP beams in Italy, composed of a reticular steel truss with an underlying steel plate which works as a formwork for the following concrete cast. The relatively light steel structure is positioned on the abutments, with only one provisional prop at midspan and bears the dead loads of the structure. In this case, fifteen 100×60 cm REP beams have been employed to form an arch 9 m wide, with a constant height of 1 m.

Each beam, spanning 45 m, is divided in three parts to facilitate transport and successively joined in situ, in order to form, after the concrete cast, a two hinged arch sustaining the deck.

The following figs. 11, 12 show the lateral elements and the central elements of the arch.

The final configuration of the structure is shown in fig. 13a and is commonly known as bow-string: the horizontal thrust produced by the arch is absorbed partly by foundations, and partly by the deck, which acts as a chain. This scheme is quite common in the case of soils with poor characteristics. Reaction coming from the dec, presents an eccentricity with respect to arch thrust and generates a couple (A), which can be balanced by the reaction offered by soil on foundations (couple B), as in our structure (fig. 13a), or by the couple of vertical forces guaranteed by the presence of a counterweight anchored to the deck (fig. 13b). These systems are quite efficient if the terminal part of the structure is sufficiently rigid to avoid relative displacements between foundations and deck, which would nullify the chain action of the latter.

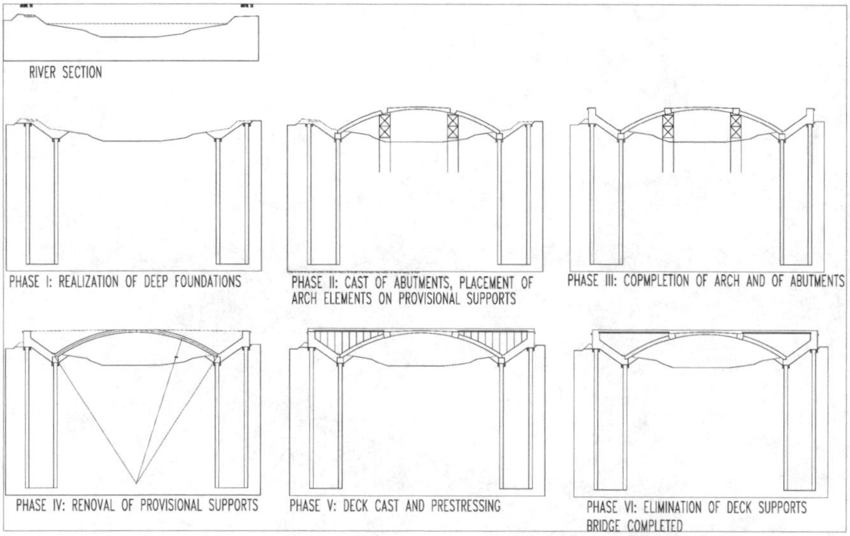

Figure 15. Erection stages of the bridge.

Figure 16. REP steel beams are set in place with just a simple provisional support at midspan. Concrete cast will follow with no formwork needed.

The deck is made in prestressed concrete cast in situ and is designed to bear live loads and the additional dead loads coming from the realization of the lateral pavements. Close to midspan arch and deck are joined, in order to reduce section's total height, as it is shown in fig. 14.

Figure 17. Final aspect of the bridge.

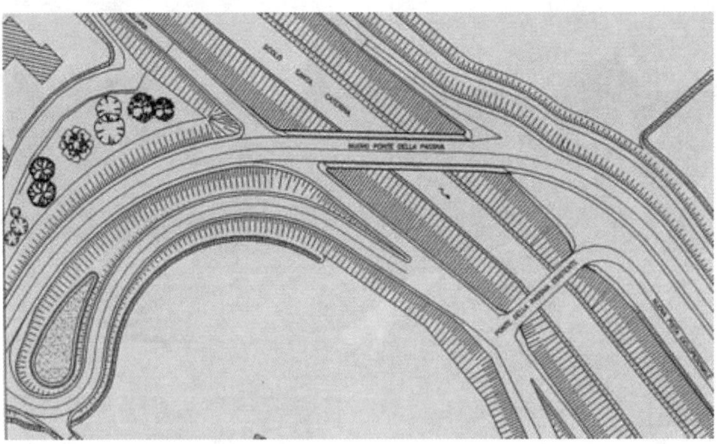

Figure 18. Plan of the area of intervention.

Concerning execution, the simplicity of the structural scheme and the use REP beams permitted to erect the arch and the deck in only 60 working days. The whole structure, with finishes and security devices, was completed in four months.

Construction phases are schematized in the following fig. 15.

The following images show the bridge during erection and the final result.

5 BRIDGE IN S. URBANO, PADUA, 1998

The bridge on the S. Caterina channel is part of the road system realized close to the city of Padua in order to guarantee an easy access to the provincial landfill. Consequently, the main type of

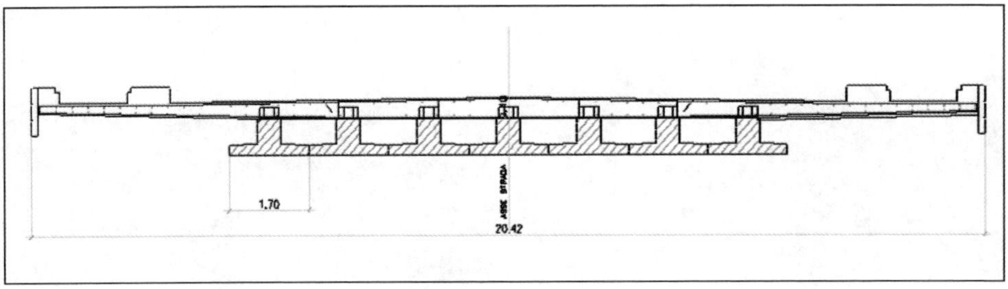

Figure 19. Cross section of the deck.

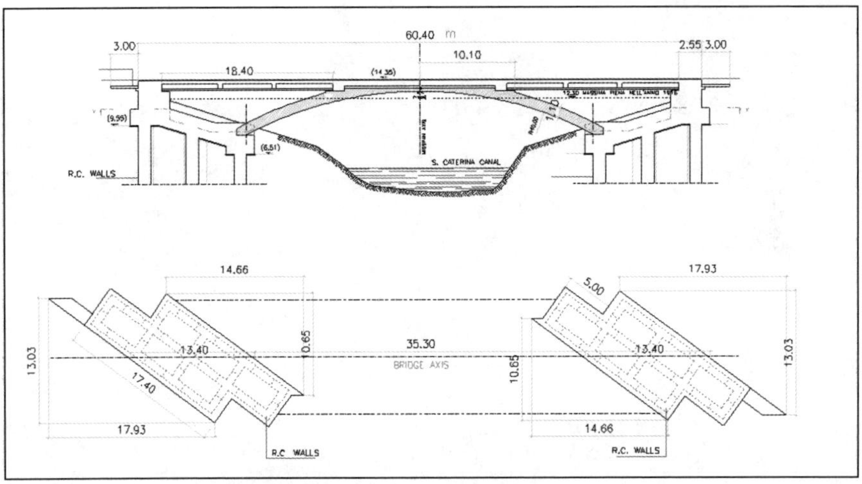

Figure 20. Longitudinal section and plan of the foundations of the bridge.

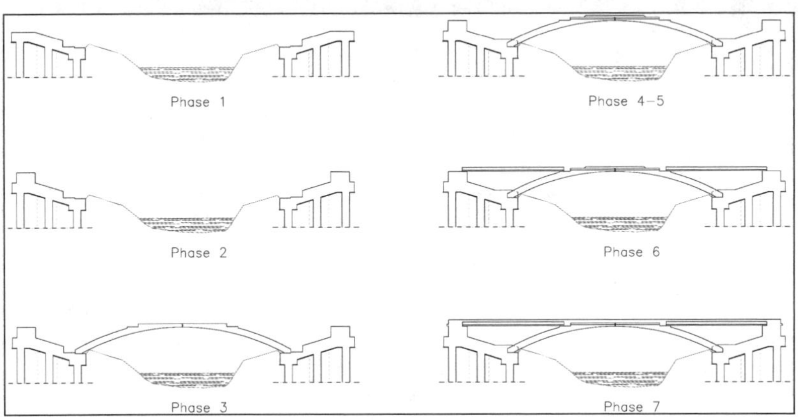

Figure 21. Building phases of the bridge.

vehicles circulating on the structure is composed of heavy load trucks. However, being the landfill quite far from the bridge, many cyclists use the structure to reach a closer park.

The bridge has two lanes and spans as a whole 60.4 m. An arch 37 m long, with a rise of 4.54 m, sustains the structure.

Figure 22. Stages of realization of the bridge.

An important complication to design was the strong obliquity of the channel with respect to the road, as it can be seen in the following fig. 18.

The deck is composed of precast T shaped beams, supported at midspan by the arch and at the edges by deep foundations. On each side of the deck a cantilever of 2.5 m allows the realization of large pavements for pedestrians and cyclists. Precast elements are made monolithic with the underlying arch and with the foundations at the abutments by means of a in situ concrete cast.

The arch has been realized with 14 + 14 REP composite beams (similar to the ones used for the bridge on the Piave Vecchia), each one spanning half of the arch length and set aside to form a 7 m wide arch. The original solution, proposed by the designers, considered the possibility to avoid completely a provisional support for the steel elements, with lateral stability guaranteed by inclined poles joined to the arch and bearing on the ground. Following this procedure, the elements form initially a three hinged arch and the horizontal thrust is borne only by the foundations, which result to be quite massive. In the final stage, after the concrete cast, the thrust is partly balanced by the deck which acts as a chain, similarly to what happens in fig. 13a.

This solution was considered as the easiest to realise, even if it needed an bigger steel truss to support the concrete cast and bigger foundations. The absence of provisional support was considered by designers as a great help to execution procedure.

Figure 23. Aspect of the bridge.

The solution proposed by designers can be schematized in seven phases, represented in the following fig. 21.

The contractor proposed a different solution, which has been chosen, with two provisional steel element supporting the arch before the complete gain of concrete's strength. This solution has been preferred because it permitted to reduce the stresses in the transitory phases of construction, allowing the realization of smaller foundations and steel elements with a reduced weight.

The following images show the sequence of construction, and the final aspect of the bridge. After the realization of the foundations and the positioning of the steel elements (fig. 22a, b) concrete has been poured directly on the steel elements with no formwork needed. It is interesting to see that the strong curvature of the arch gave some problems of concrete flow during cast, so the contractor had to use concrete with low slump in order to realize the structure properly. The strong inclination of the structure with respect to the river bed has been solved displacing longitudinally the elements of the arch, forming a "shagreened" surface, as it can be seen in fig. 22e. Figure 22f shows the static check of the bridge, compulsory by law in Italy, with heavy load trucks.

6 FINAL CONSIDERATIONS

The attention to cost reduction by means of an industrialized constructive process and the identification of a structural shape able to respond to the modern requirements of aesthetics and durability are the common thread in the design of the bridges presented in this paper.

The choice of different realization procedures, as shown, strongly influences design decisions, and vice versa. The recourse to precast elements in fact imposes a careful evaluation of the structural behaviour during the transitory stages of construction, especially with respect to the long term viscous phenomena of concrete.

Conception, design and realization cannot be considered sequentially, but have to be thought together, because any step of one implies consequences on the other.

Realization, which is the last part of the process, is actually part of the design, due to the limitations that technical capabilities impose, or offer, to the design choices.

Structural conception and realization are then indissolubly linked to form an unique design process.

REFERENCES

AITEC, 1974. Realizzazioni italiane in cemento armato precompresso 1970/74.

Alberga, G. 1953. I ponti – La pratica, Turin, UTET.

BA 42/96, 2003. Amendment No. 1, "The Design of Integral Bridges – Design manual for roads and Bridges".

Benvenuto, E. 1981. La scienza delle costruzioni e il suo sviluppo storico, Firenze, Sansoni.

England, G. L., Tsang, N. C. M. & Bush, D. I., 2000. Integral Bridges – A fundamental approach to the time temperature loading problem. Thomas Telford.

Lodigiani, V. 1991. Sperimentazione e cantiere, *R. Morandi*, Roma, Gangemi.

Varoius Autors, 2003. Rassegna: *The World of Bridges*, Editrice Compositori, Bologna.

Zanchettin, A., De la Grennelais, E. & Siviero, E. 2006. The new bridge over the river Piave in S.Donà di Piave. *2nd Fib Conference; proc. intern. symposium, Naples, June 2006.*

Zanchettin, A. & Siviero, E. 2005.Tecniche innovative per la realizzazione di ponti e viadotti a piccola e media luce. In M. Arici and E. Siviero (ed.), *Nuovi orientamenti per la progettazione di Ponti e Viadotti; proc. intern. symposium, Taormina, March 2005.*

Chapter 25

Hawkesbury Railway Bridge near Sydney, Australia

Kirti Gandhi
Gandhi Engineering Inc., New York City, USA

ABSTRACT: In January of 1886, an American contractor, Union Bridge Company of New York City, won an international competition to design and build a two-track steel railroad bridge of approximately 3,000 feet in length over the Hawkesbury River in New South Wales, about 30 miles north of Sydney, Australia. At the time it was the biggest public works project in the southern hemisphere. This paper gives a background of this project; details of 14 designs submitted by different contestants from England, France, Australia, and the US; construction methods; key individuals involved in this project; difficulties encountered during construction; and its successful completion. The caissons supporting the piers were made of iron plates, fabricated in a British shipyard from drawings sent out by the Union Bridge Company, and sunk entirely by open dredging. The depth of the caisson supporting one of the piers reached 160 feet below the high water level, a record at that time. The bridge was completed in 34 months and opened to traffic with great fanfare in May 1889. It linked the north and south regions of Australia. The bridge was strengthened several times and ultimately replaced in 1946.

1 INTRODUCTION

The subject of this paper is a bridge over Hawkesbury River (Figure 1) about 30 miles north of Sydney, Australia in the province of New South Wales (NSW). There was a scheme of connecting Brisbane in northern Australia to Adelaide in southern Australia and linking the cities of Sydney and Melbourne. The railway lines were brought to the north and south of the Hawkesbury River (Figure 2). The plan and cross-section of the crossing are shown in Figure 3.

Figure 1. Aerial view of Hawkesbury River Bridge area. (Copyright 2007 – Digital globe).

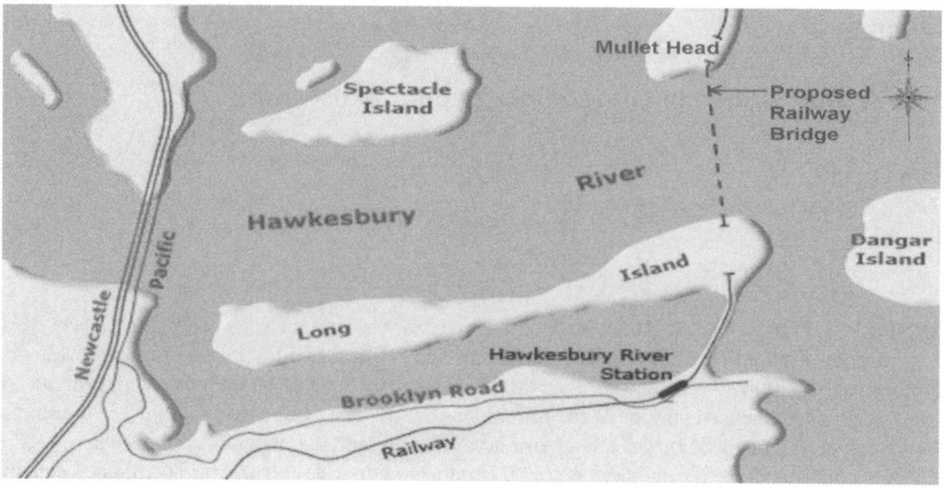

Figure 2. Location of Hawkesbury River Bridge (Google, 2007).

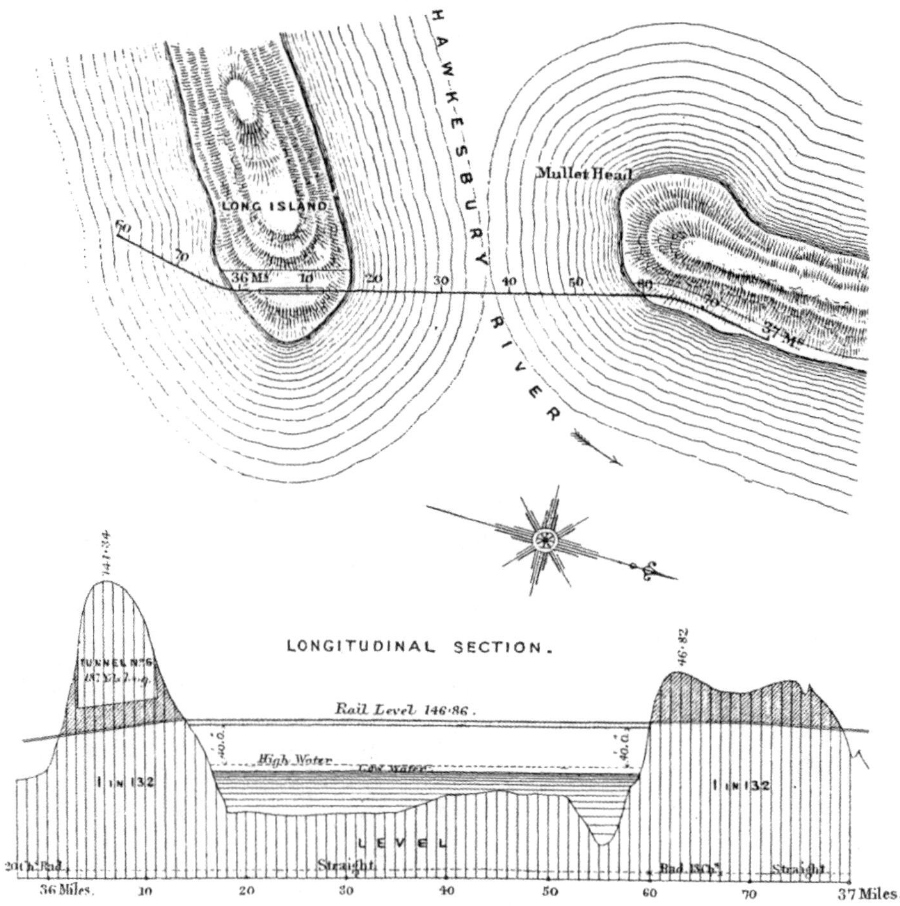

Figure 3. Plan and section of railway crossing (Burge, 1890).

NEW SOUTH WALES.

To Iron and Steel Bridge Builders.

The Agent General for New South Wales invites Designs and Tenders from Iron and Steel Bridge Builders of experience for a Steel Bridge to carry a double line of Railway across the River Hawkesbury, near Sydney, New South Wales.

The distance between the Abutments of the Bridge is about 2,900 feet. The Bridge is to be a clear height of 40 feet above high water. The River is an average depth of about 50 feet, and the foundations will have to be carried down to about 120 feet below bed of the river.

The terms and conditions, with a section of the river, and plan of site, can be obtained from the Agent General on payment of one guinea, to be sent with the application, which should be made in writing and accompanied by a list of works of similar character previously executed by the parties applying.

The Designs and the Tenders are to be delivered to the Agent General for New South Wales at the undermentioned office on or before the first day of June, 1885.

SAUL SAMUEL,
Agent General for New South Wales,
5 Westminster Chambers, S. W.,
5th February, 1885. London, England.

Figure 4. Invitation to submit designs and tenders for Hawkesbury Bridge (Scientific American, 1885b).

Soundings and borings taken by Mr. John Whitton, Engineer-in-Chief for Railways, NSW, and his staff indicated a bed of mud extending from 60 to 170 feet below high water level (HWL) overlying the sand. The foundation had to penetrate through water, mud, and soft sand down to hard gravel about 185 feet below the HWL. The difference between the HWL and low water level was about 5 feet. It was decided to place the rails 40 feet above the HWL making the height of a pier of 185 feet + 40 feet = 225 feet. Building such a deep foundation was never attempted before, and there was no assurance that it was feasible with the technology existing at that time.

Sir John Fowler & Co. was the consulting engineers to the province of NSW. Two key individuals of this firm, John Fowler and Benjamin Baker, played important roles in advocating to the government of NSW to invite international tenders for the design and construction of the Hawkesbury Bridge.

In November 1884 a circular signed by Whitton was issued by the Government of NSW inviting bridge builders in Australia, Europe and America to send to the Agent General of the Province, Sir Saul Samuel in London on or before June 1, 1885 their own detailed plans for a double-track steel bridge spanning the Hawkesbury River. The advertisement prepared by Samuel and published in a U.S. magazine is reproduced in Figure 4 (Scientific American, 1885b).

Samuel also named a board of three prominent engineers to examine and report upon the plans submitted. Two engineers Sir John Hawkshaw and Col. Douglas Galton of Britain, and W.W. Evans of the U.S. (Engineering News, 1885). Evans was one of the first American engineers to go to South America to build railroads. In 1850 he went to Chile to build Copiapo railroad which he completed in 1853. While there, he built the first pier ever built on the coast of South America. During the late 1870's, he was engaged in building and supplying a large amount of railway plant of every kind to Australia, New Zealand, and Mexico (Nason, H.B. and Young, W.H. (1887)).

2 DESIGN SPECIFICATIONS

The performance specifications prepared by Whitton and the Consulting Engineer to the Province of NSW in London, Sir John Fowler, permitted great latitude to each contesting firm in coming up with its own foundation and superstructure design (Engineering News, 1886b).

Loads and Stresses – Live load of six locomotives of 45 tons each and two trains of one ton per lineal foot each together with dead load and a wind pressure of 56 pounds per square foot of surface, shall not cause a stress exceeding 6½ tons (14,560 pounds) per square inch in tension, or 5 tons (11,200 pounds) in compression.

Quality of Material – All of the superstructure was to be of mild steel in rolled sections with an ultimate tensile strength of not less than 30 tons (67,000 pounds) and not more than 33 tons (73,920 pounds) per square inch; the test pieces were to be cut lengthwise or crosswise from the material, and they must elongate 20 per cent before breaking. When heated to a cherry red and cooled in water of 82°F, the strips must bend double without flaw or crack to a curve whose inner radius was one and one half times the thickness of the plate.

Maximum Live Load Deflection – The completed span shall be able to carry a live load of 900 tons without deflecting more than 1/1200th of its length.

Other Parameters – A bridge 2896 feet long between abutments, having a clear headway above the high water level of 40 feet, to carry a double line of railway with a gauge of 4 feet 8½ inches, and piers to be founded at an indicated depth in some cases 170 feet below water.

3 FOUNDATION INVESTIGATION

Whitton and his staff took 11 borings along the proposed alignment of the Hawkesbury Bridge spaced at 260 ft. covering approximately the middle 2,600 linear ft. of river opening between Long Island and Mullet Point. The borings were numbered from south to north and were provided to each prospective bidder.

The two-end borings were 9 ft. and 16 ft. deep below the river bed. The deepest boring was 133 ft. 6 in. Excluding the two-end borings, the average depth of the remaining 9 borings was approximately 112 ft. The records of the test borings are summarized in Table 1 below.

The bridge proposed by Union Bridge Co. is superimposed on the borings log and the resulting diagram is shown in Figure 5 (Engineering News, 1886b).

Table 1. Summary of boring results.

Boring number	Depth below river bed	Findings (distances measured from top)
1	9′-0″	3′ clay, 6′ sandstones, hard sediment
2	60′-0″	56′ black mud and sandstone, 34′ black sand, 10′ good sand
3	112′-0″	35′ softsand, mud and shells; 65′ black mud & sand, 12′ very hard dry sand
4	109′-6″	16′ sand, mud & shells; 85′ strong mud, 3′ loose sand, 5′-6″ hard sediment bored
5	105′-0″	61′ black mud, 34′ black sand, 10′ light good sand
6	114′-0″	16′ light mud, 60′ strong mud, 28′ black mud, 10′ hard sand
7	119′-0″	26′ light mud, 47′ strong mud, 44′ sand, 2′ hard sediment
8	130′-0″	27′ light mud, 56′ strong mud, 32′ black sand, 15′ very hard sand
9	133′-6′	21′-6″ very light mud, 84′-6″ strong mud, 4′ hard sandy sediment, 15′-6″ loose sand, 8′ hard sediment
10	26′-0″	31′ light mud, 87′ black mud, 8′ very hard sediment
11	6′-0″	10′ soft mud, 6′ hard sediment

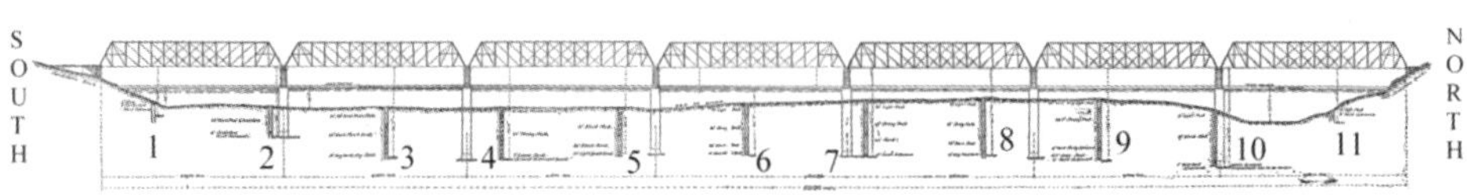

Figure 5. Location of borings with superposition of the bridge.

4 INTERNATIONAL COMPETITION

By June 1, 1885 fourteen designs were submitted to the Review Committee of Engineers slightly different than announced earlier in November 1884. Sir John Hawkshaw of Britain and W.W. Evans of the U.S. were dropped, and all three committee members were from Britain as follows:

1. W.H. Barlow, Engineer of the new Tay Bridge
2. Douglas Galton, Engineer of the Board of Trade, and
3. George Berkley, Past President of Institution of Civil Engineers.

The amounts of the tenders (excluding the Phoenix Bridge Company's offer which was for an iron bridge and not a steel bridge) ranged from £296,350 to £702,384 or US $1,440,261 to $3,413,586 (1£ = $4.86), and the time for completion from two to four years. Three tenders were submitted from the U.S., one from France, two from Australia, and eight from England and Scotland. The fourteen designs are shown in Figure 6 (Engineering, 1886a; Engineering News, 1886f).

The Committee of Engineers decided that while several of the plans for the superstructure submitted were of sufficient merit to warrant adoption, the plans of the Union Bridge Co., of New York were the only ones that were satisfactory and could be recommended. This report was then submitted to Sir John Fowler & Co. which approved and seconded the recommendation to the Province of NSW. Mr. Whitton approved the recommendation made by the Committee and Sir John Fowler & Co.

The plan of the Union Bridge Co. was for a double track steel railway bridge 2,896 ft. long between end pins, divided into five spans of 416 ft. each between pier centers, and 2 spans of 408 ft. each. The piers were to be cut stone masonry from low water to the bridge seat, and they were founded on a single iron caisson for each, 48 ft. long and 20 ft. wide with rounded ends and vertical sides. Each caisson was provided with 3 dredging wells, each 8 ft. in diameter with the concrete pockets surrounding these wells strongly braced between the wells and the shell. The cutting shoe of the caisson was 20 ft. high with an outward flare of 2 ft. all around in this height.

On January 27, 1886 the Government of NSW sent a cablegram to the Union Bridge Co. that its tender had been accepted for £327,000 or US $1,589,220, the highest of the three American bids (Engineering News, 1886a). The basic reason cited for this award was that the plan of this company for the piers was the most engineering-like of any presented and promised the greatest measure of success.

The Province of NSW had budgeted £40,000 for alterations in the project requirements after the award of the contract for a final sum of £367,000. The quantity of masonry was increased by about 700 cubic yards at the request of the railroad, and hence the contract price was increased from £327,000 to £340,000. There were some minor improvements recommended by the Union Bridge Co. for a solid floor system of built beams and buckle plates which were approved by the owner which raised the final contract price to slightly higher, but it was below the budgeted amount of £367,000.

5 REACTION OF THE PRESS AND PEOPLE

There was an apprehension in Britain that the American firm will claim "extras" to compensate for its low tender. An inquiry by "*Engineering News*" for the "extras" to the Union Bridge Co. resulted in the response that they had not heard of any, and there probably would not be any unless the Australian authorities ordered them; and specifications clearly stipulated what constituted extras (Engineering News, 1886e). In general, the press seemed to be more critical of British firms than people.

"*Engineering*" apparently was not pleased with the outcome, and noted that "in a case of this kind, where the distance was too great to expect contractors to make a personal examination of the site, a higher class of design and a greater equality in the amounts of the tenders would have been attained if the conditions had been included the alternative of either contractor's tenders on their own plans, or upon one by the Engineer-in-Chief for Railways, whose success in other works and long local experience could not fail to have been embodied in his design, giving confidence to those who no doubt under these circumstances would have tendered, and assisting others who proposed modifications or alternatives, and so enabling a selection to be made from the majority instead of the minority of firms capable of carrying out so important a work." (Engineering, 1886a).

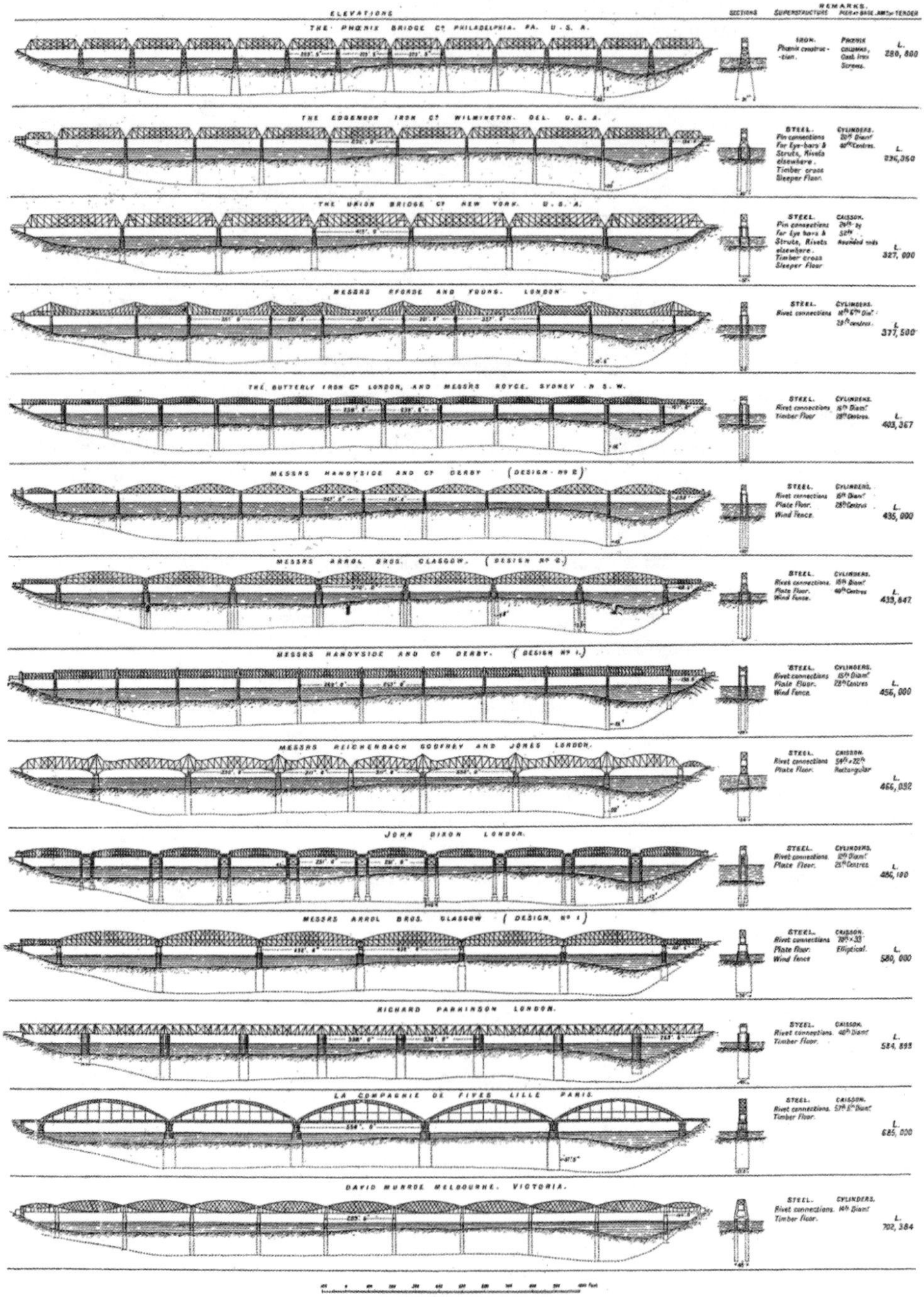

Figure 6. Competitive designs for the Hawkesbury River Bridge.

The *British Australasian* remarked that "£327,000 worth of work had been lost to the engineering trades of the mother country. It is a bad sign for the home supremacy in trade that American firms and German firms can at this juncture step in and undersell her iron and steel manufacturers in Australasian markets, and that not alone in rails and locomotives, but in bridgework as well." (Engineering News, 1886c).

The *Colonies and India* reported that "Undoubtedly, colonists would have preferred that the work should have been done by an English firm, and the fact that a New York firm is able to beat us out of a field, which almost by prescriptive right, should be our own gives rise to suggestions that are by no means pleasant or creditable to us as a competing nation." (Engineering News, 1886d).

The *Mechanical World* commented that "The Hawkesbury Bridge is being made in Glasgow to American designs. This shows that the Union Bridge Co., who secured the contract, consider they are able to obtain a cheaper bridge in Great Britain than they could at home. It is an ample contradiction to the pessimists who have sung our failing trade, but it is not very flattering to bridge designers in this country, and will no doubt by the wise be taken as evidence of a necessity for reforming our present practice." (Engineering News, 1886g).

Mr. John Dixon, a reader of "Engineering", believed that in awarding the bridge contract to an American firm, the Government of NSW was setting a questionable precedent and what he characterized as getting a "cheap and nasty" bridge; that it was impossible to build piers 100 feet in the ground and 160 feet below high water level, and that American firms were working for nothing but the name (Engineering 1886b).

A British engineer Frederick T. G. Walton reminded Mr. Dixon and others that for a new railway bridge over the Ganges at Benares in India, some of the piers were founded 140 feet below the low water level, and in one case the actual depth of ground penetrated was 145 feet (Engineering 1886e).

A foreman of a British bridge building firm identified himself as a "Guinea Pig"; and lamented that the American design was inadequate, produced a very light structure, required less than a first class workmanship, and would not be acceptable to the inspectors of Sir John Fowler (Engineering 1886c).

As if higher wages, short working hours, high railway charges, and heavy taxation were not sufficient enough to drive business away from Britain, one reader "Bridges" blamed British engineers and inspectors for their rigid attitudes, whims, and caprice for increasing cost of bridge construction by their unreasonable demands, and thereby increasing the cost of tenders submitted by British firms (Engineering 1886d).

6 CONSTRUCTION TEAM

Union Bridge Company had subcontracted most of the work to highly competent firms primarily in Britain and the U.S.

1. Union Bridge Company, New York, U.S.A.	Prime Contractor
2 Head, Wrightson and Co., Stockton-on-Tees, Britain	Fabrication of iron caissons
3. William Arrol and Co., Glasgow, Scotland	Fabrication of riveted steel superstructure
4. Colville, Glasgow, Scotland	Suppliers of steel
5. Steel Company of Scotland	Suppliers of steel for eye-bars
6. Burge and Barrow, Kent, Britain	Suppliers of Portland cement
7. Louis Samuel, Sydney, Australia	Stonework
8. Anderson & Barr, New York, U.S.A.	Foundation, sinking of caissons
9. Ryland & Morse, Chicago, U.S.A.	Erection of superstrucure

Mr. Samuel died during the construction of the bridge, and Mr. Ryland died in an accident almost at the end of the Hawkesbury Bridge erection.

7 CAISSONS AND PROBLEMS DURING THEIR SINKING

The general arrangement of pier sinking is illustrated in Figure 7 (Scientific American, 1886). This was before the actual sinking started in December of 1886, and was based on modification of the scheme John Anderson used in sinking of piers in Atchafalaya, a tributary of the Mississippi River in Louisiana in 1883 (Engineering News, 1883a).

The caissons were sunk entirely by open dredging. They were made of iron prepared in an English shipyard from drawings sent out by the Union Bridge Co. The parts when received went together with great accuracy, and the work left nothing to be desired in workmanship.

The piers were to be cut-stone masonry from low water to the bridge seat and they were founded on a single iron caisson, 48 ft. long and 20 ft. wide with rounded ends and vertical sides. The cutting edge, or shoe, of the caisson was 20 ft. high with an outward flare of 2 ft. all around in this height, making the bottom dimensions 52 ft. × 24 ft. Each caisson was provided with 3 dredging wells, each 8 ft. in diameter. Concrete pockets were formed between the outer and inner skins of the caisson all around, and between the dredging wells. The concrete used was composed of 1 part cement, 3 parts sand, and 6 parts hard broken stones. Different parts of the caisson were strongly braced against internal and external loads.

Plan, longitudinal section, and cross-section of caisson and shoe, and details of transverse cutting edge are shown in Figure 8 (Burge, 1890). The 20 ft. lowermost section of a cutting shoe of a caisson on the launching ways is shown in Figure 9. When a caisson was launched and towed

Figure 7. Sinking of caisson of Hawkesbury Bridge with store ship on left and concrete and supplies barge on right (Scientific American, 1886).

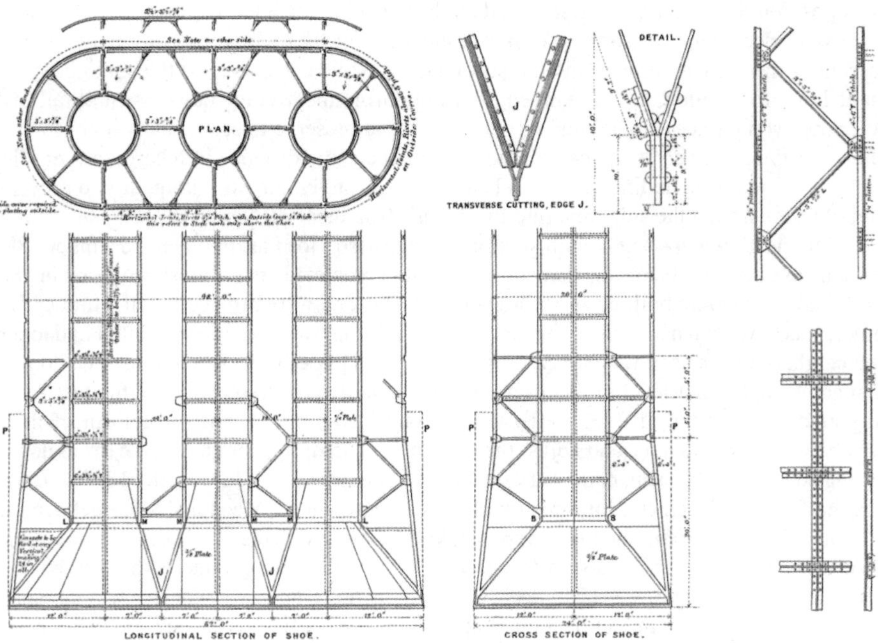

Figure 8. Plan and cross-section of caisson, shoe and details of cutting edge (Burge, 1890).

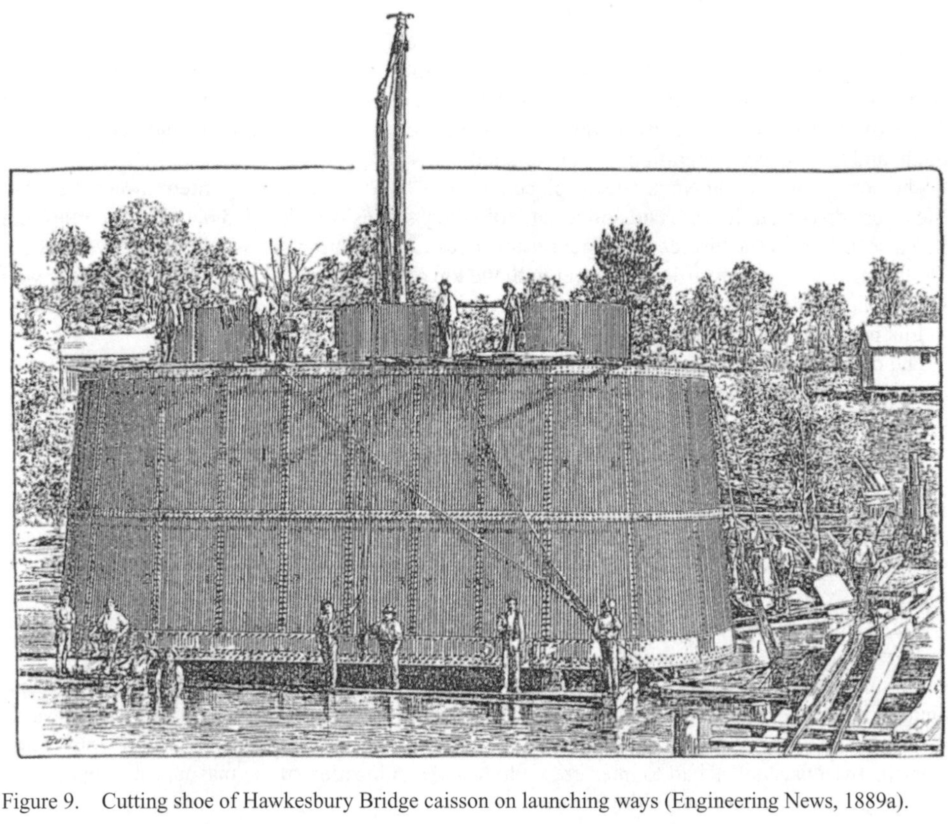

Figure 9. Cutting shoe of Hawkesbury Bridge caisson on launching ways (Engineering News, 1889a).

to its position for sinking, Anderson used six heavy ship's anchors, with 1-½ in. mooring chains. He also connected the caisson to the shore using three or four wire ropes for maintaining vertical descent of the caisson through the top soft material while concreting of the sinking pockets was carried out. In addition, to maintain equilibrium during the average daily rise and fall of 5 ft. in tide, Anderson pumped in and out sufficient water to preserve approximately a constant weight. This procedure stopped as the caisson had penetrated some distance into hard bottom. The riveting work of attaching additional panels had to be equal to boiler work in quality to prevent intrusion of water around the sides and into the 3 wells (Engineering News, 1889a).

The foundation work began with sinking of the caisson for Pier No. 5 in December 1886. This caisson gave more problems than any other and was completed after almost two years in 1888. The 20 ft. high shoes were built on shore at Dangar Island, and were provided with a timber false bottom. It was towed out and sunk to the bottom by admitting water between the outer and inner skins. The height of the caisson was increased by adding iron plates in 10 ft. sections. The concrete was kept above high water level in this caisson as the sinking progressed, and to this fact Macdonald attributed much of the difficulty afterwards found in righting it when it got out of the line. It became impossible to vary and adjust the weight; and dredging could not be carried down to the cutting edge, as the caisson, being very heavy, sank too fast (Railroad Gazette, 1890a). Burge, who was the Resident Engineer, disagreed with this assessment by Macdonald. He believed that the splay at the bottom of the shoe was responsible for these problems.

As of January 24, 1887, Caisson No. 5 was down 50 ft. in the ground with 90 ft. of connected iron work, and the water was 25 ft. deep. Preparations were underway for the next caisson, which was No. 4. Materials for the fabrication of the remaining six caissons had arrived in Sydney Harbor from England (Engineering News, 1887a).

The caisson was shifting eastward as it was sinking. At about 75 ft below the river bed, the bottom was 5 ft. to the east of the bridge centerline whereas the top was about 3 ft. The specification allowed a maximum lateral shift of 2 ft. As of April 13, 1887, the foundations for Piers 4 and 5 were sunk approximately over 80 ft. and 90 ft., respectively, below the bed of the river (Engineering News, 1887b). Excavating further down, the orange-peel bucket which weighed three tons got caught in the bottom of the middle well. A second and smaller bucket was afterward caught in the chains of the first bucket. A diver was sent down to recover the buckets, but the depth was too much, and the recovery operation was abandoned.

The bottom of Caisson No. 5 was displaced towards the east and the top tilted toward the west. Piles were driven on the east end and a mass of heavy stones was dumped between the piles and the caisson to prevent further movement toward east. The contractor installed a cribwork about 400 ft. upstream or westwards to connect with the top of the caisson, and dug into the eastern well with the hope of correcting the caisson's transverse displacement. This arrangement did not work. At this point the caisson was in sand at 144 ft. below high water level.

The contractor proposed to sink a supplementary caisson on the west end to shift the center of gravity of the combined mass towards the center line of the bridge. When this new caisson was sunk to about 106 ft., it collapsed from the pressure on the outside. So, any further sinking became impossible.

As a last resort the contractor removed the top of the supplementary caisson and side of the main caisson about 12 ft. 6 in. under the low water level with the intention to corbel with solid stones. He built a cofferdam and laid stones of 7 to 8 ft. in length with a 9 in. overhang in each course. This resulted in the center of the column of masonry coinciding with the west truss centerline, and was well within the base of the original caisson. This caisson was finally accepted.

About three months after the launch of Caisson No. 5, Caisson No. 4 was sunk without any problem to a depth of 147 ft. below high water level within the permissible tolerance.

Sinking of caisson No. 6 proved to be difficult for a different reason. At the location of the caisson the river bottom had a downward slope (Figure 10). The caisson was leaning forward towards the shore, and continued to do so as the sinking progressed. The caisson also twisted slightly, however, the twist was not as bad to interfere with the correct location of the masonry pier upon it, at right angles to the centerline of the bridge.

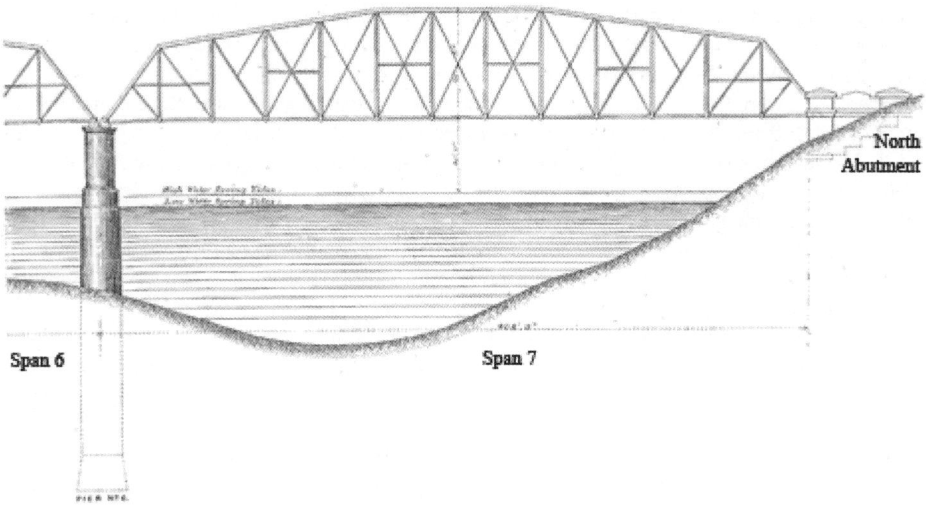

Figure 10. Downward slope of river bottom at Pier No. 6 which caused the pier to tilt to the right thereby increasing the length of Span No.6 from 416′-0″ to 420′3″ (Burge, 1890).

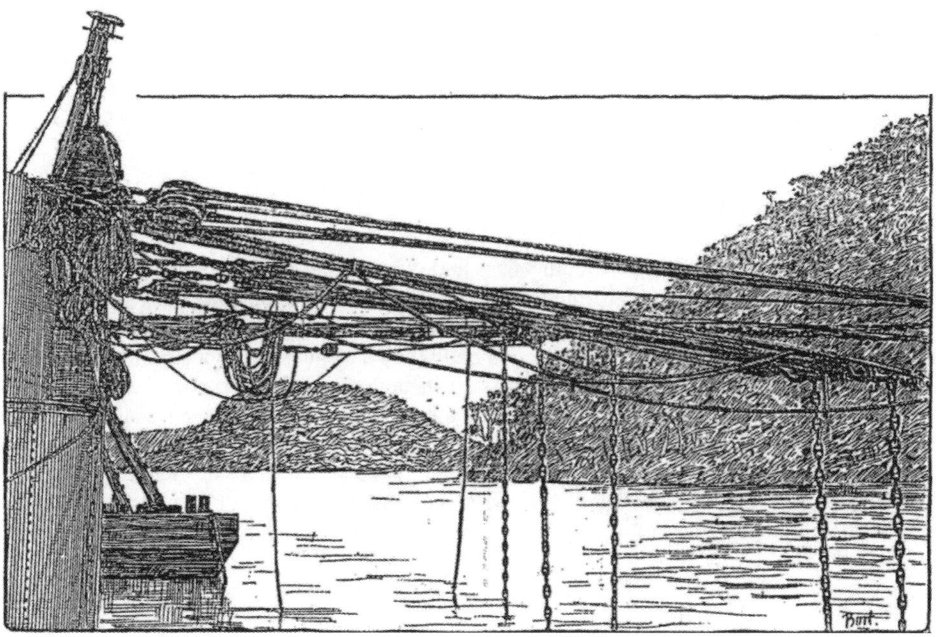

Figure 11. Arrangement to pull top of caisson No. 6 towards shore (Engineering News, 1889a).

To pull the top of Caisson No. 6 towards the shore and thus push the bottom away from the shore at the depth of about 80 ft. below the river bed, Anderson used four large iron sheaves, two on the caisson and two on the shore; and through these a wire rope 1-1/2 in. in diameter was run with strong luff-tackles attached to either end of the cable and the hauling parts of these falls led to winding engines stationed on the barges. Three chain cables of 1-1/2 in. studded round iron were also used as shown in Figure 11 (Engineering News, 1889a).

In Figure 11, the vertical parts are the loose ends of chains hanging down in the water. To further secure an away-from-shore push on the caisson and to obtain a better fulcrum for the top inward pull, Anderson dumped several thousand tons of rock on the shore side of the caisson. Even these measures were not sufficient to stop the shoreward lean of the caisson. For a pier that was over 200 ft. tall from its bottom to the top, this lean increased the distance between Piers 5 and 6 by 4 ft. 3 in. Union Bridge Company had to obtain permission to increase the span from 416 ft. to 420 ft. 3 in. from the Government of NSW.

Fortunately for the Union Bridge Co., it had not started any work on the north abutment. So the abutment location was moved by 4 ft. 3 in. north to keep the length of span No. 7 the same as originally designed. Caisson No. 6 was the heaviest among all the caissons, and the maximum pressure on its base was about 9 tons per square foot.

About 3 months after the launch of Caisson No. 6, Caisson No. 1 was readied for sinking. For sinking caissons No. 1, 2, and 3, the contractor did not use the splay plate and instead used a vertical plate outside the caisson to form a pocket which was filled with concrete from the bottom cutting edge to 24 ft. high.

Caisson No. 1 was tied to a 700-ton vessel which was used for storage and workmen's living quarters. On the night of August 18, 1887 the wind and tide were too strong for the anchorages; and the caisson, the 700-ton vessel, and derrick punt were carried away some distance down the Hawkesbury River. After a week, when the storm subsided, Caisson No. 1 was sunk without difficulty about 101 ft. below the high water level.

Caisson No. 2 was launched about 2 months after No. 1, and four months later was sunk at 155 ft. below the high water level. Two months after the launch of Caisson No. 2, Caisson No. 3 was positioned for sinking, and it took the least time, about 9 weeks, to reach a depth of 146 ft. below the high water level.

The pertinent details of sinking of caissons for various piers are summarized in Table 2 (Engineering News, 1889a).

8 FABRICATION OF SUPERSTRUCTURE

Exhaustive details of the bridge including sizes and connections of all members, and stress diagram of the truss under the worst loading conditions are covered in Engineering, 1887a,b,c,d.

Due to deep water presence of underlying soft mud it was not practical to use falsework to erect the bridge. So the Union Bridge Co. decided to build the truss spans on a pontoon which was supported on piles. On the pontoon was a trestle 50 ft. high on which the span was erected. The extreme height of the lifting apparatus used on the pontoon was 120 ft. above its deck. Even though the typical bridge span was 416 ft. long, due to shallow waters near both abutments, the length of the pontoon was reduced to 335 ft. with approximately 80 ft. of overhang. The width of the pontoon was 61 ft. to permit sufficient working space and the depth was 10 ft. to keep the pontoon floating under the 800-ton weight of the truss span. The pontoon was divided into 44

Table 2. Summary of sinking of caissons

Pier number	Date started	Date completed	Depth, HWL to river-bed	Depth below river-bed	Depth below HWL	Total height of pier
1	7/1887	10/1887	44′-6″	56′-3″	100′-9″	140′-9″
2	9/1887	1/1888	45′-0″	110′-0″	155′-0″	195′-0″
3	11/1887	2/1888	48′-0″	98′-0″	146′-0″	186′-0″
4	3/1887	6/1887	26′-0″	120′-8″	146′-8″	186′-8″
5	12/1886	10/1888	24′-6″	119′-6″	144′-0″	184′-0″
6	4/1887	7/1887	52′-0″	110′-0″	162′-0″	202′-0″

watertight compartments, and covered with galvanized iron plates on bottom and sides (Engineering News, 1888b).

The two main trusses were 410 ft. 0-1/2 in. long between centers of end pins and 58 ft. high at the center. Each truss was divided into 13 panels. The distance between the two trusses was 28 ft. on centers. The compression members were built-up members composed of plates and angles riveted together and braced using diagonal lacing bars on the open side. The tension members for both diagonals and bottom chords were pin-connected eyebars.

The floor beams were fabricated from plates and were stiffened with both lateral and cross bracing. The bridge superstructure was provided with a system of lateral bracings between both top and bottom chords. The end raking posts forming the portals for each span were stiffened laterally by cross frames latticed with angle bars, and carried down just above the height of the locomotives and cars.

The rails were supported on timber decking which in turn was supported on stringers. The ties were 24 ft. long, 9 in. wide and 8 in. high, and were spaced 16 inches on centers. A 2 inch thick planking was provided between the rails without stone ballast.

The steel for the superstructure was made by the open hearth or Siemens-Martin process with a tensile strength of between 30 and 33 tons per square inch and having an ultimate elongation of 20 per cent in a length of 8 in. (Engineering, 1887d).

When the entire superstructure was ready for erection, in order to float the pontoon, it was emptied at low tide, the valves of the watertight compartments were closed properly to prevent its filling again as the tide rose, and at high tide it was afloat. The rise and fall of tide around the Dangar Island where fabrication was carried out was about 5 ft.

9 ERECTION OF SUPERSTRUCTURE

Span No. 4 was erected first. Figure 12 shows the arrangement for moving the pontoon carrying the fully fabricated span from the pier on Dangar Island to its final location on the bridge in four different positions. Figure 13 shows position of the pontoon in third position and between third and fourth position. This being the first span erection, Ryland of Ryland & Morse, the erection subcontractor, took advantage of full moon and the resulting high tide in dragging the pontoon of its supports and floating into the river.

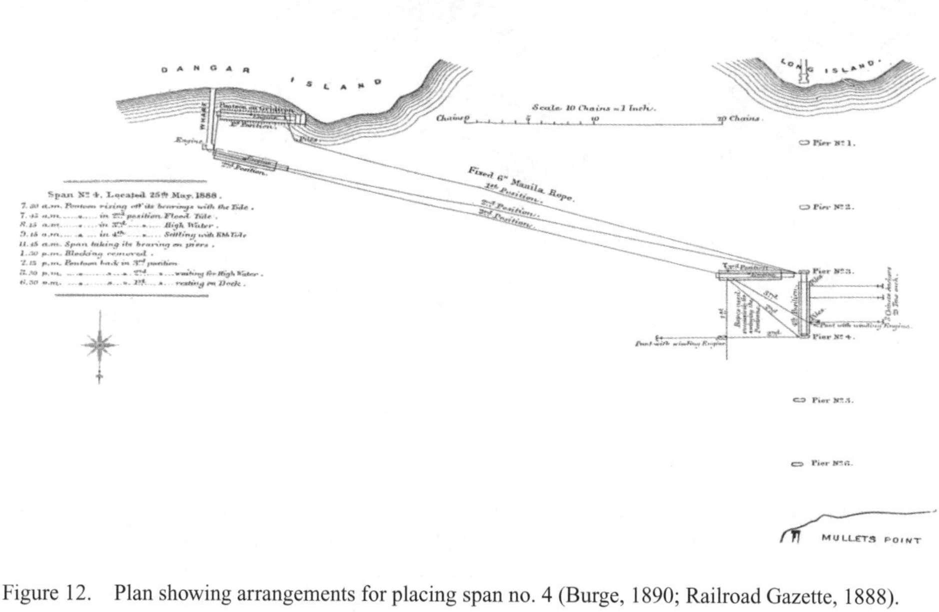

Figure 12. Plan showing arrangements for placing span no. 4 (Burge, 1890; Railroad Gazette, 1888).

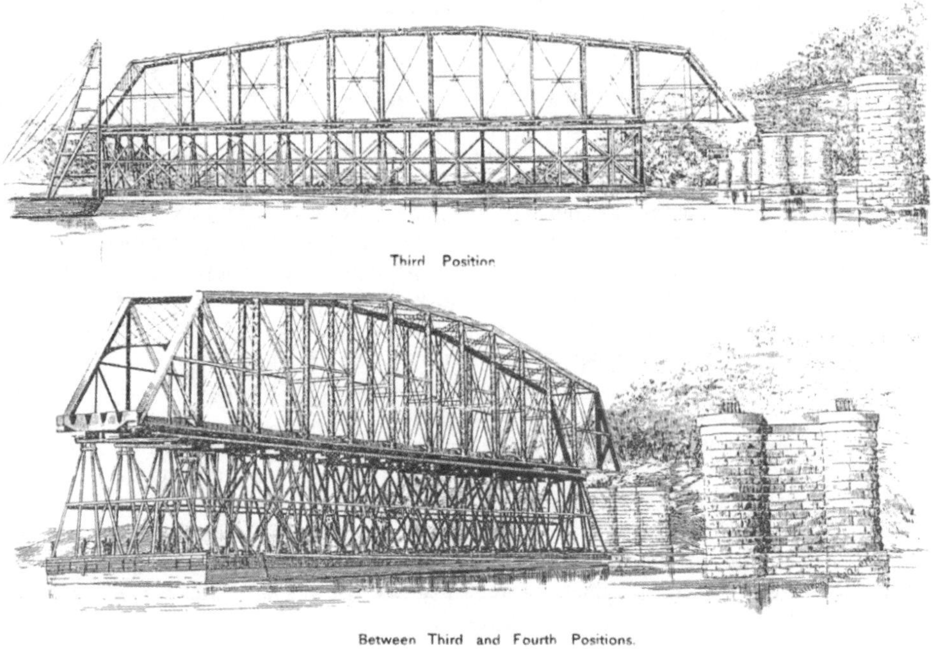

Figure 13. Erection of Span No. 4 in third position showing the overhang and between third and fourth positions (Railroad Gazette, 1888).

Figure 14. Span No. 1 being towed to its position (in the inset) and on abutment and Pier No. 1. span no. 4 in the middle of the Hawkesbury River (Engineering News, 1889a).

The pontoon was hauled out into the stream with the help of two tug boats and by tightening the 6″ manila rope which was deflected around the group of piles in the stream. When the pontoon was in water as shown in second position, the slack of the cable was taken up by the winding engines, and the pontoon was towed up to the third position by the engine and drum on the pontoon itself.

Table 3. Summary of span erection.

Span number	Date started	Date completed
1	July 11, 1888	July 12, 1888
2	September 8, 1888	September 8, 1888
3	August 16, 1888	August 16, 1888
4	May 25, 1888	May 25, 1888
5	January 29, 1889	January 29, 1889
6	March 1, 1889	March 3, 1889
7	October 5, 1888	October 8, 1888

(Engineering News, 1889a).

Here the cables from the punts with winding engines were attached to the pontoon, and the pontoon was swung into the fourth position, between Piers 3 and 4. In this position, it was secured against any movement up stream by the piles against which it was resting, and against movement in the other direction by the cables attached to the Chinese anchors.

When the position of Span No. 4 was very accurate, it was allowed to settle to its seats with the receding tide so that the bolt holes in the shoe and pedestal plates coincided. The entire operation took 11 hours, and out of these hours, 3 hours were consumed in waiting in the second position for high water to enable to brought back to the dock to its original position, ready for fabrication of another span (Railroad Gazette, 1888; Engineering News, 1888a).

It was very difficult to erect Span 1 because of violent currents and the proximity of the rocky beach. The pontoon grounded upon the rocks on a falling tide, and for some hours the pontoon and Span 1 were in danger of destruction. Ryland was able to float the pontoon off the rocks without any real damage, and the span was later safely placed between the south abutment and Pier No. 1 (Figure 14).

Spans 3 and 2 were erected without any incidence. Span No. 7 experienced difficulty due to heavy wind as the pontoon came out of the dock. The 6″ manila rope broke when the pontoon was in the middle of its journey, and Ryland thought it prudent to anchor the pontoon for the night, and request help from Sydney for more streamers which arrived the next day in the afternoon. On the third day morning during the high tide, the Span No. 7 was placed in position.

Span No. 5 was erected smoothly without any problem. For the erection of Span No. 6, the 6″ manila rope broke again during a squall of wind, and the pontoon swung and rotated due to unequal pull. It appeared certain that the Span No. 6 was going to hit the bridge already completed causing and disaster and delay. Again Ryland using dexterous maneuvers saved the span, but the two ends were reversed compared to the original plan. Since the span was symmetrical, it was installed without any superstructural problem (Burge, 1890).

Table No. 3 summarizes the erection of the seven spans.

10 LABOR PROBLEM

There were two American firms working on the construction of the Hawkesbury Bridge. Anderson & Barr was the subcontractor for the foundation work, and Ryland & Morse was the subcontractor for the erection of the superstructure. According to the Government of NSW which was building the railroad connection to the bridge, eight hours were recognized as a day's work.

The American firms insisted that the Australian workers work ten hours per day and offered extra pay to compensate for the additional time. They also wanted to complete the bridge in 30 months as per the term of their contract. Workers of Anderson & Barr refused to work extra hours and elected to strike in November of 1887. At that time Anderson & Barr had completed the foundation for four piers and the fifth one was in progress, and the caisson for the sixth and last pier was ready to be towed into position. The strike was settled in a few days, and the workers agreed to work ten hours per day with extra pay in December 1887 (Engineering News, 1887c).

Figure 15. Buckland, John L., "The Newcastle Express coming off the Hawkesbury Bridge, New South Wales" (National Library of Australia, NLA.pic-ans24768250).

11 TESTING AND OPENING OF THE BRIDGE

The contract between the Union Bridge Co. and the Government of NSW stipulated that the completed span should be able to carry a live load of 900 tons without deflecting more than 1/1200th of its length or 4¼ inches.

Henry Deane, Inspecting Engineer of the NSW Public Works Department was in charge of the load tests performed on April 24, 1889, because of the importance of this bridge, two sets of measurements were planned, one using survey instrument placed on top of the piers, and the other using a water gage. However, the water gage started leaking, and the second method was abandoned.

The 900-ton load was comprised of the following (Engineering News, 1889d):

6 freight engines of 64 tons each	384 tons
16 trucks loaded with steel rails, each weighing 32 tons	512 tons
2 brake vans, each weighing 10 tons	20 tons
TOTAL	916 tons

Deflection was measured after the load was positioned on a span for over 15 minutes, and it was determined to be 2½ inches or approximately 1/2000th of the span length. Deflection measurements were also taken when trains with locomotives were running at maximum speed across the bridge. The deflections measured were less than 1 inch in these tests.

The bridge was opened to traffic on May 1, 1889 by Lord Carrington, Governor of the Colony, in the presence of 700 to 800 invited guests including representatives from other Australian Colonies (Railroad Gazette, 1889; Engineering News, 1889c).

Figure 15 shows the Hawkesbury Bridge in use. On top of the sloping face there is a plaque, a close-up of which is shown in Figure 16.

Figure 16. Close-up of the plaque shown in Figure 15 (Sydney ieaust org au).

12 WHY DID THE UNION BRIDGE CO. WIN AND SUCCEED?

Union Bridge Company was a result of mergers of several highly reputable construction firms which not only performed the design but also did construction work. The firm was composed of Charles Macdonald, who was associated with the Delaware Bridge Company, Thomas C. Clarke, who was a part owner in the firm Clark, Reeves and Co., Col. Edmund Hayes and General George S. Field, both members of ASCE and connected with the Central Bridge Company; and Charles Kellogg and Charles Stewart Maurice of Kellogg and Maurice. The firm of Kellogg and Maurice was one of the pioneers in the building of iron bridges, and second to build those of steel.

Union Bridge Co. maintained plants in Buffalo, NY and Athens, PA. The steel for the eye-bars of the Hawkesbury River Bridge was rolled by the Steel Company of Scotland, and the eye-bars were manufactured by the Buffalo Shop. Before the eye-bars were accepted, they were annealed in special furnaces built for that purpose; and prototype tests were made in the 600-ton hydraulic testing machine also designed and built at Athens for this purpose (Macdonald, 1887). A specimen of the material and workmanship of the bar for which tests were required was sent to Baker in London.

Other noteworthy bridges built by the Union Bridge Co. were the Poughkeepsie Bridge over the Hudson River, the Cantilever Bridge over the Niagara River, the Cairo Bridge over the Ohio River, the Memphis Bridge over the Mississippi River; and several other bridges over the Mississippi, Ohio and Missouri Rivers. Union Bridge Co. submitted plans and estimates of the substructure and superstructure for a proposed suspension bridge for the Sydney Harbor in 1902. This design won a prize, but the project did not materialize.

The general impression in Britain and Australia was that the Union Bridge Co. was a prime contractor on paper only, because it had subcontracted almost the entire project to other firms. Burge (1890), who was the Resident Engineer for the NSW government, mentioned at the end of his paper that "It was a curious circumstance connected with this bridge that though the successful tender was made by an American firm, the whole of the steel and iron, except only that of the eye-bar heads, was provided by the United Kingdom, where also it was manufactured."

Thomas C. Clarke, one of the partners of the Union Bridge Co., responded by saying that because they were designing and constructing bridges for the railroads in the U.S. on a fixed price basis, they had honed their skill and ability to design for any required strength at the least cost (Clarke, 1890). As proof of this skill, he gave the summary of quantities estimated in 1885 and the final measured quantities in 1889 at the completion of the bridge:

Item	1885	1889
1. Tons of steel in superstructure	6,200	6,320
2. Tons of steel in caisson	1,600	1,668
3. Cubic yard of concrete	27,600	26,593
4. Masonry (C.Y.)	4,900	5,630
The excess of masonry was due to foundation problems of Piers 5 and 6.		
5. Time	30 months	34 months

The plans of the caissons were designed by him and his partner Charles Macdonald a year before the Hawkesbury competition, for a proposed deep foundation in the U.S. Complete plans and models were made of this, and it was adopted for the Hawkesbury Bridge without any change. As soon as the news of the international competition for the Hawkesbury Bridge were received by the Union Bridge Co., Macdonald accompanied by John F. Anderson, the foundation subcontractor, visited Australia to plan the construction of the bridge and learn the capabilities of local contractors. And, he personally went to England with the plans of the caisson and superstructure, visited several fabrication plants in England and Scotland and obtained firm prices for fabrication of the caissons and the superstructure, and delivery of cement to the job site.

After the contract was awarded to the Union Bridge Co., the caisson subcontractors Anderson & Barr assembled the foundation excavation plant and loaded it on to the ship "Anglo India" which left the New York Harbor on May 10, 1886. Thereafter, Anderson went to California, and sailed from San Francisco for Australia on June 5, 1886 to personally supervise the work on the caisson foundation.

Clarke also gave credit to his erection subcontractors, Ryland and Morse. The erection of each span weighing approximately 800 tons on a large pontoon and carrying it approximately three-quarters of a mile to the bridge site and depositing it on the bearings taking advantage of high tides was never attempted before. Although the erection scheme was designed by Macdonald and Clarke, it was the ability of Mr. S.V. Ryland to execute the erection without failures and avoiding the resulting heavy financial losses that made the project a success.

Clarke also responded to the comment by Burge that the entire bridge except the eyebars were made in England. It was a matter of economics because the prices of iron and steel were more favorable in Britain than in the U.S.; and if the fabricator William Arrol of Glasgow, Scotland had agreed to install expensive plant for the eye-bars, he would have made the eye-bars also. The reluctance of fabricators in Britain and Scotland to fabricate the eye-bars was due to the requirement that the bars should break preferable in the body of the original bar and not at any point of the head or neck. The compliance of this requirement was to be verified by testing a certain number of full-sized eye-bars to destruction.

Theodore Cooper, a prominent U.S. bridge engineer, gave the real reasons for the success of the Union Bridge Company. He noted that the partners of Union Bridge Company had built over 160,000 miles of railroads and over 3,000 miles of railroad bridges in their lifetime. This included foundations and structures of all kinds and magnitudes. As an example of the expertise of the Union Bridge Company, he cited the erection of two channel spans of Cairo Bridge over the Ohio River near its mouth at Cairo, Illinois. Each span was 518 ft. 6 in. long and weighed about 1,030 tons. It took Union Bridge Company one month and 3 days to erect the 2 spans and moving the false works, including five days of lost time while waiting for the completion of the masonry. He was not surprised that the Union Bridge Co. was a winner with the ability and experience on a task it was so well prepared to estimate and execute (Cooper, 1890).

13 PEOPLE RESPONSIBLE FOR SUCCESS OF HAWKESBURY BRIDGE

Many individuals deserved credit for this well-conceived, well-planned, and extremely well-executed project. They were from:

1. The Government of New South Wales	a. John Whitton b. Charles Ormsby Burge c. Henry Deane
2. Sir John Fowler & Co.	d. John Fowler e. Benjamin Baker
3. Union Bridge Co.	f. Charles Macdonald g. Thomas C. Clarke h. Martin Van Brocklin
4. Anderson & Barr	i. John F. Anderson
5. Ryland & Morse	j. S.V. Ryland

13.1 *John Whitton*

He was the Engineer-in-Chief of Railways when this project was planned by the government of NSW. There were private companies operating railroads in NSW, but because of labor problems caused by the gold rush and financial difficulties, they were taken over by the British Government in 1855. Whitton became the Engineer-in-Chief in 1857 and retired in 1890. It was during his services, the railroads expanded in Australia, and he provided a steady hand based on his vision and common sense. He sought guidance from Fowler, a noted British civil engineer, and advisor to the government of NSW for building the Hawkesbury Bridge. He agreed to the recommendation made by Fowler and the Board of Examiners to award the contract for the design and construction of the Hawkesbury Bridge to the Union Bridge Co. of New York.

13.2 *Charles Ormsby Burge*

He was the principal Assistant Engineer under Whitton. He started during the planning of this bridge and laying out the location of this bridge using a system of triangulation. Later, when the contract for the design and construction was awarded to the Union Bridge Co., he became the Resident Engineer for the government of NSW. He presented a paper at a meeting of the Institution of Civil Engineers in London on the Hawkesbury Bridge (Burge, 1890). He presented a paper on "Fifty years of railway and tramway construction" in Australia in 1904 which was printed by the NSW Government Printing Press.

13.3 *Henry Deane*

He was born in Britain, came to Australia in 1880 at the age of 33, and joined the Railway Department of the government of NSW. He was in charge of the load tests for the Hawkesbury Bridge. When Whitton retired a short time after the opening of the Hawkesbury Bridge in 1890, Deane succeeded him as the Engineer-in-Chief. He retired in 1906 to work in the private sector, but returned in 1912 as Engineer-in-Chief of the railway's construction division, and again left in 1914.

13.4 *John Fowler*

He was the youngest engineer to be elected the president of the prestigious British Institution of Civil Engineers in 1865 at the age of 48. His work in Egypt as a consultant to the King Ismail Pasha in designing Egypt's railway system and other civil engineering works earned him a Knighthood from the British government. Since 1850, he was considering designing cantilever bridges. After 30 years he and his younger junior partner Benjamin Baker were retained by a railway company to design the Firth of Forth Bridge in Scotland. Upon the inauguration of the bridge in 1890 by the Prince of Wales (later King Edward VII), Baker earned his Knighthood and Fowler the title of a baron.

13.5 *Benjamin Baker*

He was one of the best bridge engineers of his time. He wrote a series of articles on the design of long span railway bridges (Baker, B., 1867). He had visited the site of the Hawkesbury Bridge and estimated the cost of the bridge at £400,000 (US $1,944,000). He had also visualized that the best plan of building the bridge would be that of open-top caisson, sunk by dredging, and a steel superstructure floated out into position (Baker, B., 1890).

Fowler, Baker and Sir William Arrol, the fabricator of steel for the Hawkesbury Bridge were also associated with the construction of the first railway tunnel under the Hudson River connecting Jersey City and Manhattan in 1890. The tunnels are now used by the Port Authority of New York and New Jersey and known as PATH (Port Authority Trans Hudson) tubes.

13.6 *Charles Macdonald*

He was born in Ontario, Canada in 1837. He entered Rensselaer Polytechnic Institute in Troy, NY in 1854 and graduated in 1857 with a degree in Civil Engineering. One of his classmates was Washington A. Roebling, builder of the Brooklyn Bridge after the death of his father John in 1869. Another of his classmates, who was a member of Union Bridge Co. and participated in the design of the Hawkesbury Bridge, was Robert Escobar of Cuba. He was one of the earliest Spanish students studying engineering in the U.S.

Macdonald joined Pennsylvania Volunteer Corps in 1863 while he was employed by the Philadelphia and Reading Railroad Co. He became a prisoner in the Battle of Gettysburg until the end of the Civil War. He opened an office in New York City in 1869. He entered in an international competition to design a bridge over the east and west channels of the East River at Blackwell's Island (now Roosevelt Island) in New York City. He submitted the design of a large cantilever bridge with two spans over the two channels of 734 feet and 618 feet, and his design won the first prize in May 1876. This was the first design for a large cantilever bridge ever accepted in a competition (Engineering News, 1883b). The cost was estimated at $3,000,000 including the approaches. The bridge project was cancelled. It took 33 years to build a new cantilever bridge (Queensboro or 59th Street Bridge) in 1909 at a location south of the one proposed in Macdonald's design.

Macdonald was also one of the earliest promoters of a suspension bridge over the Hudson River at 57th Street connecting New York and New Jersey, and he had made a large model of such a bridge. He was appointed by the City of New York a Trustee of the Brooklyn Bridge which was opened in 1883. He was elected President of the ASCE from 1908 to 1909, and in 1910 he served as one of the Commissioners of the Quebec Bridge during its reconstruction.

13.7 *Thomas C. Clarke*

He was an 1848 graduate of Harvard College. He worked for the railroads before he started his own company Clark Reeves & Co. to design and build bridges. He inspected the Niagara Railway Suspension Bridge designed and built by John Roebling in 1873. In 1877, both Clark and Macdonald served as members of a three-member Commission to determine the structural integrity of the Niagara Railway Suspension Bridge (Gandhi, 2006). Clarke was a Consulting Engineer for the City of New York in building the Third Avenue and Willis Avenue Bridges over the Harlem River, and was responsible for their design and construction. He served as President of ASCE from 1895 to 1896.

13.8 *Martin Van Brocklin*

He was the superintendent of the Union Bridge Co. during the early part of construction. He was a native New Yorker, and had formal education in civil engineering. Before joining the Union Bridge Company, he was in charge of the erection of a part of the Sixth Avenue Elevated Railroad in New York City, and was one of the engineers who participated in the construction of the Oroya Railroad in Peru, South America. He died at North Platte, Nebraska in 1896. After he left the Hawkesbury Project, Charles and Frederick Macdonald assumed the responsibility of the Superintendent (Engineering Record, 1896).

13.9 *John F. Anderson*

If any one person deserved credit for the success of sinking the caissons of the Hawkesbury Bridge to unprecedented depths setting a world record in a country thousands of miles away from his home base, and making the entire project a success, it was Anderson. His partner Barr was a Dutchman. Anderson was a member of the 143rd Regiment, New York Infantry from Monticello, NY, and joined the Civil War as a Lieutenant. He was promoted to the rank of a Captain in February of 1863, and served in the Union Army until July 20, 1865.

It was Anderson's creative genius to modify an existing Milroy excavating machine used in Britain and India to sink iron cylindrical piers for the Texas and New Orleans Railroad over the Atchafalaya River, a tributary of the Mississippi River in Louisiana. With his procedure he was able to sink the foundation of the pier 125 feet below the mean water level (Engineering News, 1883a). It was the experience on this project that convinced Anderson that it was possible to go much deeper than 125 feet below the water surface using an open caisson for the Hawkesbury Bridge Project.

In 1884, Anderson obtained a patent for a bridge with movable floor system for transfer of railroad trains, wagons, and foot passengers over busy waterways, such as the Hudson River, where very heavy marine traffic prevented the use of a movable bridge, and the condition of approaches precluded the construction of a high bridge to permit the passing of ships under it (Scientific American, 1885a).

He built the lighthouse at Fourteen Foot Bank in the Delaware Bay, 22 miles from land; and he achieved a national reputation for prominent engineering works including the construction of the Chestnut Street Bridge over the Schuylkill River in Philadelphia; the building of the piers of the Washington Bridge across the Harlem River; the foundation work for Illinois Central Railroad's Cairo Bridge over the Ohio River at Cairo, IL which was opened in November 1889; the Merchants' Bridge over the Mississippi at St. Louis; the Union Pacific Bridge at Omaha, Nebraska; and piers and wharves at Havana, Cuba (New York Times, 1892).

In 1880, Anderson was a superintendent for the first Hudson River Railroad Tunnel connecting Jersey City with Manhattan. On July 21, 1880 an accident took place in the tunnel due to carelessness of a worker killing 20 workers as the tunnel was flooded and partially collapsed. In response, he developed and implemented the concept of a pilot tube 6 feet in diameter and made from 1/4 inch iron plates. This tube was carried from 30 to 60 feet in advance of the main tunnel. The radial braces were supported against the tube, which in turn supported the plates of the main tunnel tube while being assembled and put in place (Figure 17). Compared to the advancement of the main tube, the pilot tube facilitated a thorough exploration of the ground, and provided opportunity to support any weak zones thereby preventing slides and saving of lives (Scientific American, 1889). He used the same concept of pilot tube for the construction of a 12 ft. diameter relief sewer from the junction of Knickerbocker and Johnson Avenues, through Johnson Avenue and South 5th Street, to the East River in Brooklyn (Scientific American, 1885c).

In the 1880s there was a common fear in the U.S. that foreign powers may attack New York City for its wealth. The Fortification Board created by the U.S. government identified urgent need to guard 27 seaports. New York was at the top of the list with Boston and San Francisco running second and third. Anderson proposed construction of 3 forts on 3 artificial islands in New York

Figure 17. Use of pilot tube for the excavation of Hudson River Tunnel (Scientific American, 1889).

harbor between Rockaway Beach in Long Island and Sandy Hook Point in New Jersey so as to command all channels of approach and placed 12 to 15 miles away from New York City. He proposed to use the same procedure in sinking double-walled caissons with an inside diameter of 400 feet and outside diameter of 500 feet as he had used successfully for the sinking of caissons of the Hawkesbury Bridge (Figure 18). He estimated that each island could be built for about one million dollars. His plan was not implemented but received considerable attention due to its modest cost and practicality (Scientific American, 1890).

After the completion of the Hawkesbury Bridge, Anderson signed a contract in 1890 to build a lighthouse at the Outer Diamond Shoal, eight miles off Cape Hatteras, North Carolina for $450,000 based on the surveys provided by the U.S. Government. The surveys were inaccurate and unreliable, and after two years of fruitless efforts and losing more than $150,000, Anderson abandoned the project, and his firm Anderson & Barr went out of business (New York Times, 1892).

Later in his life, Anderson was connected with another history making project. In 1905, he worked as Superintendent of a marine contractor George B. Spearin, to build the Dry Dock No. 4 in the Brooklyn Navy Yard based on the drawings provided by the U.S. Government. The drawings did not show an existing 6′ diameter brick sewer crossing the site, which caused Spearin to stop the job. He requested money for the extra work, and the Government not only refused, but terminated the contract.

Spearin sued the U.S. Government, and the case went all the way up to the U.S. Supreme Court. On December 9, 1918 Justice Brandeis delivered one of the most far-reaching decisions affecting the construction industry that, "If the contractor is bound to build according to the plans and

Figure 18. Anderson's plan for the defense of New York using 500 ft. diameter caissons. Fortress (1) and Sectional elevation (2) (Scientific American, 1890).

specifications prepared by the owner, the contractor will not be responsible for the defects in the plans and specifications." . . . "Spearin was under no obligation to repair the sewer and proceed with the work while the Government denied responsibility for providing and refused to provide sewer conditions safe for work" (Engineering News-Record, 1918). Had this decision been rendered before Anderson & Barr took the Diamond Shoal Lighthouse project, they would not have been forced to go out of business due to the Government's mistakes. The firm that George Spearin founded in the late nineteenth century is still located in New York City and operating under the name "Spearin Preston and Burroughs".

13.10 *S.V. Ryland*

He was one of the best Superintendents of Erection of his time. For a number of years he was connected with the firm of Kellogg & Maurice, which became a part of Union Bridge Company. He erected the Smithfield Street Bridge at Pittsburgh designed by Gustav Lindenthal; in 1878 he erected a combination cantilever truss of 160 foot span over the Snake River at Eagle Pass, Idaho; and he was superintendent of erection of the Central Bridge Company of Buffalo, NY for the erection of the Michigan Central Cantilever Bridge at Niagara Falls in 1883. After the completion of the erection of the Niagara Falls Cantilever Bridge, Ryland was in charge of the erection of the St. John's cantilever bridge in New Brunswick, Canada.

Ryland was killed on the job almost at the end of the successful completion of the erection of the Hawkesbury Bridge. According to an article published in the Melbourne newspaper, *Argus*, of December 13, 1888, Ryland was walking along the top of one of the spans when he missed his footing and fell some 50 feet into the river below. While he was falling, a huge shark was observed immediately below. Both the shark and Ryland disappeared under the water, bringing an end to a successful career (Engineering News, 1889b).

14 POSTSCRIPT – REPLACEMENT OF HAWKESBURY BRIDGE

After the bridge was opened to traffic in 1889, there were some complaints of random vibrations resulting in irregular and frequent lateral motion of the piers, the maximum movement being about

Figure 19. Original piers of the Hawkesbury Railway Bridge after the demolition of its superstructure in 1943 (Flickr, Penny T).

1½ inches. It was explained that a small amount of movement at the top of a pier about 200 feet in height was inevitable (Railroad Gazette, 1890b). This subject was discussed in the colonial parliament; and the minister in charge said the movement was insignificant and did not affect the functioning of the bridge (Railroad Gazette, 1890c). An analysis of these random motions given by Ewald was reprinted in Engineering News, 1891.

The bridge performed admirably for about 35 years. The spans were strengthened for heavier loading between 1926 and 1931. However, the annual report of the Department of Railways of Australia in 1939 indicated the need to replace the Hawkesbury Bridge with a new bridge. The general reasons given were that the iron caissons were badly corroded, the concrete was disintegrating, the expansion bearing rollers were frozen, and the trains were much heavier than the bridge was designed for (Engineering News – Record, 1939).

To reduce loading on the bridge, only a single track was used; the speed was reduced from 35 mph to 15 mph, and finally to 5 mph. Due to the beginning of World War II, all available steel was diverted to war efforts. Even though the work on the replacement bridge about 700 feet west or upstream of the existing bridge started in July 1940, it was not completed for the next six years until July 1946; and during the WWII, the bridge carried up to 100 trains per day. The original piers of the bridge after the demolition of the superstructure in 1943 can be seen in Figure 19. The new bridge had 8 spans of 3 different lengths and the piers were carried to a depth over 180 feet. After the old bridge was demolished, its steel members were recycled and reused to build a workshop in 1954.

16 CONCLUSION

The Hawkesbury Railway Bridge was very vital to the development and economy of the Province of New South Wales, Australia. This project was an excellent example of international business and cooperation. With sound advice from its consultants, Sir John Fowler & Company of England, the Province was able to obtain specialized engineering and construction know-how of Union Bridge Company of New York to successfully build deep bridge foundations at record-setting depths, and complete the 3,300 ft. long bridge at a previously agreed upon price and schedule – a rarity in today's environment.

ACKNOWLEDGEMENTS

The author would like to thank Mr. Irving Stephens, Mr. Christopher Bessett and Mr. Quincy Collins of the Rensselaer Polytechnic Institute's Folsom Library for their help in making available the old records, and Ms. Brenda Hill, Ms. Jennifer Drieves, Mr. Stephanson Ludger, and Mr. Andre Chambers of his office in conducting research and assisting in the preparation of this paper.

REFERENCES

Baker, Benjamin (1867), "*Long-span railway bridges*", Engineering V. 3, Jan–June 1867, pp. 250, 265, 298–299, 338–9, 426–7, 441, 471, 571–2, 587–8, 611.

Baker, Benjamin (1890), "*Discussion on railway bridges*," Proc. Institution of Civil Engineers, V. 101, 1890, pp. 50–54.

Burge, Charles Ormsby (1890), "*The Hawkesbury Bridge, New South Wales*", Proc. Institution of Civil Engineers, V. 101, 1890, Paper No. 2421, pp. 2–12.

Clarke, Thomas C. (1890), "*Discussion on railway bridges*", Proc. Institution of Civil Engineers, V. 101, 1890, pp. 61–64.

Cooper, Theodore (1890), "*Discussion on railway bridges*", Proc. Institution of Civil Engineers, V. 101, 1890, pp. 65–67.

Engineering, 1886a, "*The Hawkesbury Bridge*", V. 41, April 16, 1886, pp. 367–368.

Engineering, 1886b, "*Hawkesbury Bridge*", V. 41, April 23, 1886, p. 400.

Engineering, 1886c, "*Hawkesbury Bridge*", V. 41, May 7, 1886, p. 458.

Engineering, 1886d, "*Hawkesbury Bridge*", V. 41, May 14, 1886, p. 482.

Engineering, 1886e, "*Hawkesbury Bridge*", V. 41, June 25, 1886, p. 626.

Engineering, 1887a, "*Hawkesbury Bridge*", V. 43, February 25, 1887, p. 191.

Engineering, 1887b, "*The Hawkesbury River Bridge, New South Wales*", V. 43, March 11, 1187, p. 222, p. 224 + Two page plates.

Engineering, 1887c, "*The Hawkesbury River Bridge, New South Wales*", V. 43, April 8, 1887, pp. 318–319 + Two page plates.

Engineering, 1887d, "*Foundation caissons for the Hawkesbury River Bridge*", V. 43, April 22, 1887, pp. 370–372.

Engineering News, 1883a, "*Founding piers in the Atchafalaya of Louisiana*", V. 10, May 5, 1883, pp. 208–209.

Engineering News, 1883b, "*Other cantilever bridges*", V. 10, December 15, 1883, p. 602.

Engineering News, 1885, "*A great bridge for Australia*", V. 13, June 13, 1885, p. 283.

Engineering News, 1886a, "*The Hawkesbury Bridge, New South Wales*", V. 15, January 30, 1886, p. 79.

Engineering News, 1886b, "*Hawkesbury Bridge, New South Wales*", V. 15, February 13, 1886, pp. 98–100.

Engineering News, 1886c, "*The British Australasian Says*", V. 15, April 10, 1886, p. 229.

Engineering News, 1886d, "*The New Bridge For The Hawkesbury*", V. 15, April 17, 1886, p. 250.

Engineering News, 1886e, "*The Hawkesbury Bridge*", V. 15, May 8, 1886, p. 298.

Engineering News, 1886f, "*Competitive Designs For The Hawkesbury River Bridge, New South Wales*", V. 15, May 8, 1886, 2 page insert between pages 288 and 289.

Engineering News, 1886g, "*The Hawkesbury Bridge. The Mechanical World*", V. 16, July 31, 1886, p. 79.

Engineering News, 1887a, "*Engineering News*", V. 17, March 5, 1887, p. 149.

Engineering News, 1887b, "*Engineering News*", V. 17, April 23, 1887, p. 261.

Engineering News, 1887c, "*Engineering News*", V. 18, December 10, 1887, p. 426.

Engineering News, 1888a, "*The Hawkesbury Bridge*", V. 20, August 11, 1888, p. 107.

Engineering News, 1888b, "*Engineering News*", V. 20, October 27, 1888, p. 319.

Engineering News, 1889a, "*The Hawkesbury Bridge*", V. 21, January 5, 1889, pp. 3–5.

Engineering News, 1889b, "*Engineering News*", V. 21, February 23, 1889, p. 157.

Engineering News, 1889c, "*The Hawkesbury Bridge*", V. 21, May 25, 1889, p. 477.

Engineering News, 1889d, "*Testing the Hawkesbury Bridge*", V. 22, July 20, 1889, p. 50.

Engineering News, 1891, "*The Motions Observed in the Hawkesbury Bridge*", V. 25, March 14, 1891, pp. 247–248.

Engineering News-Record, 1918, "Supreme Court fixes responsibility of contracts," V. 81, No. 26, December 26, 1918, pp. 1167–1168.

Engineering News–Record, 1939, "*Noted Australian Bridge Must Be Rebuilt*", November 23, 1939, p. 58.
Engineering Record, 1896, "Obituary" V. 33, No. 15, March 14, 1896, p. 255.
Gandhi, Kirti (2006), "*Roebling's Railway Suspension Bridge over Niagara Gorge*", 5th International Cable Supported Bridge Operators' Conference in New York City, August 28–29, 2006.
Macdonald, Charles (1887), "*The Six-hundred Ton Testing Machine at the Works of the Union Bridge Company at Athens, PA*", Trans. American Society of Civil Engineers, V. XVI, Paper No. 349, 1887, pp. 1–8 + 4 plates.
Nason, H.B. and Young, W.H. (1887), "*Biographical records of the officers and graduates of the Rensselaer Polytechnic Institute, 1824–1886.*" Benjamin Franklin Greene, D.H. Jones & Co.: Troy, NY (1855). RPI Archives and Special Collections, Troy, NY, pp. 213–214.
New York Times (1892), "Capt. Anderson gives up; Diamond Shoal Lighthouse work abandoned," April 9, 1892.
Railroad Gazette, 1888, "*Placing the Center Span of the Hawkesbury Bridge*", V. 20, August 10, 1888, pp. 516–517.
Railroad Gazette, 1889, "*The Hawkesbury Bridge*", V. 21, June 14, 1889, p. 397.
Railroad Gazette, 1890a, "*Hawkesbury Bridge Foundations*", V. 22, January 17, 1890, p. 35.
Railroad Gazette, 1890b, "*Piers of the Hawkesbury Bridge*", V. 22, June 20, 1890, p. 444.
Railroad Gazette, 1890c, "*The Hawkesbury Bridge*", V. 22, July 11, 1890, p. 496.
Scientific American, 1885a, "Bridge with movable floor system", V. 52, January 24, 1885, p. 47.
Scientific American, 1885b, "New South Wales. To iron and steel bridge builders", V. 52, March 7, 1885, p. 157.
Scientific American, 1885c, "A great sewer built by an improved method of tunneling in Brooklyn, N.Y.", V. 53, December 12, 1885, p. 367 and p. 373.
Scientific American, 1886, "*The Deepest Foundation in the World*", May 8, 1886, V. 54, p. 287.
Scientific American, 1889, "*The Hudson River Tunnel*", V. 61, September 7, 1889, p. 143 and p. 151.
Scientific American, 1890, "*Anderson's Plan for the Defense of New York*", V. 63, December 6, 1890, p. 351 and p. 356.

Chapter 26

Historic bridge replacement: A collaborative approach to context sensitive design

Thomas P. Piotrowski & Bruce Chamberlin
H2L2 Architects/Planners, LLC, Philadelphia, PA, USA

ABSTRACT: This paper presents architect's experience with a collaborative design process, aimed at balancing technical, aesthetic, and historic issues, which results in context sensitive design that has added value for all parties. The Royal Park Bascule Bridge Replacement, with prime consultant E. C. Driver in association with Kimley Horn serves as our primary case study.

Projects responsive to the purpose and needs defined by the stakeholders generate greater participation and ownership in the project. Involving the community early builds support and minimizes resistance that sometimes develops in response to change. Officials benefit by helping the public anticipate construction with the knowledge that any inconvenience will be rewarded by a structure that is responsive to the specific needs of the community. In the case of the Royal Park Bridge, a design that reflected the prevalent aesthetic and context contributed significantly to community pride by generating a positive image for residents and visitors.

1 WHAT IS CONTEXT SENSITIVE DESIGN?

A design that is "Context Sensitive" is officially defined by a number of criteria but foremost is that it results from consensus by a range of stakeholders and that it preserves environmental, scenic, aesthetic, historic and natural resource values.[1] "Stakeholders" can be the typical public officials that you would expect to manage and oversee public projects. However, more often, the general public is gaining a voice in venues that range from specially appointed civic committees composed of dozens of members, to the input of a few concerned neighbors. Community values regarding scenic, aesthetics, and history are likewise variable for each project.

Historic bridges are the greatest challenge for Context Sensitive Design (CSD). Frequently, the existing structural type and/or method of construction has become obsolete, replaced by newer

Figure 1. Deck View – Proposed Royal Park Bridge.

technologies, engineering methods, materials, and construction methods. Furthermore, it is inevitable, that due to safety regulations and various other standards, the new bridges will be higher, wider and bigger, all of which will have an impact.

To a great extent these effects can be addressed or mitigated through careful analysis of the existing structure and skillfully imbuing the new with similar characteristics. In other words, it is important to understand the contributing factors that make the existing bridge integral or special in the community and design the new structure in a manner that builds on these characteristics. While the initial impulse might be to replace with a structure in kind, modern building techniques and the cost of the original materials combined with what is typically a larger structure, may make this both prohibitively expensive or appear as a caricature of the original.

2 STAKEHOLDERS

A key component of CSD is the involvement of stakeholders. For the Royal Park Bridge, key "stakeholders" including the Landmark Commission, were relatively few, but influential. A total of four public outreach meetings were held to introduce the project to the general public and get their input. For the 1988, Columbus, Ohio, Discovery Bridge the "Community Interest Task Force" was composed of representatives of nine public or quasi-public agencies. The trend is that invited stakeholder groups are more likely to be larger and more inclusive. For example, on recent projects, we have worked with business organizations, institutions and arts groups as well as local governmental agencies. In on a five-mile highway corridor project that includes 16 bridges in Rochester, New York, the Aesthetic Committee comprised of an arts consultant, the Visitors Bureau, and the City School District.

Working with these groups requires careful planning to manage the process. This includes considering the type and means of input, as well as organizing presentations to achieve a workable and productive outcome. A skilled facilitator is invaluable in keeping meetings productive and in allowing all participants to have an equal voice. The number of outreach meetings depends on the size and complexity of the project, the level of community and stakeholders' interest. An initial meeting serves to introduce the project and define the purpose of the group as well as the goals and milestones to achieve at subsequent meetings. An overview of the project includes the opportunities and constraints as well as a brief lesson on design principles as they apply to the subject bridge. We typically prepare "precedent studies" showing creative and diverse solutions for similar situations that help participants create a vision or understanding of the project. These precedent studies include bridge and viaduct types, under bridge and deck design, urban design and river front access, pedestrian and aesthetic lighting, etc.

Interim meetings are generally workshops focusing on one or two specific issues such as the number of arch spans or the possible shapes of piers and girders, handrail design options or light fixture alternatives. The final meeting presents the completed project and serves as evidence to the community that the process is complete and that their desires, ideas and values are reflected in the final project.

3 THE CONTEXT

The Palm Beaches are affluent communities and home to many prominent full-time and seasonal residents. The Royal Park Bridge is an especially important landmark due to its location in the Palm Beach Historic District and as a dramatic and ceremonial approach to the tree lined boulevard – the Royal Palm Way. The Palm Beach Landmark Commission quickly established itself as the public advocate for the interests of its citizens and the preservation of its views. Architectural integration into the surrounding communities was accomplished as part of our collaborative approach involving the Landmark Commission, the engineers and the landscape architect. This close working relationship from the first project phases was essential to the project's success.

Figure 2. Elevation of final design.

4 THE HISTORIC ROYAL PARK BRIDGE

The Royal Park Bridge spans the intra coastal waterway linking Palm Beach and West Palm Beach, Florida. The existing bridge consists of two parallel, yet separate, two-lane structures supporting the road deck, with a bascule span at the center of the bridge. The first arch bridge was completed in 1929 and the second bridge was completed in 1957. During a routine inspection, it was discovered that wooden piles supporting the 1929 structure were severely deteriorated necessitating the emergency closure of two lanes of traffic. A major constraint to constructability and phasing was the fact that the 1929 and 1957 structures were interdependent. Work began immediately on the design of a temporary bridge as well as the design of its permanent replacement.

The existing bridge is remarkable for its simplicity, setting, and approach into Palm Beach. The bridge was aligned with Royal Park Way, the main avenue into Palm Beach, and the bridge contributed to the sequence of arrival and departure from the community. It was of a scale and character commensurate with the community; compact, solid, and restrained, and provided a simple and elegant backdrop for Palm Beach. A different set of conditions could be found at the West Palm Beach side of the bridge. The bridge terminates abruptly and unceremoniously at a boulevard that paralleled the causeway. Low and mid-rise buildings flank the west side of the boulevard, with a park along the east side adjacent to the causeway. It was important that the replacement bridge addressed both of these conditions.

5 STRUCTURE TYPE SELECTION

Technical criteria for the replacement of the bridge include navigational, structural, and roadway standards regulated by the State Department of Transportation and the US Coast Guard. A vast array of alternatives was explored for the replacement structure. Some, including a tunnel, were quickly eliminated. A key question was to determine the optimum height of the bridge; the highest alternative would eliminate the need for a bascule span but would have the greatest impact with regard to roadway approach alignment as well as visually changing the scale of the surrounding landscape. A new bridge at the same height would cause frequent openings and closings. Finally, it was decided to select a compromise elevation higher than the old bridge but low enough to require the use of a bascule span, which would open less frequently that the original.

Maintaining the existing barrel vaults, or the appearance of the vaults became a critical concern to the stakeholders and the community. A number of arch solutions were developed by the architectural and engineering team, which were evaluated for aesthetics, cost, and constructability. Photosimulation renderings of the viable options showing the effect from land and water were presented to the stakeholders. During an interactive discussion which included a lesson on structure types and nomenclature, the stakeholders decided that the preferred solutions were those alternatives closest to a true arch or barrel vault. Alternatives where arched panels were attached to beams were rejected, considered flimsy and not aesthetically acceptable. The final design solution was a segmental,

Figure 3. Mizner style.

Figure 4. Art deco style.

Figure 5. "Cracker" style.

pre-cast concrete arch girder with a barrel arch appearance. The design was dictated by the desire to maintain the look of the existing barrel vaults with the need to avoid the existing foundations.

6 ARCHITECTURAL DESIGN

The opportunity for architectural expression on the Royal Park Bridge was significant. Components consisted primarily of the control house, railings, light fixtures, texture and color, and the terminations at the east and west sides of the bridge. Of these, the most significant feature was the control house, subject of considerable study and discussion. A number of design studies were produced illustrating potential architectural approaches to the tower. Three of the earliest alternates are illustrated below. The inspiration for each alternative was derived by performing a context analysis of the project area by means of on-site visual and photographic surveys. The architectural team also studied the works of prominent local architects, including Addison Mizner, who is credited for developing the eclectic Palm Beach Style. This qualitative assessment resulted in photographic documentation and written commentary , which established distinctive contextual features and aesthetic criteria, that would become the basis for the design of the bridge's replacement and architectural components.

1. Is a style reflecting a blend of Moorish and Spanish colonial revival styles named after the architect responsible for developing the aesthetic: Addison Mizner. It is characterized by both massing and materials such as stone cladding, red clay tile roofs, ornamental tile and wrought iron. Because the Mizner style dominates the adjacent Historic District it was agreed to continue to advance the design along that theme. Final design tasks focused on detailed bridge elements such as railings, pavement, and lighting. These details were very important to the community and their design involved extensive public input and feedback.
2. This is the most original and liberal interpretation of an art deco style that is popular in parts of Florida. Although not the preferred approach the derivation did serve an important role in furthering the discussion and offered positive influence in the refinements.
3. Represents an interpretation of the indigenous "cracker" style of architecture. This treatment is typically used on utilitarian buildings in the region, and is reflective of its environment and local building materials. Characteristic is the low pitched sloped roof, placement of widows and the relationship between interior and exterior space as suggested by the veranda.

7 DESIGN TOOLS

Photosimulation or "photosim" renderings are the truest tool in exploring design options. Bridge engineers and public officials have increasingly singled this tool out as the most effective technique for expressing and communicating design intent. To produce a photosim, the site is photographed in good light from a variety of angles. It is important that the photographer be aware that it is not their mission to photograph the existing bridge – but the site of its replacement. Frequently the new bridge is higher, wider, longer and adjacent to the existing structure so the background photograph must include enough contextual information to show the proposed bridge in its entirety including approach spans. The engineer or architect then creates a 3-D wire-frame Cadd model of the alternatives to be explored. The Cadd model is manipulated to "fit" in its landscape and then "rendered" with texture, color, shadow, etc. As the design process evolved the images become more detailed and realistic. Because of their realism it is recommended that several alternatives be presented side-by-side so that the viewer does not get a false impression that they are viewing a final unchangeable design.

8 COLLABORATIVE TEAMING

Landscaping treatments prepared in conjunction with Kimley-Horn became the preferred means for dealing with the transitions at both ends of the bridge, as well as for mitigating the width of the bridge. At both ends of the bridge, designs were developed that addressed specific contextual concerns. The most extensive treatment was at the West Palm Beach side. Monumental plazas and overlooks were developed, which connected the bridge with the bicycle and pedestrian paths located in the park below that paralleled the intercoastal waterway. Large oaks of size and scale equal to the existing trees were relocated to the site. Two large stone pylons mark the entrance to the bridge at the West Palm Beach side. The design of these elements are in scale with both the bridge and the neighboring structures, and similar to the design of the 1929 original.

Of concern to stakeholders was the width of the bridge, which impacted the alignment with Royal Park Way. A landscaped median was eventually approved, consisting of, hard and soft scape material, and aligning with the existing median in Palm Beach, visually reducing and mitigating the width of the bridge. The median transitioned into a turning lane at the west end of the bridge.

For other bridge replacements, we have collaborated with lighting designers and sculptors who have played equally important roles in making bridge, viaduct and highway corridors more interesting, exciting and contextually appropriate.

Ultimately, what was important to the success of the project was the ability to produce clear concise direction for the project, based on input from the stakeholders, and to provide visualizations of the proposed solutions that would enable the shareholders to make informed decisions – as well as for the design team to try numerous visualizations to study alternate concepts.

9 CONCLUSION

Context Sensitive Design forces us to think beyond the physical limits of a bridge structure. The challenge with historic structures is to integrate modern engineering technology, open communication, and multidisciplinary collaboration to benefit and enhance our communities. Construction for the replacement of the Royal Park Bridge began January 7, 2002 and is scheduled for completion in 1,100 days.

REFERENCE

US. Department of Transportation, Federal Highway Administration, "Flexibility in Highway Design, (FHWA Pub. No. FHWA-PD-97-062).

Author index